ETHER AND MODERNITY

ETHER AND MODERNITY

The Recalcitrance of an Epistemic Object in the Early Twentieth Century

Edited by

JAUME NAVARRO

University of the Basque Country, Spain

OXFORD
UNIVERSITY PRESS

OXFORD
UNIVERSITY PRESS

Great Clarendon Street, Oxford, OX2 6DP,
United Kingdom

Oxford University Press is a department of the University of Oxford.
It furthers the University's objective of excellence in research, scholarship,
and education by publishing worldwide. Oxford is a registered trade mark of
Oxford University Press in the UK and in certain other countries

First Edition published in 2018

Impression: 1

Published in the United States of America by Oxford University Press
198 Madison Avenue, New York, NY 10016, United States of America

British Library Cataloguing in Publication Data

Data available

Library of Congress Control Number: 2018935614

ISBN 978–0–19–879725–8

DOI: 10.1093/oso/9780198797258.001.0001

Printed and bound by
CPI Group (UK) Ltd, Croydon, CR0 4YY

ACKNOWLEDGEMENTS

This project began in 2013 when I was awarded the Marconi Fellowship by the Bodleian Libraries at Oxford University. Alexandra Franklin, from the Department of Special Collections and Prof. Pietro Corsi encouraged me to organise a workshop on 'The Lure of the Ether. Physics and Modernity', which eventually took place in February 2014 at the History Faculty, Oxford. Imogen Clarke, Richard Noakes, Richard Staley, Michael Whitworth and I presented papers that were commented on by Geoffrey Cantor. In 2015, Scott Walter and I organised a special panel on the same topic at the History of Science Society meeting in San Francisco and we were joined by Connemara Doran, Linda Henderson and Richard Staley. It was in that meeting that we agreed to work towards more consistent versions of our papers and to publish this edited volume. The final papers were presented at a meeting in Donostia/San Sebastian, in March 2017, attended by all the contributors to this volume. I would like to thank the Donostia International Physics Center (DIPC) and the Basque Government for funding that meeting, as well as project HAR2015-67831-P MINECO/FEDER, EU, for funding part of the research for this book.

CONTENTS

BIOGRAPHIES OF AUTHORS

Massimiliano Badino is assistant professor of history and philosophy of science at the University of Verona. Previously, he was the Marie Sklodowska-Curie Fellow at the Universitat Autònoma de Barcelona and at the Massachusetts Institute of Technology. His research focuses on history and philosophy of mathematical physics from the Enlightenment to the twentieth century.

Imogen Clarke is an independent scholar, working in academic publishing. She completed her PhD on early twentieth-century physics at the University of Manchester in 2012, and has published articles on the history of physics with a focus on peer review, publishing practices and the relationship between the classical and the modern.

Connemara Doran (PhD, Harvard University) is a historian of science and technology, whose research focuses on the history and philosophy of the mathematical and physical sciences. She is currently a Postdoctoral Fellow at Harvard University, Department of the History of Science. Her research encompasses two areas in which the 'creative virtue' of physical and mathematical reasoning has been transformative: (1) the history of modern cosmology and (2) the history of energy resources. Her chapter in this volume draws from her book project on the intellectual adventure to empirically determine the size and shape of the universe—to conceptualise, measure and map the cosmos—during the long twentieth century. She refocuses discourse onto Henri Poincaré's odyssey to overcome the limitations of geometric empiricism by creating an *utterly new* mathematics of spatial relations.

Linda Dalrymple Henderson is the David Bruton, Jr., Centennial Professor in Art History and Distinguished Teaching Professor at the University of Texas at Austin. Professor Henderson's research and teaching focus on the interdisciplinary study of modernism, including the relation of modern art to fields such as geometry, science and technology, and mystical and occult philosophies. In addition to numerous periodical articles and catalogue essays, she is the author of The Fourth Dimension and Non-Euclidean Geometry in Modern Art (Princeton University Press, 1983; new ed., MIT Press, 2013) and Duchamp in Context: Science and Technology in the Large Glass and Related Works (Princeton, 1998). She also co-edited with Bruce Clarke the interdisciplinary anthology *From Energy to Information: Representation in Science, and Technology, Art and Literature* (Stanford University Press, 2002). In 2008 Henderson curated and wrote the catalogue for the exhibition *Reimagining Space: The Park Place Gallery in 1960s New York* (Blanton Museum of Art, The University of Texas, 2008). She is currently working on two book projects, 'The Energies of Modernism: Art, Science, and Occultism in the Early 20th Century' and 'The Fourth Dimension in Art and Culture Decade-by-Decade Through the 20th Century'.

Roberto Lalli is a Research Scholar at the Max Planck Institute for the History of Science and a Visiting Scholar in the Research Program on the History of the Max Planck Society. After having received an MSc degree in Physics, he earned a PhD in International History at the University of Milan (2011). From 2011 to 2013, he was a postdoctoral fellow in the Program on Science, Technology, and Society at the Massachusetts Institute of Technology. He is a historian of modern physical sciences whose work focuses on the interconnections between social and epistemic factors in the production and circulation of novel products in theoretical physics and in the international standardisation of scientific practices. He has published extensively on the history of relativity theories, on the transfer of quantum theory, and on the evolution of editorial practices. His monograph analyses the attempts to build an international community of general relativity experts during the Cold War. His current project concerns the development of new methodologies based on the concepts and tools of the network theory in order to jointly analyse the evolution of scientific knowledge in physics, the creation of transnational scientific communities, and the developments of scientific institutions.

Jaume Navarro is an Ikerbasque Research Professor at the University of the Basque Country. Trained in physics and in philosophy he has developed a career in the history of physics in institutions such as University of Cambridge, Imperial College and the Max Planck Institute for the History of Science. He is author of *A History of the Electron: J. J. and G. P. Thomson* (Cambridge: 2012) and co-editor of *Research and Pedagogy: A History of Quantum Theory and its Early Textbooks* (Berlin: 2015).

Richard Noakes is a Senior Lecturer in History at the University of Exeter. He has published widely on the history of physical sciences, psychical research, telecommunications and the representation of the sciences in nineteenth-century periodicals. He is the co-editor of *From Newton to Hawking: A History of Cambridge University's Lucasian Professors of Mathematics* (2003) and *Culture and Science in the Nineteenth-Century Media* (2004), and co-author of *Science in the Nineteenth-Century Periodical: Reading the Magazine of Nature* (2004). He is currently finishing a monograph, *Physics and Psychics: The Occult and British Sciences, 1850–1930.*

Arne Schirrmacher studied natural sciences and philosophy at the universities of Hamburg, Oxford and Munich. After a stay at the Max Planck Institute for the History of Science in Berlin, he became a long-term research scholar at the Deutsches Museum in Munich with projects on the history of physics, the history of twentieth century science communication as well as on the relations between science and war technologies. He was also an editor of the Hilbert Edition project at the University of Göttingen for David Hilbert's physics lectures, and a principal investigator in a research group on the history of quantum physics at the Max Planck Institute. Since 2010 he has been a senior research scholar at the Department of History of Humboldt Universität. In 2015 he won a Heisenberg Fellowship of the German Research Foundation.

Richard Staley wrote his dissertation at the University of Cambridge on the early work of Max Born in 1992, and then co-curated two museum exhibitions in the Whipple

Museum of the History of Science, 'Empires of Physics' and '1900: The New Age', before holding postdocs and visiting positions in Melbourne, Berlin and Chicago. From 2000 to 2013, he taught in the History of Science Department at the University of Wisconsin–Madison, publishing *Einstein's Generation and the Origins of the Relativity Revolution* with University of Chicago Press in 2008. He is now Rausing Lecturer in the Department of History and Philosophy of Science at the University Cambridge, conducting research on the histories of physics and anthropology, the cultural history of mechanics, and environmental physics and climate change.

Scott A. Walter teaches history and philosophy of science and technology at the François Viète Center for Epistemology and History of Science and Technology (EA 1161), University of Nantes. He serves as editor of the 'Henri Poincaré Papers' website, and has published two volumes of Henri Poincaré's scientific correspondence, devoted to exchanges with physicists, chemists and engineers (2007), and with astronomers and geodesists (2016). He has published research on the history of relativity theory in *Studies in the History and Philosophy of Modern Physics*, and he contributed a chapter on the historical origins of spacetime to the *Springer Handbook of Spacetime* (2014).

Michael H. Whitworth is a lecturer in the English Faculty, University of Oxford, and a tutorial fellow of Merton College, Oxford. He is the author of *Einstein's Wake: Relativity, Metaphor, and Modernist Literature* (OUP, 2001) and other articles and chapters on the relations of modernist literature and science. He was the co-founder of the British Society for Literature and Science. He has also written extensively on Virginia Woolf and has edited her novels *Orlando* and *Night and Day*.

Aaron Sidney Wright is a Postdoctoral Scholar at the Suppes Center for History and Philosophy of Science at Stanford University. Previously, he was a Social Sciences and Humanities Research Council Postdoctoral Fellow at the Department of History of Science at Harvard University. He received his PhD from the Institute for the History and Philosophy of Science and Technology at the University of Toronto. His research is concerned with the practice of physics over time, and what this practice tells us about science and the natural world. Particularly interested in 'historical ontology', he has made detailed studies of the work of theoretical physicists who studied the vacuum, or empty space.

1

Introduction

Ether—The Multiple Lives of a Resilient Concept

Massimiliano Badino and Jaume Navarro

1.1 ETHEREAL NARRATIVES

When did the ether disappear? Did it actually ever exist? Is it not back among some contemporary physicists? Are you trying to chase a ghost or to resuscitate a zombie? Questions like these were common in the academic events where the papers compiled in this volume were presented. Indeed, the 'Ethereal Aether', or 'the nothing that connects everything', to quote only two titles of books devoted to the history of the ether,[1] has often attracted the attention of historians of science, philosophers and physicists, and its epistemic and ontological status been the subject of much discussion and speculation. While the electron became the favourite probe for theories of scientific realism in times past, the ether was a frequent case study of theories of scientific change. For quite some time, the story of its supposed rejection via the experiments of Michelson and Morley in the late nineteenth century reassured Popperians and was used by many physics teachers. Later on, the ether became a major symbol of the incommensurability between classical and modern physics, exemplifying the Kuhnian struggle between a young generation of relativists and quantum physicists and the stubbornness of conservatives and crackpots. More recently, social historians of science have moved away from the categories of belief and unbelief in determining the nuances of the abandonment of the ether in the early twentieth century, bringing in elements such as technical or pedagogical use.[2]

In any case, there is a major consensus in thinking that the ether was mostly erased from any central role in physics and in culture after World War I. The best and most often quoted work on the history of the ether, the 1981 volume *Conceptions of Ether*, took for granted this timeline by limiting the study of *the History of Ether Theories* to the period 1740–1900.[3] The recurrent question 'what did Einstein *really* think of the ether' is only an exception to this rule, and even then the view is that the ether qua theory disappeared *in spite* of (some of) Einstein's opinions.[4]

Badino, M. and Navarro, J., 'Introduction: Ether—The Multiple Lives of a Resilient Concept' in *Ether and Modernity*, edited by J. Navarro. © Oxford University Press 2018. DOI: 10.1093/oso/9780198797258.003.0001

Certainly, in the narratives of the death, rejection or demise of the ether, Einstein and the theory of relativity play a central role. But, as the chapters in this book show, far from killing the ether off, special and general relativity (and, to a lesser degree, quantum physics) caused an explosion of ether narratives into different directions, and only a few of them can be traced back to a Manichean dichotomy between classical and modern. These stories cross each other; they share contexts, aims and actors, thus making it difficult to disentangle them into simple narrative strands. Thus, the absence of a chapter specifically devoted to 'Einstein and the ether' should not be understood in terms of Edmund Whittaker's old strategy to diminish Einstein's role in the development of relativity but as a way of emphasising the complexities of the ether in the early twentieth century.[5]

Rather, this book is a snapshot of the multiple lives of the ether in the first decades of the last century. From developments in pure mathematics to wireless technologies, from modernist art to spiritualism and from popular to alternative views of physics, the chapters in this book present us with an array of narratives that develop along several lines of fracture, none of which can give us a single coherent picture. Thus, far from consolidating the traditional dichotomies between classical and modern, between British and Continental, between material and spiritual and between 'true' and popular physics, this book challenges the explanatory role of these fissures. Indeed, these dichotomies are as many ways to tell the story of the ether: however, they fail to capture the whole story and it is in their failure that one can catch a glimpse at the peculiar features of this concept. Thus, the goal of this introduction is to follow some of these lines of fracture and to unearth the dynamics underlying the place of the ether in the early twentieth century.

1.2 CLASSICAL AND MODERN

The most obvious way to tell the story of the ether in the twentieth century is perhaps along the line of fracture that separates classicality and modernity. It is very tempting to make use of the revolutionary watershed, which includes both relativity and quantum theory, in order to safely locate the ether within the realm of 'classical physics' as opposed to a newer and advanced 'modern physics'. The narrative of the ether would thus unfold as the battle of some diehards for hopelessly old-fashioned scientific practices and values and against the unstoppable tide of progress. But, as some historians of science have shown, rather than mere labels, 'classical' and 'modern' are extremely fluid, largely co-constructed and continuously renegotiated categories which transcend the boundaries of physics and belong to the public sphere at large.[6] For this reason, the central argument of this volume concerns the comparison of images of the ether in science and in culture with the explicit goal of tracking the different dynamics, timescales, and meanings with which the term 'ether' was appropriated by different communities. Only in this way does it become possible to map out the complex conceptual clusters hidden under the deceivingly simple notions of classicality and modernity. What is normally used as a sharp dichotomy reveals so many intersections, entanglements and overlaps that it was not inconsistent to regard something like the ether as classical and modern at the same time, although for different reasons in different places and by different actors.

Let us begin at the beginning, that is, with physics. In the early years of the past century, Einstein's relativity theory was viewed by many more as an opportunity to rethink the ether than a reason to reject it. In Chapter 5, Scott Walter argues that the foundations of electron theory rested on a plexus of dynamical and kinematic problems. On the one hand, the description of electromagnetic phenomena in different states of motion called for a redefinition of the terms 'space', 'time', 'speed of light' and 'reference frame'. On the other, the bare fact that electric charges were involved required a medium of propagation for electromagnetic waves. For all its cumbersomeness, the ether allowed for a unification of kinematics and dynamics: it was the supporting medium and the absolute frame of reference and it was the storage device of energy. The British versions of electron theory, described by Richard Noakes in Chapter 6, made good use of these features. In addition, the ether proved useful even when combined with the most advanced mathematical tools as shown by Poincaré's combination of group transformations and Langevin's velocity waves to obtain the Lorentz contractions. By contrast, Einstein only offered 'a pair of postulates, the logical consistency of which was suspect' and the hypothesis of the relativity of time, a suggestion that even a flexible mind like Poincaré's looked upon as uninviting.

In this context, it is scarcely surprising that some physicists were reluctant to abandon the ether completely and tried instead to rethink its role under the kinematic constraints of Einstein's relativity. This quest for a novel way to understand the dynamical foundations of the electrodynamics of moving bodies generated what Walter calls the 'conceptual drift of the concept of ether': a series of attempts at filling up with a renewed ether the void that Einstein's epistemological austerity had left behind. Deprived of kinematic properties, the ether simply became a *substratum*, the loftiest metaphysical concept a physicist's conscience could bear and deal with. It is within this framework that one has to look at the reception of Minkowski's work, which, slowly but steadily, became the veritable vehicle of acceptance of special relativity. Connemara Doran's novel step-by-step analysis of the road from Poincaré to Minkowski's geometrisation tools, provided in Chapter 4, helps understand this evolution.

Efforts to make relativity and the ether happy together continued during the 1910s and 1920s in Britain. In Chapter 12, Aaron Wright describes this process as a sort of 'domestication' of the ether, an approach towards modern physics which marked Dirac's physical education and eventually fed his understanding of the role of the ether. Einstein's 1905 famous conclusion that the ether was 'superfluous' contained a positive message: after all, relativity had not pronounced the ether non-existent. This alleged compatibility between the ether and relativity was one of the main weapons on the ether front. If the ether could be neither observed experimentally nor undermined theoretically, the only remaining option was to characterise it through its functions. This was the rationale behind many of the 'relabelling strategies' so clearly described by Wright in the opening sections of the chapter.

This functional reinterpretation of the ether is key to understanding its vicissitudes. Fortunately, relativity, particularly general relativity, offered plenty of conceptual space. Its unprecedentedly complicated mathematical architecture was far from being self-explanatory. On the contrary, several corners in its symbolic maze had to be interpreted and many of them were open to alternative interpretations. Einstein himself admitted this point in 1916, and in 1918 he explicitly stated that, according to general relativity, one had to ascribe physical properties to the empty space via specific components of the gravitational potential. As Richard Staley explains in Chapter 10, for Einstein, this mathematical prescription created the site for a new role of the concept of ether. This is just one instance, albeit possibly the most illustrious, of a line of thought that became popular especially in the British scientific and public discourse: relativity had not declared the ether dead and, in fact, it might even vindicate it eventually. Thus, for example, Ebenezer Cunningham drew a very straightforward conclusion from special relativity: if there are multiple inertial frames, then there should be multiple ethers. On the contrary, as partly seen in Chapter 2, by Imogen Clarke, and Chapter 8, by Jaume Navarro, Arthur Eddington was happy to assume only one collection of point events, and he saw in this symbolic representation of the physical world the same connecting function formerly accomplished by the ether. Another interesting instance is Poincaré's 'structural' reinterpretation, discussed by Doran in Chapter 4. According to Poincaré, the existence of the ether could be reduced to 'a natural kinship between all the optical phenomena'. From this point of view, which could safely be defined as ontic structural realism *avant la lettre*, the ether hypothesis was no longer true or false—it was simply more or less convenient.

These examples illustrate two deeply entwined points. First, the development of modern physics, relativity as well as the quantum, did not happen as a gestalt jump out of the classical world. Quite the contrary—new theories, concepts and perspectives were conceived in constant dialogue with tradition, which never really died out. It is by exploring the consistency of new ideas with regards to classical wisdom that the full extent of their revolutionary potential can be appreciated.[7] Second, and related, modern physics is not a complete overcoming of the classical one, as the narrative mentioned at the beginning of this section would have it.

The conceptual space for the ether lies at the very heart of the dialectic between classic and modern: in what modern science does not tell us—in what it does not bother to specify, determine and interpret. Dirac, as discussed in Chapter 12, represents another illuminating example of this entanglement between the classic and the modern. For him, a reasonable relativistic or quantum theory of a system should be based, via analogy, on a corresponding classical theory of the same system, a point critically reiterated in his 1951 theory of electron.[8] Precisely this intuition led him from a new classical theory of the electron to formulating an ether for quantum electrodynamics as a field of velocities that a charge sitting in a certain point would pick up. Once again, there is a specific mathematical object, the electromagnetic potential, signalling the presence of the ether, but, as Wright perceptively observes, Dirac did not confine himself to carving out a space for the ether: he showed how a classical concept could still contribute to solve the riddles of modern science.

For many experts, relativity and the ether could coexist peacefully and even fruitfully but, for others, they were mutually exclusive. In Chapters 7 and 10, Arne Schirrmacher and Richard Staley, respectively, explore in different ways the politically tinged support for this claim in Germany and we shall return to this issue later in this introduction, in Sections 1.3 and 1.4. For the moment, however, we would like to focus on the interesting case of Dayton C. Miller, thoroughly discussed by Roberto Lalli in Chapter 9. The conceptual flexibility of the 'modern ether of relativity', so to speak, contrasted with stricter experimental demands, namely a very precise value—a round zero—for the ether-drift experiments. As Miller promptly noted, no such exact result had ever been attained, thus leaving some space for the ether. Moreover, the very possibility of performing Michelson–Morley-like experiments presupposed the existence of the ether. These two points made a powerful effect on the American physical community, which had built its fortune especially on physical astronomy and recognised the ether as an indispensable tool to make sense of their daily practices. This explains why the ether-drift experiments, promptly if not fully satisfactorily accommodated in classical physics, were paradoxically resuscitated by relativity.

This experimental branch of the story turned the survival of the ether into an all-or-nothing affair. But even in this case of head-on confrontation, the actors were not rejecting modernity as a whole. Instead, they were denouncing relativity as a wrong track, both scientifically and culturally. This suggests that the dialectic of modernity should not be viewed as a clash between blocks, but rather as the result of crafty political and rhetorical tactics aiming at formulating the most effective alignment of concepts, values and emotions.

One can find excellent examples of this tactic in the public discourse on the ether. The first key factor to understanding the dynamics underlying the debate on the ether in the public sphere is the particular conjuncture of European society and culture in the early decades on the twentieth century. New social actors and new political ideas shook the very foundations of the international order patiently put together during the nineteenth century. The appalling carnage of World War I and the tumultuous upheaval happening in Russia took the entire continent aback. Behind the lights and the dances of the belle époque, fear and uncertainty were the dominating feelings among Europeans. In this period of emotional instability, Oliver Lodge's plea for the ether could find more than one interested ear.

As Imogen Clarke and Michael Whitworth duly stress in Chapters 2 and 3, respectively, Lodge's communication prowess was almost unmatched. In his capacity as scientific expert and public figure, he managed to make his points heeded both at physics meetings and in the social arena. More importantly, he knew how to get at the layman's heart. In an age in which the cleavage between the common man and official science was rapidly increasing, Lodge was able to create a sense of proximity between people and the ether. The ether was still with us—relativity had never really expunged it and it could still be seen at work as he was broadcasting his speeches on the BBC. The ether embodied all those cognitive values of intelligibility, visualisability and intuitiveness that the mathematical abstruseness of modern physics wanted to obliterate. It rested on a perfectly logical analogy: in the same way that water waves need water, so electromagnetic waves need an electromagnetic medium to propagate. Furthermore, the ether was indispensable for the good face of modernity: as Chapter 8 shows, the association of the ether with wireless technologies gave the former a clear seal of modernity and the latter a reassuring sense of stability and continuity. While relativity and quantum physics were useless for understanding technology, let alone improving it, the ether, by contrast, played a central role. The magic of the wireless, which allowed people from the opposite ends of the earth to speak as if together in the same room, made perfect sense once the electromagnetic waves were pictured as perturbation of an all-pervasive medium. As the output of a type of physics now in jeopardy, the wireless technology literally embodied scientific as well as cultural values, both modern and traditional. It is unsurprising, therefore, that the wireless community became instrumental in sustaining the concept of the ether well beyond its survival in official science.

In Lodge's capable hands and with the cooperation of the wireless practitioners, the ether became a highly symbolic notion. It obviously related to electromagnetic theory, a chapter of science in which Britons had been unquestioned masters, from Faraday to Maxwell, William Thomson, Joseph J. Thomson and, why not, Lodge himself. And it also related to the wireless technology and Britain's imperial supremacy in electrifying and communicating the world. The ether, in sum, symbolised the cherished values of the classical British civilisation: stability, intelligibility and common sense. And, while it opposed all the uncomfortable novelties of the new century—moral uncertainties, social changes, Bolshevik revolutions, Freudian abysses and cerebral mathematical theories—it was modern in a good way: it expressed sensible progress and dynamism; it had a future in science as well as in society. The public image of the ether was neither classical nor modern but a cunning combination of both.

1.3 KNOWLEDGE, ETHOS AND AUTHORITY

One of the most serious flaws of the standard narrative on the ether is that it sets the entire issue in terms of a choice between scientific theories. By making the question depend on the intrinsic cognitive qualities of the two options, relativity theory and the ether hypothesis, it regroups all the nuances of the debate into two broad categories: the former were necessarily *right*, and the latter, *wrong*. By contrast, this book shifts the focus from the theories to the actors of this debate, to retrieve their distinctive voices, backgrounds and agendas. In this way, we can now appreciate elements previously neglected by historians. One major example is the theme of authority, which permeates several articles.

In order to be an influential contributor to a debate, one needs to gain sufficient authority; but—and this is the particularly intriguing aspect—in the ether debate, authority is a peculiar mixture of knowledge and ethos. Dirac, a Nobel laureate, was wittingly provoking the scientific community in 1951 by connecting his well-respected work with the anachronistic notion of the 'æther', in his battle for a new formulation of quantum electrodynamics. Although the attempt did not meet with great acceptance, Dirac not only used his authority and knowledge to bring back some form of ether but also used the authority of the ether, so to speak, to create a debate on the foundations of quantum electrodynamics.

Oliver Lodge and Arthur Eddington are two major examples of this entanglement between ethos, knowledge and authority. In Chapter 3, Whitworth describes with commendable precision Lodge's techniques to approach his public in order to foster not only a sense of expertise but also moral trust. By showing wisdom rather than just knowledge and by helping the reader (or the listener) to co-create the space for the ether, Lodge becomes the personification of those ideals of stability, reasonableness and Christian patience which were associated with the Victorian culture and of which the ether was an expression. And as Clarke argues in Chapter 2, Lodge and Eddington were very effective in convincing the general public that there was a genuine scientific discussion on the ether and that his was an active voice in this discussion. Although different on most points, Lodge's and Eddington's argumentative strategy included using their scientific authority to promote a critical attitude towards the infallibility of science. By claiming that, on the ether problem, the jury was still out, they achieved multiple results: they presented themselves as humble men with no pretension to know everything, they challenged some mainstream interpretations of relativity and they sympathetically nodded to their privileged audiences—the wireless community in the case of Lodge, mathematical idealists in the case of Eddington, and a broad array of liberal Christians in both.

This interplay of ethos and knowledge is key to understanding the complex dynamics of scientific and popular culture in the early decades of the twentieth century, but it also helps clarify the tectonic shifts within the scientific culture itself, particularly along both the geographical and the disciplinary fracture lines. During the nineteenth century, these differences were increasingly painted in critical terms and, often, related to specific *mentalities, spirits, cultures* and *ethos*. The criticism levelled by William Thomson and Peter Guthrie Tait against the Continental abuse of mathematics, for instance, and Pierre Duhem's famous distinction between the French and the British 'minds' are examples of this train of thought. The debate on the ether was not immune from these geographical undertones. As Noakes argues in Chapter 6, Lodge blamed Germany's materialist, decadent and ultimately dangerous morals on a philosophy that neglected, among other things, the ether. After all, the infection of modern science had started in Germany, the fatherland of both relativity and quantum theory. Given the peculiarities of the German thought and ethos, this genesis was not surprising, and the result had to be properly domesticated before any British attempt of appropriation.

But, on the other side of the Channel, they thought it differently. Notoriously, Philip Lenard exposed relativity as contrary to the German—to wit, Aryan—spirit, using his authority as a Nobel laureate to challenge Einstein's views and even to accuse him of plagiarism; he also fought a bitter struggle against J. J. Thomson and George Simpson, whom he regarded as careless and unfair experimenters. On several occasions, as Schirrmacher describes in Chapter 7, Lenard contrasted his rigorous and systematic approach to experimental physics with the sloppy attitude of the Britons. Much to his dismay, the latter were also more successful in terms of publication policy and accessibility, a feature that only made Lenard more furious.

The case of Lenard shows how national differences often intersect disciplinary boundaries. But the issue was more complex. Relativity and its mathematical maze were not born within the experimental domain. With the new strange player in the scientific community, the theoretical physicist, doubts could be legitimately raised that the theoretician's mathematical dreams could simply lead too far. Unlike those of the experimenter, the theoretician's constraints were vague and shaky: for the latter, common sense was a burden, a relic of a past scientific ethos. And the ether played on both sides of the divide: for some, it was a theoretical figment of the imagination, but for others it was the most experimentally grounded object in the fight against the excesses of modern theoretical physics. Miller rebelled against this drift of modernity, but his story reveals an ironic twist. His painstaking efforts to obtain precise checks of the predictions of relativity, brilliantly described by Lalli in Chapter 9, eventually came to a close via the experiments of Georg Joos, who was not only a German but also a representative of a new—and modern—experimental physics. Eventually, the ultimate application of the experimenter's ethos made the experimenter himself a dispensable player in the scientific game.

1.4 MATTER AND SPIRIT

In nineteenth-century science, the ether served two main functions: it was the supporting medium for electromagnetic radiation and it was the storage for the energy no longer available for human usage. Each function required the ether to be a material entity of a very peculiar nature. It had to be rigid enough to allow the extremely fast propagation of transversal waves, thin enough not to slow down the planets moving through it and capable somehow of keeping the dispersed energy until, at some point in time, it could be released. None of these formidable properties was ever made consistent with the others: they were all justified by conceptual requirements internal to the electromagnetic theory. But, for many, the unresolved tension between the several properties of the ether was more of an opportunity than a hindrance. An entity of undefinable, amphibious features, the ether could not simply be an ordinary part of the physical domain: rather, it could be the bridge between our world and the ultramundane. The ether became an obvious resource to the resurgent spiritualism of the second half of the nineteenth century.

A late convert to this trend, Oliver Lodge, became a powerful advocate of this culture, placing himself in the British tradition of natural theology. Lodge could easily inscribe his own agenda in this unfinished project as the rhetorical strategies unfolded by Clarke, Whitworth and Noakes show. Lodge's favourite argument that science does not have all answers was meant in the broadest way possible. The verdict against the ether was not carved into stone, not only because science always reassesses its results but also because there were superior goals in intellectual inquiry, goals which science could not avoid. Thus, the reminder that the universe was still for the better part mysterious and 'unseen' was not only the expression of a healthy epistemological modesty but also the evocation of the spiritual subtext accompanying the scientific discourse. It is precisely within this framework that Lodge could parallel electron theory and psychic research: both had been mysterious at one time but were now 'concrete and tractable'.

On this emotional front, World War I also played an important role. As Lodge, Lenard and many others had lost their beloved ones in the trenches, the need to believe that the cruel carnage was not the ultimate end grew stronger. This powerful thought, which Lodge condensed in his book *Raymond*, could not find a more sympathetic audience: the storage power of the ether worked for electricity and the still-mysterious soul equally well.

In Britain, the connection with the project of natural theology gave to the discourse on the ether a characteristic psychical and almost religious tinge. By contrast, in other areas of Europe, the tension between matter and spirit presented itself in cultural, political and artistic varieties. Lenard's political usage of the ether as a brand of *Deutsche Physik* was part of a powerful neo-Romantic movement whose declared aim was to recover German *Naturphilosophie* and a more authentic relation with the spiritual side of nature. Aggressively opposed to the ruthless industrialisation and the blind scientificity of the modern world, the upholders of this movement sang the praise of a mystical connection between natural environment, people and spiritual tradition. This was the notorious concept of the *Volk*, a profound combination of blood, soil and an omnipresent 'ether' connecting individuals in a spiritual unity.[9]

The connection between occultism and radical political ideas is old. In his book, *Mesmerism and the End of the Enlightenment in France*, Robert Darton superbly described how mesmerism—a practice later supported by the Society for Psychical Research—got mixed with Rousseau's ideas of a primeval human energy and eventually became a vector of radical politics during the Enlightenment.[10] In the 1920s, many German intellectuals moved along a similar track. Inspired and supported by the indefatigable Eugen Diederichs, the publisher of a massive edition of Meister Eckart's works, the self-proclaimed Free German Youth organised mountain walks and gatherings at which Madame Blavatsky's theosophy was discussed in a lively manner. What the Free German Youth had in common with other spiritualistic movements with which it was in contact was its struggle against the damaging aspects of modernity, together with a firm belief in an all-pervading medium. One Herbert Reichstein even claimed in the 1920s that the first Aryan was created by a shock from an electro-spiritual ether. However, the Free German Youth was unique in that it stressed the connection between science and art. In its pretension to explain everything, modern science progressively neglected the inward eye and turned man into a cosmopolitan producer of material goods. The path to counter this decline was forcefully outlined by Julius Langbehn—one of the most influential minds in the Free German Youth—in his *Rembrandt als Erzieher* (also published by Diederichs):

Germans must be turned into artists, and science must be turned into art. Only through an artistic relation with the world would it become possible to recover the 'spirit' and 'possession of such a spirit meant recalling that which was truly genuine, the Germanic past, as opposed to modern and evil rationalism'.[11]

Science, spiritualism and art were thus intertwined in a project of cultural renewal. However, the relations with modernity could be wavery. For the intellectuals of the Free German Youth, art had primarily an antimodernist meaning. Fidus (the pseudonym of Hugo Höppener), arguably its most representative artist, declared overtly that his work originated from a connection with the ultramundane world; but, in terms of content, he was committed to retrieving the legacy of traditional Germany. By contrast, in Chapter 11, Linda Henderson shows very clearly that Umberto Boccioni was attempting a very daring synthesis between spiritualism and modernity via a reconfiguration of the notion of matter. For Boccioni—not by chance an enthusiastic reader of theosophical writings himself—the storage function of the ether, on which Lodge's occultist arguments rested, became the vehicle to cross the borders between ether and matter. Unlike cubism, in this case the mixture of past, present and future was not in the memory but in the bodies themselves: matter was ether condensed, and ether was matter vaporised. Hence, matter embodied the elan of the spirit, to use an expression of Bergson (whose works were published in German, once again, by the energetic Diederichs).

Although the appropriation of modernity changed importantly from one current to another, a common pattern seems recurrent, to wit, the rejection of purely rational thought and the search for alternative ways of accessing reality, possibly a deeper reality. In this regard, some chapters in this book demonstrate that sensory perception (at times opportunely extended) and aesthetic intuition maintained a central role. The aesthetic approach to science, as much as the aesthetic approach to life, aspired to attain a superior form of knowledge, one that could not be easily conceptualised, but could serve as an effective guide for daily tasks, scientific or otherwise. Doran, Staley and Wright insist, with different nuances, on this theme. It is less relevant that, for Poincaré, Mach and Dirac, the path to the unity and beauty of this world view passed through the symmetries of mathematics or those of sensorial experience; what is relevant is that this form of knowledge was immediate, intuitive, irreflexive and, in some sense, spiritual.

1.5 EPISTEMOLOGY AND EMOTIONS: A PLEA TO PLURALISM

Unsurprisingly, we have circled back to knowledge. The philosophical stories we began with insisted that the ether was the idle wheel of electromagnetic theory.[12] From the point of view of an epistemology that focuses exclusively on what leads to successful predictions, this claim is hardly questionable: the ether as such certainly did not directly contribute to Maxwell's equations. This epistemology, in turn, whispered in the historians' ear the leitmotiv we see at work in many standard narratives: once Einstein awoke the scientific world from its ethereal slumber, the ether remained an option only for the nostalgic, the passé, the left behind. However, the chapters in this book show that the methodological dichotomies these narratives traditionally hinged on—classic versus modern, progress versus conservatorism, continuity versus discontinuity—should not be taken to be as clear-cut as they used to be, because the discourse on the ether crosses these distinctions transversally. It is not by chance that Einstein is more evoked than dealt with in this book, for the really interesting question is not 'Why were people so stubborn to stick to an outdated and blatantly false idea?', but rather 'What made the ether such a resilient concept?'. This novel question forces us to jettison the hegemonic positivistic picture of knowledge and assume a pluralistic point of view. The answer is to be found in the complex epistemological landscape and in the structure of feeling, as Raymond Williams used to call it,[13] that allowed the ether to survive in the wake of relativity.

If we explore this landscape more carefully, we are immediately struck by the richness and variety of the discourses on scientific knowledge in which the ether can be found. For example, as Whitworth shows in Chapter 3, Lodge's literary ingenuity helped him create a sense that the ether was physically present—and made good use of it in wireless telegraphy—and no abstruse mathematics could convince us to the contrary. The idea that we might have a sense for the ether, present among some British writers, might be traced back to William Thomson's presidential address at the Birmingham Midland Institute (3 October 1883), 'The Six Gates of Knowledge'.[14] Famously, Thomson argued that knowledge always entered through the senses which, supplemented with the 'sense of force', accounted for the six gates of the title, and he did not discard the possibility that other senses—perhaps even a sense of the ether—might at some point be discovered.

If we expand our gaze beyond the Channel, we find that both in *Der Analyse der Empfindungen* (1886) and in *Erkenntnis und Irrtum* (1905), Ernst Mach repeatedly used his own body as a probe to explore the external world, although he added an important twist to the common-sense empiricism of the British tradition. For Mach, experience was constituted by 'elements', a notion importantly larger than just sensorial data:

> Perceptions, presentations, volitions, and emotions, in short the whole inner and outer world, are put together, in combinations of varying evanescence and permanence, out of a small number of homogeneous elements. Usually, these elements are called sensations. But as vestiges of a one-sided theory inhere in that term, we prefer to speak simply of elements.[15]

Mach was opening up a world in which the transcendental subject of the Enlightenment progressively lost its role of legislator of nature.[16] The boundaries between subject and object became as blurry as those between matter and spirit and the ways to access knowledge expanded to positively encompass emotions, desires and volitions. From this perspective, mere experimental undetectability mattered very little. As Staley and Schirrmacher show, the permanence of the ether in the Germanic culture was related to its capability to serve epistemological functions much higher than that served by the lab.

Ultimately, it is the adaptability of the ether to multiple cultural contexts, and the epistemological pluralism that underlies this capability, that the rich diversity of the essays in this volume capture so effectively. Thus the obvious question is: what is it in the epistemological fabric of the concept of ether that allowed it to live so many lives well beyond its early proclaimed redundancy? A preliminary answer, which is more a collection of thoughts for further research than a fully fledged response, is that the ether may be a sort of *interstitial concept*. Let us close this introduction by elaborating this thought.

The reorganisation of knowledge, leading to a deep scientific revolution, always leaves epistemic interstices. One obvious example is the famous phenomenon of the 'Kuhn losses'.[17] Thomas Kuhn argued that, when a new paradigm takes over, it might happen that problems previously well understood in the old paradigm suddenly become intractable. In other words, paradigms always leave epistemic gaps that must be filled with resources external to them. By the same token, relativity made the ether kinematically superfluous, but it left behind the problem of making sense of electromagnetic events occurring in the vacuum, as Dirac would point out. It is hardly surprising, then, that many authoritative physicists tried to fill this interstice by reconfiguring the concept of ether.

Analogous interstices can crack open at the interface between science and culture. The chapters in this book provide numerous stories of appropriation of the ether by different communities. The wireless practitioners, the occultists, the Italian futurists, the German experimental physicists, and so on, had questions about which modern science only had useless, irrelevant or even unpleasant answers, if at all. By contrast, the ether could comfortably live in all these spaces made available by the imperfect interconnection between science and culture. As an interstitial concept, the ether was plastic and pliable enough to be adapted to diverse contexts, because it was no longer a specific object but rather a multidimensional concept able to serve a number of epistemic, symbolical, social, political, emotional, moral and even scientific functions, some of which, in contrast, were perfectly in tune with modernity. But such an extreme flexibility had, of course, a downside. Sitting at the interstices between multiple discourses, the ether was not integral to any of them and was not autonomous. Hence, it had to be sustained by the continuous effort of authoritative figures energetically acting in the public sphere. This, in turn, generated a complex dynamics of alliances, negotiations and strategies reaching out a considerable variety of debates. It is precisely its interstitial nature that makes the ether such an effective entry point in that thorny juncture of modernity that was the beginning of the twentieth century.

NOTES

1. Lloyd S. Swenson, Jr, *The Ethereal Aether. A History of the Michelson–Morley–Miller Aether-Drift Experiments, 1880–1930* (Austin: University of Texas Press, 1972); Joe Milutis, *Ether. The Nothing that Connects Everything* (London: University of Minnesota Press, 2006).
2. Andrew Warwick, *Masters of Theory. Cambridge and the Rise of Mathematical Physics* (Chicago: Chicago University Press). See also Graeme Gooday and Daniel J. Mitchell, 'Rethinking "Classical

Physics"', in Jed Z. Buchwald and Robert Fox, eds., *The Oxford Handbook of the History of Physics* (Oxford: Oxford University Press, 2013), 721–64.

3. Geoffrey N. Cantor and Michael J. S. Hodge, *Conceptions of Ether. Studies in the History of Ether Theories, 1740–1900* (Cambridge: Cambridge University Press, 1981).
4. Ludwik Kostro, *Einstein and the Ether* (Montreal: Apeiron, 2000).
5. Edmund Whittaker, *A History of the Theories of Aether and Electricity*, vol. 2 (London: Longmans, 1953).
6. On this point, see Richard Staley, 'On the Co-Creation of Classical and Modern Physics', *Isis*, 96 (2005): 530–58.
7. On this process of exploration of the relation between classical and modern ideas in physics, see Jochen Büttner, Jürgen Renn and Matthias Schemmel, 'Exploring the Limits of Classical Physics: Planck, Einstein, and the Structure of a Scientific Revolution', *Studies in History and Philosophy of Modern Physics*, 34 (2003): 37–59. For a pedagogical take on the same issue, see Massimiliano Badino and Jaume Navarro, 'Pedagogy and Research: Notes for a Historical Epistemology of Science Education', in Massimiliano Badino and Jaume Navarro, eds., *Research and Pedagogy: A History of Quantum Physics through its Textbooks* (Berlin: Edition Open Access, 2013), 7–30.
8. On the role of analogy in the emergence of quantum physics see Olivier Darrigol, *From c-Numbers to q-Numbers: The Classical Analogy in the History of Quantum Physics* (Berkeley: University of California Press, 1992).
9. See e.g. George L. Mosse, 'The Mystical Origins of National Socialism', *Journal of the History of Ideas*, 22 (1961): 81–96, and George L. Mosse, *Nazi Culture: Intellectual, Cultural, and Social Life in the Third Reich* (Madison: University of Wisconsin Press, 1966).
10. Robert Darton, *Mesmerism and the End of the Enlightenment in France* (Cambridge, MA: Harvard University Press, 1968).
11. Mosse, 'The Mystical Origins of National Socialism', p. 88.
12. See e.g. Philip Kitcher, *The Advancement of Science* (Oxford: Oxford University Press, 1993), pp. 143–9.
13. Raymond Williams, *Marxism and Literature* (Oxford: Oxford University Press, 1977).
14. William Thomson, 'The Six Gates of Knowledge', in *Popular Lectures and Addresses*, vol. 1 (London: Macmillan, 1891), 260–306.
15. Ernst Mach, *The Analysis of Sensation* (New York: Dover, 1959), p. 22.
16. On this point, see Jürgen Habermas, *Knowledge and Human Interests* (Boston: Beacon Press, 1971), pp. 81–90.
17. Thomas S. Kuhn, *The Structure of Scientific Revolutions* (Chicago: University of Chicago Press, 3rd edn, 1996), pp. 103–10.

2

The Ether at the Crossroads of Classical and Modern Physics

Imogen Clarke

The matter of which I have been speaking so far is the material which builds up the earth, the sun, and the stars, the matter studied by the chemist, and which he can represent by a formula; this matter occupies, however, but an insignificant fraction of the universe, it forms but minute islands in the great ocean of the ether, the substance with which the whole universe is filled.

J. J. Thomson, Address of the President of the British Association for the Advancement of Science, 1909.

2.1 INTRODUCTION

On the evening of 25 August 1909, inside the Walker Theatre in Winnipeg, Canada, the esteemed British physicist J. J. Thomson opened the seventy-ninth meeting of the British Association for the Advancement of Science. Thomson's presidential address, an overview of current developments in physics, focused largely on the ether, which he declared to be 'not a fantastic creation of the speculative philosopher' but rather 'as essential to us as the air we breathe'.[1] Some three decades later, in his 1942 biography of Thomson, Lord Rayleigh criticised this approach, noting that his subject had not taken the 'fashionable view about ether' but instead displayed a 'robust confidence in its reality and necessity' in the face of mounting evidence to the contrary.[2]

Looking back on the physics of the first half of the twentieth century, the twenty-first century reader is inclined to agree with Rayleigh's assessment. Against the backdrop of relativity theory, quantum mechanics and radioactivity, Thomson's commitment to a decidedly nineteenth-century theory certainly appears outdated. And yet, at the time of the Winnipeg speech Thomson was Director of Cambridge University's Cavendish Laboratory, an institution renowned for its extensive role in the development of a new style of modern physics during this period. Under Thomson's helm, Cavendish physicists moved away from a focus on precise measurement of physical constants and towards investigations into the fundamental structure of matter, theoretical speculation and the

Clarke, I., 'The Ether at the Crossroads of Classical and Modern Physics' in *Ether and Modernity*, edited by J. Navarro. © Oxford University Press 2018. DOI: 10.1093/oso/9780198797258.003.0002 favouring of visual assessments of experiments over metrical precision. Where previous studies of matter had investigated it in relation to the ether, there was here a move from the macroscopic to the microscopic, as Cavendish researchers attempted to uncover the internal structure of rays and atoms. Thomson himself was famed for his turn-of-the-century 'discovery' of the electron, a steadfast symbol of modern physics. Nevertheless, he maintained throughout his life a commitment to the ether, a cornerstone of classical physics.[3]

How do we reconcile these two aspects of J. J. Thomson's life, the classical and the modern? The complexities of Thomson's career have been discussed and picked apart, with it now being established that Thomson himself did not conceptually connect the 'corpuscles', for which he found experimental evidence, with existing theories of the electron.[4] But Thomson could also be well served by a broader shift in how we frame our histories of early twentieth-century physics, a move away from a traditional approach that divides physics and physicists into classical and modern.[5] This dichotomy obscures much of the complexities of the period, often resulting in an incomplete picture, with only half the story being told. Indeed, Rayleigh's account of Thomson's life and career is a prime example of such failings, with the ether scarcely mentioned in the biography of a man who saw it as fundamental to his life's work.[6] With recent scholarship challenging the dichotomy of classical/modern, now is an opportune time to re-evaluate the ether's position in the twentieth century.[7]

In this chapter, I aim to liberate the ether from its historiographical assignment to classical physics, and consider its role in debates surrounding the future of the discipline. I begin by looking at the discussions underway in professional spaces between 1909 and 1914, and suggest that a physicist's commitment to the ether does not classify them as a 'classicist' but rather as an advocate of continuity in the discipline. I then examine the ether's 'popular' life following the well-publicised 1919 eclipse expedition, and the subsequent expository efforts by the 'classical' Oliver Lodge and 'modern' Arthur Stanley Eddington. In doing so, I move beyond the classical–modern divide, and suggest a more substantial role for the ether in professional and popular early twentieth-century British physics.

2.2 THE ETHER AND DISCONTINUITY, 1909–14

When thinking about the rapid changes that took place in physics in the early years of the twentieth century, it seems as though a more helpful dichotomy than classical and modern is that of continuity and discontinuity. The great continuous medium was under threat by a new conception of nature as formed of discrete matter and energy. With large changes potentially taking place in the discipline, the broader meaning of continuity was also in play. Would physics progress in continuous steps, or through a discontinuous shift resulting in the abandonment of long-held concepts? And within this, what would be the fate of the ether? There was a continuum of positions on the nature and existence of the ether, and a range of approaches to the role of the old physics in the changing discipline.

Perhaps the most vociferous British opponent of the ether in these early years was the Cambridge-trained experimental physicist Norman Campbell. Campbell's attitude towards physics, developed while working in Thomson's Cavendish from 1902, involved the rejection of ideas that could not be experimentally verified.[8] His 1907 textbook *Modern Electrical Theory* characterised the Cambridge traditions of mathematical and experimental physics as retrospectively 'old physics' and 'new physics'.[9] While mathematical physicists clung to old ideas, hindering the prospect of progress, Campbell believed the new experimentalists held no such deep commitments and were thus best placed to move forward. Campbell became interested in relativity theory within the context of his ongoing project to describe the electrodynamics of moving bodies without recourse to either the ether or mathematics. In 1909 he wrote in the *Proceedings of the Cambridge Philosophical Society*:

> The trend of modern theory is everywhere to replace by discontinuity the continuity which was the basis of the science of the last century. Any method which is especially applicable to discontinuous processes is certain to be fruitful of results in every department of investigation, and any considerations which can be advanced in the elucidation of such a method are not devoid of value.[10]

The idea of the physical world as fundamentally continuous, crucial from an ethereal point of view, was further challenged in the coming years. In 1911, Ernest Rutherford published a paper detailing his nuclear model of the atom, with the model consisting of a central charge surrounded by a cloud of electrons. The world was now dramatically filled with empty space, although it was entirely conceivable that the particles of matter could still be connected by the ether. Two years later, however, Niels Bohr incorporated the concept of discontinuous energy into Rutherford's model by proposing that a small, positively charged nucleus was orbited by electrons that could jump from a higher energy orbit to a lower one, emitting a quantum of discrete energy as it did so. The concept of quantum radiation had now been applied to a theory of atomic structure, combining discontinuous matter and discontinuous energy.

Campbell praised Bohr's theory in a 1914 *Nature* article on the structure of the atom, supporting it on the basis of its accordance with experimental results. For Campbell, Bohr's theory represented a sharp divide between two groups of physicists:

> There are only two alternatives open to the modern theoretical physicist: he may either suppose that the principles of the older mechanics are true, and that all the brilliant results which have followed from the application of the conceptions of Planck and Einstein to the most diverse phenomena are illusory and devoid of evidential value; or he may suppose that they are not true. Bohr's theory offers him the choice in its most striking form.[11]

Campbell had also recently published a second version of his *Modern Electrical Theory*, which again argued against the need for an ether in physical theories.[12] This was reviewed in *Nature* by the chemist and radioactivity researcher Frederick Soddy, who noted the following:

> Physical theories at the present moment are so shaky at the foundations that the doubt arises sometimes whether the superstructure is not being built up too rapidly. The difficulties, now ten years old, in reconciling the undulatory and corpuscular types of radiation in one theory, the hopeless confusion that prevails as to the necessity for the existence of an ether, and the modern discrete or quantum theory of energy, seem to call for a more drastic reconsideration than we find here of many of the simplest physical conceptions and their experimental basis.[13]

For Soddy, Campbell's rejection of many physical principles was still too conservative. As a chemist, he held no deep commitment to physical theories and was instead concerned with analysis of chemical properties and changes. He was also not one to bow down in the face of tradition more generally, and campaigned for economic, social and institutional reform in the 1920s and 1930s.[14] There was perhaps here a connection between his attitude towards rejection of long-held theories in physics, and rejection of tradition more generally. For Soddy, where a system was no longer working, be it economical, social or physical, the answer often lay in dramatic reform. Progress could not necessarily be achieved through small measures but rather required profound conceptual shifts.

Of the many physicists who opposed such profound conceptual shifts regarding the status of the ether, Lodge was perhaps the most vocal. He also benefited from a considerable platform, as the president of the 1913 British Association for the Advancement of Science. As Thomson had done in 1909, Lodge used the event as an opportunity to set forth a defence of the ether, in front of a large audience in his home territory of Birmingham, and an even larger readership of the subsequent published account.[15] Speaking for an hour and a half, Lodge criticised what he called 'modern tendencies' in science, including the current 'irresistible impulse to atomise everything', which he saw as not only a problem in physics, with subatomic particles and quantum radiation, but also in biology, with the emergence of Mendelian heredity. He also accused modern science of denying the existence of anything which could not be readily sensed or measured. To make his point absolutely clear, Lodge titled his talk 'Continuity', here referring to continuity of matter and energy, but also continuity of thought, a link between the past and the present, a physics that would hold onto the ether and Newton's laws.[16]

The problem of continuity and discontinuity with regard to quantum developments was further addressed two days later in a debate on radiation. James Jeans, a recent convert to quantum theory, led the debate, framing the discussion in terms of continuity versus discontinuity, and placing himself firmly on the side of discontinuity. He declared that perhaps the 'boldest and simplest attempt at reconciliation between the conflicting theories lies in abandoning the ether altogether, and relying on some purely descriptive principle, such as that of relativity'.[17] This was also the view taken elsewhere by O. W. Richardson, who concluded that, if relativity was indeed a 'Universal Principle', then it followed that the ether was a 'superfluous hypothesis'.[18] Jeans was committed to a scientific method that began with certain premises and from them deduced valid knowledge.[19] He would not pragmatically adopt a theory simply on the basis of its current usefulness in explaining experimental results. Thus, he was more than willing to discard the old, providing the new had solid and viable foundations.

Other British physicists in the room took an opposing view. The mathematician Augustus Love refused to accept that 'existing theories of dynamics and electrodynamics need to be supplemented by the theory of the quanta', instead proposing that recent results could be interpreted within 'ordinary theories'.[20] There was a contribution from Joseph Larmor, whose book *Aether and Matter* had been influential in early twentieth-century Cambridge pedagogy.[21] He commented on the new work relating to specific heats at very low temperatures, suggesting that 'there is nothing in it that is destructive to the principles of physics which have led to so rich a harvest of discovery and synthesis in the past'.[22] He looked for reconciliation between the old and new ideas, continuing his ongoing search for interactions between the ether and electrons. Furthermore, he used the rhetoric of destruction, clearly placing the new ideas in the context of a dramatic revolution, and emphasising the nature of discontinuity. Meanwhile Lodge stayed quiet, but did invite the committee of Section A of the British Association to continue the debate at his house the following Sunday.[23]

The discussion was written up in *The Times* by E. E. Fournier d'Albe, then an assistant lecturer in physics at Birmingham.[24] D'Albe was a member of the Society for Psychical Research and had fairly radical views on what the 'new' physics could teach about the nature of the soul. He believed that the discrete nature of matter, as revealed by the electron, provided evidence for continuity of life after death.[25] D'Albe was also deeply committed to the ether, and had opposed relativity theory on the basis of a perceived threat to it. He thus had much in common, intellectually and institutionally, with Lodge. He did not, however, share Lodge's opposition to quantum theory, hoping instead that 'the investigation of this fascinating problem will teach us a great deal about the inter-stellar aether which conveys the messages'.[26]

In *The Times*, D'Albe described the debate as a 'pitched battle between the adherents of the doctrines of Young and Fresnel, Maxwell and Hertz on the one hand, and the revolutionary followers of Planck, Einstein, and Nernst on the other'. Ultimately, it was an 'old controversy' between continuity and discontinuity. While the battle was still 'raging', d'Albe noted that opinion seemed to be in favour of the quantum theory. Detailing the views of the opposition, he described Love as fighting 'with conviction for the older and more conservative view', while Larmor was 'somewhat pathetically seeking the way of salvation through the falling *debris* of cherished views'.[27] As d'Albe gleefully reported on a battle between conservatism and revolution, he did not, despite being an adherent to the ether, appear too threatened by the choice being made. Indeed, d'Albe, who omitted a comment by James Jeans about abandoning a dynamical approach, did not necessarily see a fundamental change occurring. With the ether and the quantum theory entirely compatible in his understanding of physics, d'Albe could support both the 'classical' and the 'modern'.

Not all 'modern' physicists were as embracing of change as d'Albe. A colleague of both Lodge and d'Albe, Samuel McLaren was a graduate of Trinity College and worked as a mathematics lecturer at Birmingham University from 1906 to 1913. McLaren was keenly interested in the 'new physics', as he was able to grasp the complex mathematics underpinning it, and also struck up a friendship with Niels Bohr during the Danish physicist's 1911 visit to England.[28] McLaren was thus not unreceptive to new ideas, but he did take issue with the more destructive consequences of some interpretations. Writing in the *Philosophical Magazine* (for which Lodge was an editor), McLaren accused 'Einstein's idea of the Quantum' of being 'destructive of the continuous medium and all that was built upon it in the nineteenth century'. McLaren's desire to retain the continuous medium of the ether was more than simply a commitment to a physical principle. He began his article by declaring that 'the unrest of our time has invaded even the world of Physics, where scarcely one of the principles long accepted as fundamental passes unchallenged by all'. The problem was not simply the discontinuity of energy, but rather the discontinuity of progress, of physics proceeding not by gradually building upon the work of those who had gone before, but by tossing old theories aside and replacing them with wildly different ones. And this predicament was not exclusive to physics. Indeed, quite the opposite was true: McLaren believed that a more general 'unrest of our time' had infected physics, and referred to a 'spirit of revolution'.[29]

Through McLaren and Lodge, we see a strong connection between discontinuous matter and energy, and discontinuous progress. In addition, McLaren's reference to the 'spirit of revolution' reveals a sense that physics was not alone in facing the challenge of discontinuity. Indeed, the years surrounding 1913 saw equivalent developments in art and literature, with cubist paintings and stream-of-consciousness novels breaking up time and space. Here, as with science, long-standing authorities were being challenged (see Chapter 11).[30] And it was within this age of revolution that threats to the ether would start to be more widely discussed. As the twentieth century advanced, conversations about the ether, and about the nature of matter, the universe and physics more generally, began to take place in more popular spaces. In British popular physics in the 1920s, two key actors would appear: Arthur Stanley Eddington and Oliver Lodge. With these physicists later coming to personify modern and classical physics, respectively, their shared and divergent popular communications about the ether shed considerable light on the nature of this apparent divide.

2.3 EDDINGTON VERSUS LODGE: THE ETHER AND MODERN PHYSICS

In May 1919, two teams of astronomers travelled to the Brazilian municipality of Sobral and the island of Principe, off the west coast of Africa, in order to measure the deflection of starlight during an eclipse. As the well-known story goes, the astronomers set out to obtain observational evidence of the general theory of relativity, proposed by Einstein in 1916. The results, in favour of Einstein's theory, were presented at a joint meeting of the Royal Society and Royal Astronomical Society in November 1919, widely reported in the British press and led to increased popular interest in physics. The Cambridge astronomer Arthur Stanley Eddington, whose overwhelming support for Einstein's theory has subsequently led to (now largely debunked) criticisms of his scientific objectivity, played a major role in planning the expedition and carrying out the measurements, as well as promoting the endeavour more widely.[31] Eddington and other members of the Joint Permanent Eclipse Committee (JPEC) framed the expedition as a crucial experiment, with a potential 'trichotomy' of results: a full deflection of 1.74″, confirming Einstein's predictions; a half-deflection, in line with Newton's theory of gravitation; or no deflection at all, requiring an entirely new understanding altogether.[32] Using connections at *The Times*, the JPEC were able to lead a successful 'publicity campaign' during and after the expedition, culminating in *The Times*'s announcement of the results under the heading 'Revolution in Science'.[33]

In the tale of the eclipse, Lodge plays a bit part, that of an antagonistic sceptic of Einstein, one who storms out of the joint meeting when it becomes clear that Einstein has emerged victorious. (Lodge subsequently maintained he had a 'long-standing engagement and a 6 o'clock train'.)[34] I would argue, however, that Lodge played a significant role in how both the expedition and physics were subsequently popularised. Take, for example, one of the more esoteric newspaper accounts of the eclipse expedition, published in the *Aberdeen Press and Journal*, a daily Scottish newspaper, on Tuesday, 11 November 1919. Headlined 'A Great Discovery. May Reduce Physical Labour To A Minimum', the article announced a 'sensational discovery' that 'will, it is believed, give the clue to the nature of ether, and that, in turn, may provide the clue to the way of utilising the tremendous energy of ether, which would give mankind control of forces of power and energy undreamt of'. Predicting a future with physical labour reduced to a minimum, the author noted that Lodge, while urging caution against 'generalising as to the great possibilities founded upon the discovery', had admitted that 'a great and splendid result has been achieved, capable of far-reaching possibilities'.[35]

It seems likely that the author of this report had paid less attention to the initial announcement in *The Times* than to Lodge's letter to the paper the following day. Under the heading of 'The Ether of Space. Sir Oliver Lodge's Caution', Lodge described the eclipse result as 'a great triumph for Einstein', but issued 'a caution against a strengthening of great and complicated generalisations concerning space and time on the strength of the splendid result: I trust that it may be accounted for, with reasonable simplicity, in terms of the ether of space'.[36] Nowhere in this short letter did Lodge refer to hitherto undreamt of stores of energy, but he had suggested this in an article in *Philosophical Magazine* earlier that year.[37]

There was little mention of the ether in the promotional reports surrounding the eclipse expedition. Eddington did not see general relativity as having implications for the ether, but rather viewed the theory as an exciting development that would create new opportunities for physical theories to be tested by astronomers and open further avenues for research into the large-scale structure of the universe.[38] But it took Lodge little more than a single letter to *The Times* for the entire endeavour to be reframed (in one Scottish newspaper anyway) as being about the ether (and about Lodge). And the recently retired Lodge did not shy away from media attention in the subsequent years, working to centre the ether in discussions about the new physics. In the magazine *Nineteenth Century and After*, Lodge warned the physics world to not take the implications of the results too far and to not 'be revolutionary to a rash and hasty extent', arguing for an interpretation of the new results in terms of a 'generalisation' of the old theories, a generalisation that would hold on to the ether.[39] Writing in the *Fortnightly Review* in 1920, in an article unambiguously titled 'The Ether Versus Relativity', Lodge insisted that his late nineteenth-century experiments needed to be repeated and that 'extraordinary and expensive means' were required to detect the extremely slow speed of the ether stream. Lodge saw 'no reason why a National Laboratory should not undertake such an experiment'.[40]

Meanwhile Eddington was now in the position of trying to temper the narrative of revolution that he had partially created, while also promoting the success of Einstein's theory. Writing in the *Contemporary Review* in late November, he mocked the hyperbolic headlines that had appeared in newspapers: 'REVOLUTION in Science—Newton and Euclid Dethroned—Bending of Light—The Fourth Dimension—Warping of Space!' He accused such judgements of being perhaps 'too hasty', but admitted that the 'fundamental nature of the change has not been exaggerated'.[41] He also attempted to lay to rest any claims of Newton's overthrow, arguing that Newton had in fact predicted, in his *Opticks*, that light could bend. Furthermore, he ended by declaring that it was 'not necessary to picture scientists as prostrated by the new revelations, feeling that they have got to go back to the beginning and start again. The general course of experimental physics will not be deflected, and only here and there will theory be touched.'[42]

By the time the 1920 British Association meeting came about, relativity theory had been discussed at length. Perhaps as a result, Eddington used his presidential address to Section A to talk about an entirely different topic, the internal constitution of stars.[43] The astronomical journal, the *Observatory*, later lamented an absence of relativity theory at the meeting, suggesting this was 'because those chiefly concerned had become a little jaded with the strenuous conflict'.[44] Lodge, however, does not appear to have suffered from this, and a large crowd gathered to hear him deliver a 'Controversial Note on Relativity'.[45] Again admitting that Einstein's equations were supported by experimental observations, Lodge went on to argue that some interpretations of these equations were 'threatening to land physicists in regions to which they had no right of entry', into metaphysical reasoning 'beyond their ken'.[46] He disagreed with any attempts to 'build up on an equation an elaborate metaphysical structure', arguing that such equations were open to numerous interpretations. Perhaps most damaging to those physicists who were trying to disassociate themselves from revolution, Lodge ended his talk by suggesting that relativists should perhaps 'be regarded as Bolsheviks and pulled up'.[47] The press duly responded to Lodge's successful courting of controversy, and the Bolshevik sound bite was reported in *The Guardian*, the *Daily Telegraph* and the *Daily News*.[48]

Eddington did tackle relativity theory at the 1921 British Association meeting, in a highly anticipated talk attended by apparently nearly 2,000 members of the association.[49] *The Times*'s account of the event reported Eddington labelling the ether as an 'idle hypothesis, unsupported by experiment and giving explanations of nothing'. The article ended with Lodge's contribution to the discussion:

> Sir Oliver Lodge, proposing a vote of thanks to Professor Eddington for his delightful address, said that he was not yet prepared to abandon the ether. Professor Eddington, he added, was so immersed in the doctrine of relativity that he thought it was self-evident. He was like the cricketer who on being asked to explain a 'yorker' retorted that it was just a yorker. Einstein's relativity theory was of great mathematical interest, but the general interest taken in it throughout the world was entirely due to *The Times*.

As a result of Lodge's final remark, this report was titled 'Relativity or Ether?' and subtitled 'Sir Oliver Lodge and *The Times*', allowing the ether enthusiast to hijack Eddington's address. With neither physicist presented as more authoritative than the other, the framed question of 'Relativity or Ether?' had no clear answer in this report.

The Times was a long-standing faithful reporter of the annual British Association events, and no doubt had sent a scientific reporter to cover this meeting, one already aware of ongoing discussions relating to the impact of relativity theory on the ether. Local newspapers, however, reported the event quite differently, with little mention of the continuous medium. A variety of similar reports began by noting Eddington's notoriety as the only physicist who could intelligibly explain Einstein's theory, and ended with a brief mention of Lodge. However, Lodge's explicit mention of the ether was absent in these reports, as was Eddington's clear dismissal of the medium. These newspapers referred to Eddington's suggestion that there were some physicists who believed velocity through the ether did not exist, but this remark was not highlighted as important, nor accompanied by any suggestion that the ether itself was in question.[50]

Wherever Eddington could be found vigorously championing relativity theory, it seems that Lodge was rarely far behind. A well-known public figure, even a minor contribution from him was likely to be reported, injecting scepticism into any discussions of the new physics. As a result, he was able to entirely reframe how lectures were reported, steering the conversation onto the ether. In addition to 'hijacking' Eddington's lectures, Lodge was also starting his own conversations elsewhere, and was a prolific popular writer and lecturer throughout the 1920s. In 1924, Ernest Benn published Lodge's book *Atoms and Rays*. Here, Lodge detailed 'current' knowledge about matter, discussing the structure of the atom, quantum theory and the nature of energy. Throughout, he described the ether as the fundamental 'cementing substance' that held everything together and was responsible for the transmission of energy.[51] In 1925, Lodge wrote the opening chapter of *Phases of Modern Science*, a publication produced alongside the second showing of the Royal Society's 'Pure Science' exhibit at the 1924 British Empire Exhibition. In Lodge's chapter 'Radiation', he informed the reader of the 'most usual view of radiation' as waves in the ether.[52]

That same year, Lodge delivered his sweeping account of science in the modern age, in a series of talks on 'Atoms and Worlds', broadcast over the wireless in October and November 1925. Subsequently published in Benn's Sixpenny Library series as *Modern*

Scientific Ideas: Especially the Idea of Discontinuity, the book, resembling a small pamphlet, came to only seventy-nine pages, a factor that, coupled with its affordable price, may have attracted readers daunted by heftier tomes. Within the pages, Lodge put forward his interpretation of modern science, which he saw as in flux. He made clear, in the introduction, that 'continuity remains the fundamental idea to which scientific philosophy will in the last resort return', but spent most of the book detailing the new and many ways that matter and energy had been found to be discontinuous. Thus, 'it must be admitted that modern science is in a rather complicated, though very interesting, stage'. Scientists had not yet 'attained full knowledge' and were 'encountering a number of facts the full explanation of which will need some generations of work on the part of our leaders'.[53] Lodge's book was, in his mind, the presentation of a nascent and unfinished subject, science so cutting edge that we did not yet know the answers.

Lodge's views on the ether were not unchallenged by his peers. *Atoms and Rays* was reviewed in *The Observer* by Edward Andrade, Professor of Physics at the Artillery College in Woolwich. Andrade, who had written his own book on the *Structure of Atom* (a treatise lacking any reference to the ether), advised his readers that Lodge was 'rather unorthodox... in his constant reference of everything back to the ether'. He remarked that physicists had barely any knowledge at all about the ether, and knew simply that, as Einstein had shown, it 'has not got any mechanical properties, which rather spoils its usefulness'. However, overall, Andrade recommended Lodge's book, providing the reader be careful to differentiate 'the certain' from 'the less certain'. He praised Lodge's 'freshness, charm and polished simplicity of style', and described the 'great skill and enthusiasm' with which Lodge discussed the quantum theory.[54]

In *Discovery*, a popular science magazine, *Atoms and Rays* received a very positive anonymous review, with the declaration that to 'the student of Physics, as well as to everyone who is interested in Physical Science, the appearance of a new publication by Sir Oliver Lodge is always a memorable event'.[55] *Modern Scientific Ideas* was also favourably reviewed, in *Nature*, with reference to Andrade's contribution to the same series, on *The Atom*, in which the ether-sceptic scientist depicted nature as increasingly more atomic and discontinuous.[56] The reviewer noted that the two books covered similar ground, and that it would thus be 'of considerable interest to note the varying manner of treatment of the same material by two decidedly individualistic writers'.[57] Notably, Andrade's book was not depicted as being more accurate, or even 'better', than Lodge's, but simply different. In a review in *Discovery*, the two books were presented as complementary.[58] The author V. E. Pullin, Director of Radiological Research at the Royal Arsenal in Woolwich (an institutional neighbour of Andrade's Artillery College), described *Modern Scientific Ideas* as 'an excellent preamble' to these more specialised books, providing an overview of modern physics. He declared that to 'acclaim Sir Oliver Lodge as an expounder of modern science would be to gild the lily'.[59]

Furthermore, the mere appearance of Lodge's chapter in *Phases of Modern Science* was a sign of tacit approval for his views, placing them at the very beginning of a reference work accompanying an ambitious, state-funded celebration of the British Empire.[60] Indeed, Lodge was very deliberately chosen for this role, appointed Vice-Chair of the organising committee (despite making it very clear he would be too busy to attend any meetings) so that he could contribute to the book.[61] Similarly, Lodge was invited to contribute to *Nature*'s special relativity issue in 1921, appearing alongside such experts as Eddington, Dyson, Crommelin, Jeans, Lorentz and even Einstein himself.[62] He was afforded an additional air of expertise by his acquaintanceship with biologist and science journalist Peter Chalmers Mitchell, who wrote a weekly column in *The Times* on the 'Progress of Science'. Mitchell, who corresponded frequently with Lodge on the subject of spiritualism, evidently viewed the older physicist as a reliable authority on physics, with Lodge's name cropping up regularly in Mitchell's column.[63]

Outside of *The Times*, smaller papers followed suit. Lodge's 1923 Silvanus Thompson memorial lecture, 'The Origin or Basis of Wireless Communication', was reported in *The Western Daily Press*, with the header 'The Ether Indispensable'. The paper quoted Lodge's warning not to be misled 'by any apprehension of the theory of relativity into supposing that that theory dispenses with the ether merely because it succeeds in ignoring it'. Lodge went on to say that 'the leaders in that theory were well aware that for anything like a physical explanation of light or electricity or magnetism, or cohesion or gravitation, the ether was indispensable'.[64] In 1925, the *Dundee Evening Telegraph* reported on Lodge's radio talk 'Vibrations and Waves and What They Signify'.[65] Notably, the chosen headline for this report was 'Deep Secrets about Ether'; the existence of ether was not questioned in the article. The *Nottingham Evening Post* reported Lodge's 1928 Kelvin Lecture on 'The Revolution in Physics', in which Lodge noted that he personally saw the current situation as the beginning of 'the real theory of the ether'.[66]

While the lead-up to the 1920s began with concerted efforts by Eddington and others to promote their expedition and the theory of relativity, which they believed it had decisively proved, these efforts were somewhat derailed by a rogue Lodge. For, while Eddington was often introduced as the primary British expert on Einstein, he did not usurp Lodge as the expert on *all* physics, including (crucially) modern physics. Thus, Lodge was able to repeatedly inject the ether into any conversation about relativity, before more widely promoting his views on the place of the ether in modern physics. As a result, the ether may have started the 1920s as a classical medium under threat of obsolescence at the hands of modern physicists but, by the end of the decade, it could still be interpreted as part of the future of the discipline. Lodge's popularisation work was not viewed as a desperate defence of classical physics but rather an authoritative account of the new. The ether maintained a healthy popular existence, and a sense of relevance, up until the end of the decade. And with Lodge's voice among the most widely heard, the nuances between different kinds of ethers were likely lost to a popular audience.

2.4 CONCLUSION

In 1928, Eddington's Gifford Lecture of the previous year was published as the bestselling *Nature of the Physical World*, which by 1944 had sold 80,000 copies worldwide.[67] Now with arguably a larger platform than even Lodge, Eddington was able to set forth his views on the ether to a wide popular audience. His first chapter was entitled 'The Downfall of Classical Physics', apparently the first attempt by a British scientist to actually define the category of 'classical physics'.[68] Here he compared Einstein and Minkowski's drastic altering of our concept of time and space, and Rutherford's nuclear model of the atom, proposing that the latter was in fact the more dramatic. Eddington found it strange that physicists would accuse relativists of Bolshevism (perhaps a reference to Lodge's 1920 British Association talk), but happily accept 'the dissolution of all that we regard as most solid into tiny specks floating in void'.[69] Furthermore, this most revolutionary development was entirely compatible with the concept of an ether, for, thanks to Rutherford, 'we now realise that the aether can slip through the atoms as easily as through the solar system'.[70]

While Eddington was 'inclined to think that Rutherford, not Einstein, is the real villain of the piece', his definition of classical physics placed the former, not the latter, in the category. However, Eddington tackled the impact of Einstein (and thus 'modern physics') on the ether in depth in a later discussion of relativity theory. He noted that with the theory of relativity 'evidently bound up with the impossibility of detecting absolute velocity', the ether was called into question, as motion with respect to the ether is equivalent to absolute motion. Furthermore, with no ethereal frame having been found, the notion of velocity through the ether is meaningless. However, he emphatically stated: 'This does not mean that the aether is abolished. We need an aether.'[71]

Without the ether, Eddington explained, the physical world would be analysed into isolated particles of matter or electricity with featureless interspace. The role of the ether was thus to provide this interspace with character, much as 'we postulate matter or electricity to bear the characters of the particles'. Abolishing the ether would be equivalent to abolishing matter. Certainly, the ether could no longer be conceived of as a kind of matter, but Eddington suggested that this view had ceased to be orthodox at some point in the nineteenth century, and certainly before the advent of relativity theory. While our conception of the ether had changed, in particular the idea of velocity through the ether, the ether itself was 'as much to the fore as ever it was'.[72]

Eddington's ether was, of course, not the same ether as that proposed by Lodge. However, the two men held a shared belief, that an ether played an important role in modern physics. This they defended passionately to popular audiences, with the form of their ether differing considerably, but the extent of their belief largely on a par. As with Lodge, Eddington's ether was vital in maintaining continuity in physics, providing a medium that connected together the increasingly discrete building blocks of nature, while also forming a bridge between the past, present and future of the discipline. Throughout the early twentieth century, adherence to the ether displayed not a 'classical' but a continuous approach to physics.

NOTES

1. J. J. Thomson, 'Address', in *Report of the Seventy-Ninth Meeting of the British Association for the Advancement of Science* (London: John Murray, 1910), 3–29, p. 15.
2. Lord Rayleigh, *The Life of Sir J. J. Thomson* (Cambridge: Cambridge University Press, 1942), p. 161.

3. On Thomson and the Cavendish, see Jaume Navarro, *A History of the Electron: J. J. and G. P. Thomson* (Cambridge: Cambridge University Press, 2012); Isobel Falconer, 'J J Thomson and "Cavendish" Physics', in Frank A. J. L. James, ed., *The Development of the Laboratory*, (London: Macmillan, 1989), 104–17; S. B. Sinclair, 'J. J. Thomson and the Chemical Atom: From Ether Vortex to Atomic Decay', *Ambix* 34 (1987): 89–116.
4. George E. Smith, 'J. J. Thomson and the Electron, 1897–1899', in Jed Z. Buchwald and Andrew Warwick, eds., *Histories of the Electron: The Birth of Microphysics* (Cambridge, MA: MIT Press, 2001), 21–76; Isobel Falconer, 'Corpuscles, Electrons and Cathode Rays: J. J. Thomson and the "Discovery of the Electron"', *The British Journal for the History of Science* 20 (1987): 241–76.
5. See e.g. Helge Kragh, *Quantum Generations* (Princeton: Princeton University Press, 1999); David M. Knight, 'Classical Physics', in *Public Understanding of Science: A History of Communicating Scientific Ideas* (London: Taylor & Francis, 2006), 167–181; Jochen Büttner, Jürgen Renn and Matthias Schemmel, 'Exploring the Limits of Classical Physics: Planck, Einstein, and the Structure of a Scientific Revolution', *Studies in the History and Philosophy of Modern Physics* 34 (2003), 37–59, p. 37, in their response to Thomas S. Kuhn, *Black-Body Theory and the Quantum Discontinuity, 1894–1912* (Oxford: Oxford University Press, 1978).
6. On Thomson and the ether, see Navarro, *History of the Electron*.
7. Imogen Clarke, 'Negotiating Progress: Promoting "Modern" Physics in Britain, 1900–1940' (PhD dissertation, University of Manchester, 2012); Richard Staley, 'On the Co-Creation of Classical and Modern Physics', *Isis* 96 (2005): 530–58; Graeme Gooday and Daniel Mitchell, 'Rethinking Classical Physics' in Jed Z. Buchwald and Robert Fox, eds., *The Oxford Handbook of the History of Physics* (Oxford: Oxford University Press, 2013), 721–64.
8. AndAndrew Warwick, 'Cambridge Mathematics and Cavendish Physics: Cunningham, Campbell and Einstein's Relativity 1905–1911. Part II: Comparing Traditions in Cambridge Physics', *Studies in History and Philosophy of Science* 24 (1993): 1–25.
9. Norman Robert Campbell, *Modern Electrical Theory* (Cambridge: Cambridge University Press, 1907).
10. Robert Norman Campbell, 'The Study of Discontinuous Phenomena', *Proceedings of the Cambridge Philosophical Society* 15 (1909): 117–36, p. 117.
11. Robert Norman Campbell, 'The Structure of the Atom', *Nature* 92 (1914): 586–7, p. 586.
12. Norman Robert Campbell, *Modern Electrical Theory* (Cambridge: Cambridge University Press, 1913).
13. Frederick Soddy, 'Modern Physical Ideas and Researches', *Nature* 92 (1913): 339–40, p. 339.
14. Jeff Hughes, '"Divine Right" or Democracy? The Royal Society "Revolt" of 1935', *Notes and Records of the Royal Society* 64, Supplement 1 (2010): S101–17. For Soddy in general, see Linda Merricks, *The World Made New: Frederick Soddy, Science, Politics, and Environment* (Oxford: Oxford University Press, 1996); George B. Kauffman, ed., *Frederick Soddy (1877–1956): Early Pioneer in Radiochemistry* (Dordrecht; Lancaster: Reidel, 1986).
15. In addition to the official British Association report, Lodge's talk was also published in book form by J. M. Dent.
16. Oliver Lodge, 'Continuity', in *Report of the Eighty-Third Meeting of the British Association for the Advancement of Science* (London: John Murray, 1914), 3–42.
17. 'Friday, September 12. Discussion on Radiation', in *Report of the Eighty-Third Meeting*, 376–86, p. 380.
18. O. W. Richardson, *The Electron Theory of Matter* (Cambridge: Cambridge University Press, 1914), p. 325.

19. Matthew Stanley, 'So Simple a Thing as a Star: The Eddington–Jeans Debate over Astrophysical Phenomenology', *British Journal for the History of Science* 40 (2007): 53–82.
20. 'Friday, September 12', pp. 383–4.
21. AndAndrew Warwick, *Masters of Theory: Cambridge and the Rise of Mathematical Physics* (Chicago: University of Chicago Press, 2003).
22. 'Friday, September 12', p. 386.
23. British Association. 'Problems of Radiation. Modern Universities and the State. Electric Heating and Cooking', *The Times*, 13 September 1913.
24. D'Albe wrote two anonymous reports of the discussion: 'British Association. Improvement of British Canals. Incubation of Eggs in Egypt', *The Times*, 13 September 1913, and 'British Association. Problems of Radiation. Modern Universities and the State. Electric Heating and Cooking', *The Times*, 13 September 1913. In both cases, d'Albe wrote only the paragraphs relating to the radiation discussion (information from *News International* archivist).
25. Richard Noakes, 'The "World of the Infinitely Little": Connecting Physical and Psychical Realities Circa 1900', *Studies in History and Philosophy of Science Part A* 39 (2008): 323–34, p. 328.
26. D'Albe's views on relativity theory and the quantum were expressed in Edmund Edward Fournier d'Albe, 'The Radiation Problem', *Nature* 92 (1914): 689–91, p. 690.
27. 'British Association. Improvement of British Canals. Incubation of Eggs in Egypt'.
28. McLaren is discussed in Alex Keller, 'Continuity and Discontinuity in Early Twentieth-Century Physics and Early Twentieth-Century Painting', in M. Pollock, ed., *Common Denominators in Art and Science: The Proceedings of a Discussion Conference Held under the Auspices of the School of Epistemics, University of Edinburgh, November 1981* (Aberdeen: Aberdeen University Press, 1983), 100–2.
29. Samuel B. McLaren, 'The Theory of Radiation', *Philosophical Magazine* (Series 6) 25 (1913): 43–56, p. 43.
30. Peter Gay, *Modernism: The Lure of Heresy: From Baudelaire to Beckett and beyond* (W. W. Norton, New York, 2008); Robert Wohl, 'Heart of Darkness: Modernism and its Historians', *Journal of Modern History* 74 (2002): 573–621; Christopher Butler, *Early Modernism: Literature, Music, and Painting in Europe, 1900–1916* (Clarendon Press, Oxford, 1994); Arthur I. Miller, *Einstein, Picasso: Space, Time and the Beauty that Causes Havoc* (Basic Books, New York, 2002).
31. Alistair Sponsel, 'Constructing a "Revolution in Science": The Campaign to Promote a Favourable Reception for the 1919 Solar Eclipse Experiments', *British Journal for the History of Science* 35 (2002): 439–67; Matthew Stanley, '"An Expedition To Heal the Wounds of War": The 1919 Eclipse and Eddington as Quaker Adventurer', *Isis*, 94 (2003): 57–89; John Earman and Clark Glymour, 'Relativity and Eclipses: The British Eclipse Expeditions of 1919 and Their Predecessors', *Historical Studies in the Physical Sciences* 11 (1980): 49–85.
32. The trichotomy is discussed most comprehensively in Earman and Glymour, 'Relativity and Eclipses'.
33. Sponsel, 'Constructing a Revolution'; 'Revolution in Science', *The Times*, 7 November 1919. While the 'Revolution in Science' article was not written or commissioned by the JPEC (and was instead the work of biologist and science journalist Peter Chalmers Mitchell), it was the indirect result of several months of promotion.
34. Oliver Lodge, 'The Ether of Space', *The Times*, 8 November 1919.
35. 'A Great Discovery. May Reduce Physical Labor to a Minimum', *Aberdeen Press and Journal*, 11 November 1919.
36. Oliver Lodge, 'The Ether of Space. Sir Oliver Lodge's Caution', *The Times*, 8 November 1919.

37. Oliver Lodge 'On a Possible Means of Determining the Two Characteristic Constants of the Æther of Space', *Philosophical Magazine* (Series 6) 37 (1919): 465–71.
38. Warwick, *Masters of Theory*, pp. 464–5. Eddington also saw this as an opportunity to foster internationalism, as a representative of Quaker pacifism: Stanley, 'An Expedition to "Heal the Wounds of War"'.
39. Oliver Lodge, 'The New Theory of Gravity', *Nineteenth Century and After* 86 (1919): 1189–201, p. 1196.
40. Oliver Lodge, 'The Ether versus Relativity', *Fortnightly Review* 107 (1920): 54–9, p. 58.
41. Arthur Stanley Eddington, 'Einstein's Theory of Space and Time', *Contemporary Review* 116 (1919): 639–43, p. 639.
42. Eddington, 'Einstein's Theory', p. 643.
43. A. S. Eddington, 'Address to the Mathematical and Physical Science Section', in *Report of the Eighty-Eighth Meeting of the British Association for the Advancement of Science* (London: John Murray, 1920), 34–49; for Eddington's astrophysical work, see Stanley, 'So Simple a Thing'.
44. 'Notes', *The Observatory* 44 (1921): 312–24, p. 321.
45. 'British Association', *Yorkshire Post and Leeds Intelligencer*, 27 August 1920.
46. Oliver Lodge, 'Popular Relativity and the Velocity of Light', *Nature* 106 (1920): 325–6, p. 325.
47. Oliver Lodge, 'Popular Relativity', p. 326.
48. 'British Association: Go-As-You-Please Schools. Sir Oliver Lodge and Einstein's Theory', *The Manchester Guardian*, 27 August 1920; 'Sir Oliver Lodge on Einstein's Theories', *Daily Telegraph*, 27 August 1920; *Daily News* quoted in 'Notes', *The Observatory* 43 (1920): 375–6.
49. 'Relativity or Ether? Sir Oliver Lodge and "The Times"', *The Times*, 13 September 1921.
50. 'Einstein's Theory', *Western Morning News*, 13 September 1921; 'The British Association', *Northern Whig*, 13 September 1921; 'Einstein's Theory Defined', *Western Daily Press*, 13 September 1921; 'Einstein's Theory', *Lichfield Mercury*, 16 September 1921; 'Einstein's Theory Defined', *Yorkshire Post and Leeds Intelligencer*, 13 September 1921; 'The Value of the Potato', *Aberdeen Press and Journal*, 13 September 1921; 'British Association', *Belfast News-Letter*, 13 September 1921.
51. Oliver Lodge, *Atoms and Rays: An Introduction to Modern Views on Atomic Structure and Radiation* (London: Benn, 1924), p. 12.
52. Oliver Lodge, 'Radiation', in Royal Society, *Phases of Modern Science: Published in Connexion with the Science Exhibit Arranged by a Committee of the Royal Society in the Pavilion of His Majesty's Government at the British Empire Exhibition, 1925* (London: Royal Society, 1925), 1–21, p. 3.
53. Oliver Lodge, *Modern Scientific Ideas* (London: Benn's Sixpenny Library, 1927), p. 13.
54. E. N. da C. Andrade, 'Books of the Day: The New Physics', *The Observer*, 10 August 1924.
55. 'Reviews of Books: Atoms and Rays', *Discovery* 5 (1924): 228.
56. Edward Neville da Costa Andrade, *The Atom* (London: Benn, 1927).
57. 'News and Views', *Nature* 119 (1927): 824–9, p. 827.
58. V. A. Pullin, 'Benn's Sixpenny Library: First Scientific Titles', *Discovery* 9 (1928): 163–5.
59. Pullin, 'Benn's Sixpenny Library', p. 165.
60. On the exhibition, see Donald R. Knight and Alan D. Sabey, *The Lion Roars at Wembley: British Empire Exhibition, 60th Anniversary 1924–1925* (New Barnet: D. R. Knight, 1984).
61. Lodge's appointment (and the article he is to write) is referred to in a letter from Thomas Martin (Secretary of the British Empire Exhibition) to F. E. Smith, 25 February 1925, File: F. E. Smith, British Empire Exhibition 1924 Correspondence, Royal Society Archives; The decision to appoint him came out of the Publications Subcommittee File: Publications Subcommittee, British Empire Exhibition 1924, Correspondence, Royal Society Archives; Lodge is listed as

Chair in *Phases of Modern Science.* Lodge's inability to attend meetings is referred to in Thomas Martin (Secretary of the British Empire Exhibition) to Lodge, 21 October 1925, British Empire Exhibition 1924, box 2, Royal Society Archives.

62. *Nature* 106 (1921).
63. On Mitchell, Lodge and *The Times* see Clarke, *Negotiating Progress*, pp. 134–5.
64. 'The Ether Indispensable', *Western Daily Press*, 2 February 1923.
65. 'Deep Secrets about the Ether', *Dundee Evening Telegraph*, 21 January 1925.
66. 'Epoch-Making Discovery in Physics', *Nottingham Evening Post*, 26 April 1928.
67. Michael Whitworth, 'The Clothbound Universe: Popular Physics Books, 1919–39', *Publishing History* 40 (1996): 55–82, pp. 67–8. Loren R. Graham, *Between Science and Values* (Vancouver: Columbia University Press, 1981); Matthew Stanley, *Practical Mystic* (Chicago: University of Chicago Press, 2007).
68. Gooday and Mitchell, 'Rethinking Classical Physics'.
69. Arthur Stanley Eddington, *The Nature of the Physical World* (Cambridge University Press, 1928), p. 1.
70. Eddington, *Nature of the Physical World*, pp. 3–4.
71. Eddington, *Nature of the Physical World*, p. 31.
72. Eddington, *Nature of the Physical World*, p. 32.

3

Transformations of Knowledge in Oliver Lodge's *Ether and Reality*

Michael H. Whitworth

3.1 INTRODUCTION

Oliver Lodge's *Ether and Reality*, published in 1925, presents an intriguing anomaly. Published in an era when popular writing on physics was dominated by attempts to come to terms with the new physics, particularly special and general relativity, it placed at its centre an epistemic object that had been disputed for many years. Moreover, by the end of the book, the ether remains a shadowy entity. Although *Ether and Reality* ostensibly presents itself as a popular science book, it aims to provide other kinds of knowledge. At points, particularly in Lodge's tracing analogies between natural phenomena, it seems quietly to continue the tradition of natural theology. A reading of it needs to be alert not only to scientific contexts, but to literary and biblical ones. In reading it, we need to ask how the reader is involved in producing the knowledges it conveys, and we need to ask how Lodge uses literary tropes and figures such as allusion and parallelism to shape his readers' responses.

It may be no accident that the ether remains elusive at the end of *Ether and Reality*. In his 1913 British Association lecture, which was subsequently published as the book *Continuity*, Lodge had encapsulated something paradoxical in his understanding of it: 'Matter it is not, but material it is; it belongs to the material universe and is to be investigated by ordinary methods.'[1] As Richard Noakes has remarked, Lodge wishes the ether to be something 'vastly different from ordinary matter' yet not wholly immaterial.[2] This paradoxical quality creates a problem of exposition. Ordinarily, exposition depends on relating unfamiliar concepts to more familiar ones, and the more familiar are usually material things that exist at a human scale and are perceivable by the unaided human senses. To explain the immaterial ether in this way is to risk reducing it to ordinary matter.

Ether and Reality began as a series of seven radio broadcasts on the BBC, on Tuesday evenings 9.30 to 10 pm, fortnightly in January to March 1925. As such they potentially reached an audience of over a million.[3] Four of the lectures were reprinted in the BBC's

Whitworth, M.H., 'Transformations of Knowledge in Oliver Lodge's *Ether and Reality*' in *Ether and Modernity*, edited by J. Navarro. © Oxford University Press 2018. DOI: 10.1093/oso/9780198797258.003.0003 listings magazine, the *Radio Times*, and in this form reached a readership of between 610,000 and 750,000.[4] The book was published by Hodder and Stoughton in May 1925 in their 'Broadcast Library' series at a price of 3s 6d: in the year to 31 March 1926, it sold 17,129 copies in the UK, and in the following ten years, assisted by a cheaper 2s 6d edition from 1929 onwards, it sold a further 7,295 copies.[5] Even if the book had not achieved this significant numerical reach, it would be important to study, not least because what it presents as scientific orthodoxy differs so greatly from the other influential and bestselling popular texts of the 1920s and early 1930s: A. N. Whitehead's *Science and the Modern World* (1925), A. S. Eddington's *The Nature of the Physical World* (1928) and James Jeans's *The Mysterious Universe* (1930). The nature of the exposition and the problems it entailed were very different from theirs. Moreover, from the outset Lodge was aiming at a large audience, whereas the works by Whitehead, Eddington and Jeans originated in elite academic lectures: the Lowell lectures, the Gifford lectures and the Rede lecture.

Martin Eger's account of the popular scientific book enables us to begin thinking of it as a genre, although Eger sees it as a late twentieth-century development and does not touch on its early twentieth-century antecedents. Eger claims that works in the genre have a certain structural 'rhythm' in common, one in which scientific exposition is sandwiched between philosophical discussions:

> In its pure form, the book begins with a philosophical survey, sometimes only an introduction, in which certain human or social problems are posed and related to some aspect of science; that science is then expounded in a lengthy technical or quasi-technical section, and the work ends with the original philosophical problems to which solutions or responses are now offered *on the basis of the scientific content just presented.*[6]

The structure of *Ether and Reality* anticipates this form, but the book articulates the 'human or social problems' less clearly than do Eger's examples. The clearest articulation comes in the preface, where Lodge notes the existence of 'material and spiritual worlds'; by proposing the ether as 'the connecting link' between them, he implies that the problem is one of disjunction. It should also be noted that the scientific filling of Lodge's sandwich is not as purely technical as those in the works described by Eger. Although Lodge presents what might be mistaken for a technical account of the ether and its functions, he eschews technical vocabularies and intersperses them with remarks about 'Higher Things' and the 'Unseen Universe'. In his conclusion, he returns explicitly to the 'Higher Things' and, in a final allusive phrase to which we shall return, he describes the ether as 'the living garment of God' (p. 179).[7]

Although this essay focuses primarily on the text and on Lodge's persuasive techniques, his text did not exist in a vacuum, and a high percentage of Lodge's readers would have been aware of his other works and would have had heard him broadcast. In a memoir, the BBC announcer Stuart Hibberd recalled that Lodge 'could talk simply, directly, effortlessly, and with great humility, in non-technical terms, about complex scientific subjects, giving the impression that he was taking you into his confidence and that you and he were sharing secrets together'.[8] As another characterisation of his style put it, 'he persuaded rather than lectured, shared secrets rather than imparted information'.[9] Building on the trust earned through his broadcasts, throughout *Ether and Reality* Lodge aims to build what several critics have termed 'ethos': if certain aspects of an exposition are to be taken on trust, it is important for the writer to establish his or her trustworthiness.[10] Part of that trustworthiness derives from his humility, but part derives from his recognised credentials as a scientist. I have noted elsewhere the extraordinary presentation of Lodge's credentials on the title page: twelve dense lines of type listing his degrees, honorary degrees and membership of scientific societies.[11] The credentials given are scientific: the title page tells us that Lodge was formerly President of the Physical Society, but not that he was formerly President of the Society for Psychical Research. The creation of ethos does not end with the title page and, throughout the text, Lodge is at pains to present himself as a man who is patient, cautious, experienced and wise; and as a man who is forgiving of others' errors and misapprehensions.

3.2 INVISIBLE ETHER AND TANGIBLE ENTITIES

In the central expository section, Lodge repeatedly connects abstract entities to concrete ones, familiar from everyday experience. One means by which he makes the connections is the redescription of the familiar world in less familiar terminology such as the language of 'vibrations' and 'fields'. In a long and lyrical passage in Chapter 1, immediately after casting doubt on the usefulness of 'our most delicate [scientific] instruments', Lodge invites the reader to put his or her hand in front of a 'blazing fire':

> It is not hot air that we feel; the air remains cool: it is not heated by radiation. What we feel is due to Ether vibrations: they excite the nerves of the skin, and give us the tingling sensation which we call heat or temperature. (p. 28)

A 'quiver of the Ether' can cause sunburn and sunstroke. The vibrations blacken photographic plates; plants photosynthesise because of it. Lodge leads us through their growth and death and their reuse as fuel: 'when you make a wood or coal fire you liberate the stored or dormant energy, the ether recovers it and gets it back again in perceptible form' (p. 29). A version of this text formed part of the second BBC broadcast in the series, on 20 January 1925, and a significant proportion of the audience would have been listening to Lodge in homes heated by open fires.[12]

A little further on he continues to relate the ether to tangible uses: 'We are using it now in radio telegraphy: but before that we used it to drive our streetcars and machinery: in America they use it for judicial killing' (p. 30). In all these cases, the abstract entity of the 'Ether vibration' is embodied in something familiar and, in many cases, homely. To say that we are using ether 'now' in radio telegraphy is to speak not only of the 'now' of the modern world but also of the 'now' of the original broadcast. As, in the idiom of the time, radio broadcasts were heard 'over the ether'; it was a stroke of simple genius to use the medium to persuade listeners of its reality. Another instance of the shifter 'now' appeared in the original broadcasts but was revised in the book. In *Ether and Reality* the passage reads thus:

> The waves emitted by a radio-aerial will get to the Antipodes, say New Zealand, in the 1/14th part of a second. (p. 44)

But, in the *Radio Times*'s text and, it is safe to assume, in the original broadcast, the sentence begins with 'The waves that we are now using will get to the Antipodes'.[13] Although Lodge does not labour the point, and did not need to, there are several other references to broadcasting (e.g. p. 164, p. 178). Although Lodge could not perform experiments in front of his audience, the 'now' of the original broadcast implies the medium between speaker and audience and, like the blazing fire, reconnects abstractions with the audience's immediate physical situation.

In one respect, Lodge's use of radio as an everyday illustration is anomalous: more typically, his examples do not reflect the visible modern world of the early twentieth century. Expositors of special and general relativity often drew their illustrative examples from distinctly modern technologies: elevators, trains and indoor cycle tracks.[14] Indeed, as early as 1920, J. W. N. Sullivan was able to speak mockingly of the trains in one exposition of general relativity: 'we are always meeting these trains in expositions of Relativity'.[15] By contrast, the everyday world of Lodge's illustrations would have been familiar in the mid-nineteenth century, if not centuries earlier. In his illustration of a particle disturbing its medium, a boat sends out ripples, and a bell vibrates the air (p. 40); in his illustration of one body acting upon another, a horse pulls a cart, the earth pulls down an apple, a golf club hits a ball, and a gun shoots a bird (p. 65). His examples of action at a distance are somewhat more modern: a letter or telegram; a railway signalman; the construction of the Forth Bridge; the leadership of World War I; and mosquitoes preventing the construction of the Panama Canal (pp. 80–1). The example from World War I is the most recent but, insofar as its organisation by bureaucracies remote from the front line resembles that of previous wars, the median date in these instances is still somewhere in the late nineteenth century.

3.3 THE IMPLIED READER OF *ETHER AND REALITY*

Recent work in relation to the popularisation of science has emphasised the active participation of the audience or readers: as Ralph O'Connor notes, audiences 'received the proffered knowledge in different ways and put it to their own uses, sometimes drawing—and in turn promoting—meanings far removed from the popularisers' original intentions, reshaping the science in the process'.[16] Indeed, what in 1990 Hilgartner called the 'dominant view' of popular science, that of an active elite scientist simplifying science for transmission to a passive, ignorant audience, can no longer be considered dominant within the history of science.[17] The difficulty comes in finding evidence for the complex interactions whereby science is reshaped in the wider culture. *Ether and Reality* was reviewed in some of the sixpenny weeklies and in other literary journals, although less widely than might be expected; in any case, reviewers are not typical readers, and reviews are a peculiar kind of reception evidence.[18] If *Ether and Reality* was actively reshaped in the literary sphere, such appropriations have been neglected, but the rectification of the omission is beyond the scope of the present essay. While we cannot observe Lodge's actual readers in the act of reading, it is possible to look at the interaction of Lodge with his *implied* reader. We can discern what the implied reader of *Ether and Reality* is expected to know and what mental processes they are expected to undertake. Judging by the analogies that Lodge introduces, the reader is expected to know about the behaviours of various material objects in the world, with a particular emphasis on those that cause vibrations or facilitate communication: things such as boats, bells and horse-drawn carts. The reader comes to the text with a knowledge of the earth's atmosphere and, by extension, our living in a medium of which we are usually unaware; the reader understands phenomena such as magnetism and wave motion.

At times, the evidence is ambiguous, partly because of Lodge's tact when it comes to reminding the reader of things he or she ought to know. When Lodge describes the speed of light and the immense scale of the universe, he concludes that 'all this is well and even popularly known' and that the difficulties lie elsewhere (p. 43). His use of 'popularly' is interesting, because it suggests that the reader might position him- or herself as someone who does not need to be reminded of 'popular' knowledge. In other words, the idea of 'popular' knowledge is sometimes disavowed in the texts of what we now call 'popular science'; the disavowal allows the reader to believe that he or she is joining the elite. Similarly, when Lodge considers the relative densities of mercury and water, of water and air and of lead and iron, as well as those of other substances, he concludes, 'All these things are familiar; but I remind you of them because they are of the causes that regulate the rate of propagation of waves through these different substances' (p. 51). To say that the facts 'are familiar' is to remind the reader that he or she ought to have known them. It suggests that Lodge believed many of his readers would know at least some of those densities, and that he believed his readers were familiar with the general concept of 'density' but needed it to be related to more concrete examples.

A passage early in Chapter 2 of *Ether and Reality* serves to indicate some of what the implied reader knows, and the limits of his/her knowledge. Lodge explains the apparent discontinuity of matter:

> Fire an infinitely long-range projectile into the sky, and the chances are it will not hit anything: Lord Kelvin reckoned that the chance of hitting anything by such a projectile was about the same as the chance of hitting a bird if you fired a gun at random. That is one of the first things to realise about matter: there are great gaps between its particles. (p. 36)

Several details might be noted: the reader is expected to take in his/her stride the concept of infinity and the thought experiment of 'an infinitely long-range projectile'; more trivially, perhaps, he/she is expected to have some knowledge of the density of the bird population, and is expected to recognise the name Lord Kelvin and to respect the accuracy of his estimate. The final sentence in the quotation suggests that the reader needs to have the fundamental point reiterated, but that he / she will essentially already understand the idea about large-scale discontinuity. When, in the next paragraph, Lodge turns to the small-scale discontinuity between particles, he anticipates and paraphrases the reader's thoughts:

> That is all very well for the sky and the stars and the planets; but what about the earth? What about a piece of rock, or furniture, or any solid object? Do you mean to say that the particles of a body like that are widely separated, with great spaces between them in proportion to their size, and that a straight line might penetrate them deeply without encountering a particle? Yes, I do: that is what is meant by the discontinuity of matter. It is discontinuous on a small scale as well as on a large scale. (p. 36)

That Lodge presents the reader as someone asking doubting questions suggests he anticipates a reader less familiar with small-scale discontinuity than with large-scale discontinuity, but also that he anticipates a reader capable of learning by analogy. Moreover, it suggests that he wishes to encourage readers to ask such questions: at some level, he is discouraging reader passivity. A reader-response account of *Ether and Reality* cannot determine whether readers did position themselves as Lodge wished or process his information and analogies in accordance with his ideas, but it does indicate a basic outline of what a sympathetic reader might have done. Some of the text's more oblique methods of exposition also require an active reader.

3.4 THE RIDDLE

As the 'blazing fire' experiment suggests, in *Ether and Reality* Lodge is particularly concerned with the trustworthiness of the senses; but, because we have no sense organ for the ether, he needs to transform them and at times to render them unfamiliar. What we might call 'the riddle' is a frequently recurring feature of his rhetoric: a descriptive passage in which Lodge describes something in terms that make it recognisable, but in which he withholds its name until the end of the passage. (One might alternatively see the riddle as 'delayed decoding', following Ian Watt's well-known analysis of Joseph Conrad's representational techniques.)[19] The riddle is not uncommon in popular philosophical and popular scientific writing in the period. The common pattern is to describe the properties of an object in detail and only then to reveal its name, allowing the processes of perception and cognition to be slowed down and examined.[20]

Unlike his contemporaries, Lodge also uses the riddle form in relation to the senses themselves. This was something he had already done in *The Ether of Space*, which he published in 1909 and where he had confronted the problem of how he could assert the existence of the ether without material evidence. There, Lodge remarks that humans are not restricted to the sense of touch, which he implies belongs to an earlier phase of evolutionary development:

> We are conscious of matter by means other than its resisting force. Matter acts on one small portion of our body in a totally different way, and we are said to *taste* it. Even from a distance it is able to fling off small particles of itself sufficient to affect another delicate sense. [21]

Having dealt with hearing in a similar way, he goes on to ask whether there are 'any more discoveries to be made', and answers in the affirmative.[22] The effect is to remake the familiar, automatic processes of sensory perception into a more generalised account; this account then opens a space for the existence of other, as yet unnamed senses. This leads to an earlier version of the 'blazing fire' experiment:

> You are now as directly conscious as you can be of the ethereal medium. True the process is not very direct. You cannot apprehend the ether as you can matter, by touching or tasting or even smelling it; but the process is analogous to the kind of perception we might get of ordinary matter if we had the sense of hearing alone. It is something akin to *vibrations in the ether* that our skin and our eyes feel.[23]

Having established the analogy, Lodge has prepared the ground for the more outrageous proposal, worthy of scientific romance, that 'we are growing a new sense'.[24] Although he immediately retracts the literal suggestion and explains that he means that instruments such as electroscopes, galvanometers and telephones are acting as prosthetics, he sows the seed of the idea that we might possess a sense organ specific to the ether. By encouraging his audience to entertain the idea of an ether sense, Lodge encourages them to believe in the reality of ether. In his reference to scientific instrumentation, Lodge preserves the late nineteenth-century position that the advance of science was primarily dependent on the development of more sensitive measuring devices.[25] However, the suggestion that *Homo sapiens* is developing new senses recalls the less orthodox position developed by William Crookes in his 1897 presidential address to the Society for Psychical Research. Crookes had led his audience through a series of thought experiments about the nature of the human body under different conditions of gravity and scale, and about the different notions of physical law that such bodies might develop.[26] Lodge was a member of the council of the society in 1897, and would certainly have been familiar with this argument.

In *Ether and Reality*, Lodge presents a miniaturised version of the riddle—only one sentence long. It comes immediately after the 'blazing fire' passage: 'Every part of the skin is sensitive to the Ether quiver; but a particular region is localised in nearly all animals so as to be especially sensitive, and it is called an eye' (p. 29). The passage reimagines the senses as a sensitivity of the skin, reimagines the eye as a patch of skin and reimagines human beings as animals. Before the last six words, the reader is allowed momentarily to imagine that humans and other animals might possess a previously unrecognised ether sense. A similar gesture appears later in the book, in a passage where Lodge again tries to persuade the reader that there might be sensory evidence for the ether. In a series of syntactic parallelisms, he attempts to connect perception and cognition:

> We can see pieces of matter moving. When we see the gold-leaves of an electroscope diverge, we know that they are electrically charged: when we see a galvanometer deflect, we

> know that there is an electric current in its neighbourhood: when we see two pieces of iron or steel jump together we know that they are magnetized. (pp. 116–17)

'We see . . . we know' and 'we see . . . we know': Lodge omits intermediate steps of theorisation, and the possibility that what we see might complicate what we know. Lodge continues:

> We have a further most extraordinary sense. When we look at the flame of a fire or the filament of a lamp, or at the moon or a star or a landscape or indeed any visible object, we know—at least some of us have learnt to know—that there is an electromagnetic disturbance in the Ether, travelling at a known pace, reaching our eye, and in some curious way stimulating the optic nerve. (p. 117)

While 'we look' gives a large hint that the sense under discussion is the familiar one of sight, until we reach 'our eye', we cannot be certain that Lodge is not about to introduce a new sense. The riddle as a form enables Lodge to do conventional popularising work of representing familiar entities in a new, scientific vocabulary, thus making the reader see them from a new angle; but it also allows him to insinuate that there might be other, previously unrecognised entities in the universe.

3.5 CASTING DOUBT ON SCIENCE

Throughout *Ether and Reality*, Lodge casts doubt on the certainty of our scientific knowledge. Of course such a gesture would not be unexpected in a popular science book, especially in the opening phases: many writers undermine the reader's certainties with the expectation that the exposition will reconstruct them in accordance with a new idea. But Lodge goes beyond the conventional gesture. He cautions that gravitation 'is not fully understood' (p. 38) and that the mode of connection between matter particles and ether 'is not known; it is earnestly being sought' (p. 39). Spinning loops of ether 'are a reality' but 'what is going on in them is a speculation' (p. 131). Related to this tactic is Lodge's use of leading questions such as the following:

> Do an electron and proton ever clash together and obliterate each other in a pulse of radiation? Can matter, as we know it, radiate itself away and cease to be matter, until at some future time perhaps it is reconstituted? We do not yet know for certain—we only suspect it. (p. 144)

Alongside the leading questions, the speculative reference to a 'future time' opens doubt in the reader's mind about the permanency of physical law. The phrase hints at a millenarian transformation. In the following paragraph, Lodge considers the length of time which the sun has been radiating away its energy, and asks further questions:

> How can we account for all that radiation? Whence comes the energy which a body like the Sun is constantly emitting, and which some stars are emitting thousands of times faster? And what has become of it all? Here are two questions. We are not yet ready to answer the second; but Astronomers believe they can answer the first. (p. 145)

Although in the following paragraph Lodge gives a brief account of the process of atomic disintegration that lies behind the energy release, he also casts doubt on it: 'I do not say that it is true; no one positively asserts as yet that it is true; but if it were true, it would account for the energy' (p. 145). The specific concerns in the quoted passage about the origin and destination of energy echo those of *The Unseen Universe*, which was published in 1875 and in which Balfour Stewart and Peter Guthrie Tait notoriously claimed that the energy seemingly 'lost' through entropy had only gone to another universe.[27] Later, in *Ether and Reality*, Lodge remarks that the human attributes of mind, consciousness, memory and affection, which 'do not seem to belong to the material or physical universe' in fact 'belong to the Unseen Universe, the universe which makes no appeal to our senses' (p. 163). While the allusion to Tait and Stewart's work would not have been recognised by every reader, to some readers, given that the book had been a bestseller in its own time, the allusion would have been very clear. Here, as elsewhere in the book, Lodge asks leading questions, having encouraged his reader to answer them with 'the ether'.

3.6 THE RHETORIC OF SIGNIFICANCE

Throughout *Ether and Reality*, Lodge repeatedly ascribes *significance*, or potential significance, to physical theories. Lexically, he modulates up from 'knowledge' and its cognates to terms such as 'understanding', 'meaning' and 'significance'. Lodge's assertions about 'deep' meaning encourage the reader to speculate about what will ultimately be revealed, and one obvious conclusion is that the deeper meaning will concern the ether. Another aspect of the purpose of the rhetoric of depth and significance appears when Lodge is speaking of the length of the evolutionary process which has produced 'the life and mind which is now' and 'the mind which is still to come'. In doing so, he asserts that 'there is some far-seeing design, some lofty meaning in it all' (p. 148). The reader is encouraged to believe that the deep and significant meanings touch upon God.

However, to reduce this aspect of Lodge's writing to the religious function would be to oversimplify. Also at stake is his quarrel with the descriptionist school of Ernst Mach and Karl Pearson. Courtney Grean Raia has traced this dispute back to an 1895 exchange of public letters with Pearson over psychical research, and has noted further manifestations of it in Lodge's *Life and Matter*, which was published in 1900, and *My Philosophy*, which was published in 1933.[28] Early in *Ether and Reality*, Lodge asserts that the aim of science is not to produce 'mere formularies or convenient modes of expression' (p. 15). While he does not gloss 'convenient modes of expression', the informed reader would recognise that it engages with Mach and Pearson. In *The Grammar of Science*, Pearson had spoken of scientific laws as formulae that economically summarise a wide range of facts, and of scientific conceptions such as force as being merely 'convenient' modes of expression.[29] Lodge's opposition to descriptionism, an opposition founded on the belief that matter and force were fundamental realities, dated back as far as 1893.[30] The belief that physics ultimately relates back to fundamental bodily experience runs throughout Lodge's writings on ether and its relation to relativity.[31] Lodge's treatment of symbols as tools for 'work[ing] out things we do not fully understand' and his assertion that 'when we fully understand a thing, we can express it in words' (p. 58) is consistent with those views. Lodge apparently believes that when we have a proper understanding of a scientific truth, it relates back to an immediate bodily experience (of motion or force), and can therefore be expressed in natural language rather than a specialised mathematical formalism.

Lodge's references to 'the mind which is still to come', when coupled with his self-presentation as a patient student of the facts, tell a story about science: it is a discipline in which we should trust facts rather than our prejudices or generalisations, but be aware of the limitations of our bodies as entities that have evolved in particular physical conditions. It is a discipline in which we need faith that conventional empirical proof may not be available for many years; it is a discipline in which patience is rewarded.

It is in the context of Lodge's beliefs about patience that we should see the surprising cluster of biblical allusions at the end of Chapter 1. Throughout *Ether and Reality*, Lodge quotes from and alludes to literary works, particularly poetry, and to the Bible, but the appearance of five biblical allusions in 220 words of prose calls for particular comment. It may appear to be only an instance of the general tendency of popular science writing in the 1920s and 1930s to take some form of religious truth as its ultimate justification. *Ether and Reality* ends by claiming that the ether is 'the living garment of God' (p. 179); in *The Mysterious Universe* (1930), James Jeans was to claim, notoriously, that the 'Great Architect of the Universe' was a pure mathematician.[32] As I have noted elsewhere, the generic pattern was sufficiently established by the early 1920s that Bertrand Russell's refusal to discuss 'the Creator' in the concluding sentences of *The ABC of Atoms* (1923) appears to be an atheist's disavowal of it.[33] Lodge's biblical allusions in *Ether and Reality* were likely to persuade his readers that the book was moving towards a higher truth. In the prologue, he had justified his work on the grounds that 'our present physical knowledge, when properly grasped and accepted, constitutes a beneficent source of power, a fertilising influence, a body of illustration and parable, which can be drawn upon and used by those whose business it is to deal with still Higher Things'; he goes on to refer to their 'congregations' (p. 21). The allusions may also be seen as part of Lodge's technique for creating ethos; biblical echoes and allusion commonly function this way in Victorian non-fictional prose.[34]

But the allusions at the end of Chapter 1 also imply something important about the as yet unproven truths that Lodge attempts to expound:

> We are constantly closing our minds to Reality, or rather we are living among realities of which we have no apprehension. Open our eyes that we may see! (p. 32)

The argument here is part of a battle between prejudice and facts. The phrase 'Open our eyes that we may see' draws on either 2 Kings 6:17 ('I pray thee, open his eyes, that he may see') or Psalm 119:18 ('Open thou mine eyes, that I may behold wondrous things out of thy law'). The biblical allusion turns an epistemological failing into a spiritual one.

More significantly, it suggests that a higher power might one day make a deeper knowledge possible.

The four other allusions all fall in the final paragraph of the chapter:

> Revelation has not ceased, though it takes many forms; and those to whom perception has come must relieve their minds by utterance, and not expect to be fully understood in their day and generation. Their own knowledge, so far as it is sound, is but a step in advance; and they must have faith that humanity will take it in due time. Meanwhile they must possess their souls in patience; they need not strive nor cry; in quietness and confidence shall be their strength. (p. 32).

The first of these is the most glancing allusion: not being understood in one's own day and generation calls to mind Jesus's saying, as recorded in all four Gospels, that 'a prophet hath no honour in his own country' (John 4:44; see also Matthew 13:57, Mark 6:4 and Luke 4:24). The implication is that even those who have defeated prejudice may not be understood at the present time, but that at some future millennial moment their worth will be recognised. That they must 'possess their souls in patience' (cp. Luke 21:19, 'In your patience possess ye your souls') further emphasises that the scientific truth about ether will be known, but only in the future. The allusion to Matthew ('he shall not strive, nor cry', Matthew 12:19) is less important, but the final sentence, taken more or less directly from Isaiah 30:15, returns us to the topos of prophets not understood in their own time. Allusion here works as another way of drawing analogies. The scientist—or, at least, the open-minded scientist who is willing to accept the evidence for the ether's existence—is compared to a biblical prophet, and the qualities of patience required of a scientist are identified with those of a prophet not honoured in his own time.

Scientific knowledge is not the only form of knowledge that Lodge's readers are expected to bring to the text and, if we ask what kind of knowledge *Ether and Reality* allows the reader to produce, we must recognise that it is not, or not purely, scientific knowledge. It is clear from the biblical allusions at the end of Chapter 3, and elsewhere throughout the book, that the reader is expected to be familiar with Christian texts; it is clear from the references to life and mind surviving death (p. 120, p. 168) that the reader brings an interest in mortal questions, whether understood in a Christian framework or not. The opening pages gloss 'Reality' as signifying far more than physical reality. In the epilogue, Lodge returns both explicitly and implicitly to questions of man's place in a mysterious universe. He again makes use of literary allusion. For example, he suggests that the fact that the physical universe consists entirely of 'positive and negative electrons' brings us close to 'the mystery of existence', and goes on to say:

> We grope in a kind of helplessness with our few animal senses, and we live our short animal lives encouraged by a faith and hope that we are something more than appears, and that in the deep roots of our being we belong to another order of things, which is associated with this physical order only for a time. (p. 176)

The vocabulary here echoes that of Tennyson's *In Memoriam A. H. H.*:

I stretch lame hands of faith, and grope,
And gather dust and chaff, and call
To what I feel is Lord of all,
And faintly trust the larger hope.[35]

The allusion should not be surprising. Lodge had taken the epigraph to the Epilogue to *Ether and Reality* from Tennyson's 'Ulysses', and had quoted Tennyson in other works.[36] Indeed, in *Reason and Belief*, published in 1910, he had explicitly addressed the question of whether the 'student of science' had 'the right to quote poets or men of letters in support of his hypothesis'.[37] He concluded that 'the utterances of poets' were not 'part of the facts that can be appealed to in support of a thesis', but claimed that 'their evident reasonableness carries with them a conviction of truth' and that 'the intuitions of genius must not be ignored'.[38] The same work had provided a five-page list of its references to its literary quotations. However, quotation is not allusion, and the echo of *In Memoriam* is, as allusions go, a subtle one; it is possible that Lodge accidentally echoed Tennyson's vocabulary without intending to allude. Either way, the connection suggests that *Ether and Reality* belongs not only in the tradition of popular science writing but also in a tradition of literary writing about death, consolation and a higher order of things.

In the final paragraph of *Ether and Reality*, Lodge turns to the sense of touch. He argues that atoms are never truly in contact, but are always cushioned by ether: 'Hence even our apparently most material sense is dependent on this omnipresent medium'. The ether, he claims, is 'the primary instrument of Mind', the 'vehicle of soul' and the 'habitation of Spirit'. In his final sentence, he goes further, saying that we may call it 'the living garment of God' (p. 179). The phrase derives ultimately from a passage from Goethe's *Faust, Part One,* which was published in 1808 and in which the Earth Spirit describes his weaving as '*der Gottheit lebendiges Kleid*'. The phrase had been translated in a wide variety of ways, and Lodge's rendering may owe something to Thomas Carlyle's *Sartor Resartus,* published in 1833–4 and where, before quoting a translation of the relevant passage, Carlyle had written of '*the living visible Garment of God*'.[39] An anonymous translation of the phrase as 'the living garment of God' had appeared in 1872.[40] Lodge had previously alluded to Goethe's 'garment' in the peroration to *Continuity,* which was published in 1913. There he had argued that religion makes no appeal to the senses, and that

> we are deaf and blind therefore to the Immanent Grandeur, unless we have insight enough to recognise in the woven fabric of existence, flowing steadily from the loom in an infinite progress towards perfection, the ever-growing garment of a transcendent God.[41]

The original passage in Goethe is rich with implications for the relation of the visible material world to the invisible world of God, and of the relation of academic knowledge to deeper or more authentic forms; it is not clear which of them—if any—Lodge wishes to invoke in *Ether and Reality.* It may be that Goethe simply provides a resonant phrase. The idea in *Continuity* that the garment is 'ever growing' is replaced with the vaguer notion of one that is 'living'. The phrase could be seen as a gesturing not so much to Goethe as to Lodge's own 1913 lecture and its claim that some form of non-testable 'insight' or faith provides the foundation to empirical science. What the Tennyson and the Goethe allusions share is a concern with touch and, by implication, the mediation of presence through the ether. *Ether and Reality* is not quite a theodicy: even though 'God' is literally the last word of the book, the Christian deity is not prominent throughout, and Lodge does not seem directly concerned to justify his ways to men. In many respects, 'God' stands for a form of knowledge that is more authentic than any that science can attain.

Lodge's aim in *Ether and Reality* seems to be to communicate not what the ether is, but rather to establish the conditions for its possibility and to evoke what it does, understood in terms available to the unaided human senses. Moreover, in communicating these things, Lodge very often insinuates a 'spiritual' dimension to physical entities, whether by association or analogy. By assembling entities which might plausibly resemble the ether, by creating strings of analogies and by engaging his readers in thought experiments and riddles, in *Ether and Reality* Lodge creates a space for the ether which the sympathetic reader is allowed to fill. Furthermore, he simultaneously encourages his readers to make use of sensory evidence, while suggesting that the human body may possess senses as yet unknown and unnamed. By creating an image of himself as a thoughtful, experienced thinker and by implicitly comparing the scientist to a biblical prophet, Lodge creates conditions for scientific knowledge in which the ether should be believed in as a reality not yet fully revealed.

ACKNOWLEDGEMENTS

I presented earlier versions of this paper at a workshop on the ether in Oxford (organised by Jaume Navarro; February 2014), at the annual conference of the British Society for Literature and Science (organised by Greg Tate at the University of Surrey; April 2014), at a seminar on Oliver Lodge at the Science Museum, London (organised by Jim Mussell; July 2015) and at a colloquium in Donostia/San Sebastián, (organised by Jaume Navarro; March 2017). For their comments at those events, I would like to thank Sarah Hanks, Linda Dalrymple Henderson, Jaume Navarro, Richard Staley, Emilie Taylor-Brown and Aaron Wright; I am also grateful to Bernie Lightman for comments on an earlier draft.

NOTES

1. Oliver Lodge, *Continuity* (London: J. M. Dent, 1913), p. 60.
2. Richard Noakes, 'Ethers, Religion and Politics in Late-Victorian Physics: Beyond the Wynne Thesis', *History of Science* 43 (2005): 415–55, p. 423.
3. By the end of 1924, 1,129,578 licences for radio receivers had been sold in Britain: Asa Briggs, *The Birth of Broadcasting*, vol. 1, *The History of Broadcasting in the United Kingdom* (Oxford: Oxford University Press, 1995), p. 175.

4. Oliver Lodge, 'The Mystery of the Ether', *Radio Times*, 30 January 1925; 'Solving the Ether's Secrets', *Radio Times*, 13 February 1925; 'The Marvels of Magnetism', *Radio Times*, 27 February 1925; 'Life, Mind, and The Ether', *Radio Times*, 10 April 1925. For circulation figures for the *Radio Times*, see Briggs, *Birth of Broadcasting*, p. 278.
5. Hodder and Stoughton archives, London Metropolitan Archives, CLC / B / 119 / MS16312 / 037 (covering 1925–9) and CLC / B / 119 / MS16312 / 012 (covering 1930–7).
6. Martin Eger, 'Hermeneutics and the New Epic of Science', in Murdo William McRae, ed., *The Literature of Science* (Athens: University of Georgia Press, 1993), 186–209, pp. 189–90.
7. All references to Oliver Lodge, *Ether and Reality* (London: Hodder and Stoughton, 1925), are given in the text.
8. Stuart Hibberd, *This—Is London* (London: Macdonald and Evans, 1950), p. 26.
9. Briggs, *Birth*, p. 261.
10. George P. Landow, *Elegant Jeremiahs: The Sage from Carlyle to Mailer* (Ithaca: Cornell University Press, 1986), pp. 154–88; Sarah Tinker Perrault, *Communicating Popular Science: From Deficit to Democracy* (Basingstoke: Palgrave Macmillan, 2013), pp. 96–112.
11. Michael H. Whitworth, *Einstein's Wake: Relativity, Metaphor, and Modernist* Literature (Oxford: Oxford University Press, 2001), p. 53.
12. It formed part of 'Our Sense of Ether: Vibrations and Waves and What They Signify', broadcast on 20 January 1925, 9.30–10 pm, reprinted as Lodge, 'Solving the Ether's Secrets'.
13. Lodge, 'The Mystery of the Ether', p. 242.
14. A. S. Eddington, *Space, Time and Gravitation* (Cambridge: Cambridge University Press, 1920), p. 64; Albert Einstein, *Relativity: The Special and the General Theory*, trans. Robert W. Lawson (London: Methuen, 1920), pp. 9–31; Charles Nordmann, *Einstein and the Universe*, trans. Joseph McCabe (London: T.F. Unwin, 1922), p. 141.
15. J. W. N. Sullivan, 'An Abstruse Theory', *Athenaeum* 4724 (1920): 621–2, p. 621. Signed 'S': The authorship of unsigned and initialled *Athenaeum* articles has been confirmed by reference to the editor's marked copies, City University, London.
16. Ralph O'Connor, 'Reflections on Popular Science in Britain: Genres, Categories, and Historians', *Isis* 100 (2009): 333–45, pp. 334–5.
17. Stephen Hilgartner, 'The Dominant View of Popularization: Conceptual Problems, Political Uses', *Social Studies of Science* 20 (1990): 519–39.
18. Ronald Campbell Macfie, 'Ether and Reality', *Bookman* [London] 68 (1925): 197–8; G. W. de Tunzelman, 'Modern Theories of the Ether', *Nation and Athenaeum* 37 (1925): 348, 350; Algernon Blackwood, 'A Thriller', *Time and Tide* 6 (1925): 592–3; 'Ether and Reality', *Times Literary Supplement*, 14 May 1925 (the *TLS Historical Archive* identifies this review as being by J. W. N. Sullivan).
19. Ian Watt, *Conrad in the Nineteenth Century* (Berkeley: University of California Press, 1979), pp. 175–9.
20. Whitworth, *Einstein's Wake*, pp. 97–9.
21. Oliver Lodge, *The Ether of Space* (London and New York: Harper and Brothers, 1909), p. 20; Lodge's emphasis.
22. Lodge, *Ether of Space*, p. 20.
23. Lodge, *Ether of Space*, p. 21.
24. Lodge, *Ether of Space*, p. 26.
25. H. Otto Sibum, 'Science and the Changing Senses of Reality circa 1900', *Studies in History and Philosophy of Science* 39 (2008): 295–7, p. 295.

26. William Crookes, 'Address by the President', *Proceedings of the Society for Psychical Research* 12 (Appendix to Part 31) (1897): 338–55.
27. Greg Myers, 'Nineteenth-Century Popularizations of Thermodynamics and the Rhetoric of Social Prophecy', *Victorian Studies* 29 (1985): 35–66; Jaume Navarro, 'A Victorian Theory of the Multiverse', in Jaume Navarro, ed., *Science and Faith within Reason: Reality, Creation, Life and Design* (Farnham: Ashgate, 2011), 83–102.
28. Courtney Grean Raia, 'From Ether Theory to Ether Theology: Oliver Lodge and the Physics of Immortality', *Journal of the History of the Behavioral Sciences* 43 (2007): 19–43, pp. 34–5.
29. Karl Pearson, *The Grammar of Science* (London: Walter Scott, 1892), pp. 93, 144.
30. Lodge, *Ether of Space*, p. 15. Cp. Lodge 'The interstellar ether', *Fortnightly Review* 53 (1893): 856–64.
31. Oliver Lodge, 'The Geometrisation of Physics', *Nature* 106 (1921): 795–800; and see Whitworth, *Einstein's Wake*, pp. 203–4.
32. James Jeans, *The Mysterious Universe* (Cambridge: Cambridge University Press, 1930), p. 134.
33. Bertrand Russell, *The ABC of Atoms* (London: Kegan Paul, Trench, Trubner, 1923), p. 171; Whitworth, *Einstein's Wake*, p. 49.
34. Landow, *Elegant Jeremiahs*, pp. 163–4.
35. Alfred Tennyson, *In Memoriam A. H. H.*, part 55, stanza 5.
36. John Holmes, '"The Poet of Science": How Scientists Read Their Tennyson', *Victorian Studies* 54 (2012): 655–78.
37. Oliver Lodge, *Reason and Belief* (London: Methuen, 1910), p. 196.
38. Lodge, *Reason and Belief*, p. 197.
39. Thomas Carlyle, *Sartor Resartus*, ed. Kerry McSweeney and Peter Sabor (Oxford: Oxford University Press, 1987), p. 44 (Carlyle's italics).
40. 'Love Lane, from Kensington to Fulham', *Cornhill Magazine* 26 (1872), 272–385, p. 286. The article quotes this translation from an anonymous manuscript. The *Wellesley Index* attributes the article to Charles Kegan Paul, and it seems likely that the manuscript was his own; however, his published translation of *Faust*, which appeared the following year, renders the phrase as 'The garment of life which the Godhead wears': *Faust: A Tragedy*, trans. C. Kegan Paul (London: Henry S. King and Co., 1873), p. 24.
41. Lodge, *Continuity*, p. 93.

4

Poincaré's Mathematical Creations in Search of the 'True Relations of Things'

Connemara Doran

Can science teach us the true relations of things?

Henri Poincaré, *The Value of Science*, 1905.

4.1 INTRODUCTION

'Mathematical physics and pure analysis... mutually interpenetrate and their spirit is the same.'[1] Addressing (in absentia) the first International Congress of Mathematicians (ICM) in Zürich in 1897, Henri Poincaré (1854–1912)—French mathematician, mathematical physicist and philosopher—explained his vision of the mutual needs and shared spirit of these worlds, a perspective shaped by his deep conceptual work at that interface for almost two decades. 'Mathematics have a triple aim', he argued. It is 'not enough' that mathematics aims to 'furnish an instrument for the study of nature' so that the physicist could 'know it better', an aim notably exemplified by advances 'already rough-hewn' in celestial mechanics and mathematical physics. '[Mathematics must also] have a philosophical aim, and, I dare maintain, an aesthetic aim. They must aid the philosopher [and physicist] to fathom the notions of number, of space, of time.'[2]

Pure analysis—a battlefield long engaging the profound subtleties of infinitesimals and infinity, differentiability and continuity, number and the mathematical continuum—helped secure those advances in understanding nature (notably, Poincaré's new methods of celestial mechanics), but it also thereby deepened the needs and promises within mathematical physics. Mathematics, he asserted, must probe these depths with the 'aesthetic purpose' of enabling philosophers and physicists to grasp, articulate and expose such subtleties and relations within nature:

> The physicist cannot ask of the analyst to reveal to him a new truth; the [analyst] could at most only aid [the physicist] to foresee it.... All laws are therefore deduced from experiment; but to enunciate them, a special language is needful; ordinary language is too poor, it is besides too vague to express relations so delicate, so rich, and so precise. This therefore is

Doran, C., 'Poincaré's Mathematical Creations in Search of the "True Relations of Things"' in *Ether and Modernity*, edited by J. Navarro. © Oxford University Press 2018. DOI: 10.1093/oso/9780198797258.003.0004 one reason why the physicist cannot do without mathematics: it furnishes him the only language he can speak.[3]

Poincaré's mathematical corpus attested to the richness of conceptual advance, pure and applied, that is possible within the intertwined fields of analysis and physics. Some among his audience would soon come to understand aspects of his vision and help advance it. Poincaré was just beginning another surge in activity at that interface, creating an array of utterly new mathematics (and associated philosophical discourse) that would ground his response to the radical new challenges within fin-de-siècle physics, a crisis within theory that greatly increased the stakes regarding our notions of number, space, time and, therewith, the physicists' ether.

In 1900, three international Congresses met in Paris—the second ICM, the first International Congress of Physicists (ICP) and the first International Congress of Philosophy (ICPHIL)—pushing the boundaries of open questions in the respective disciplines and their interconnections. In mathematics, emphasis on a fully rigorous axiomatic, strictly deductive, formal approach to proof advanced in tandem with a drive for a mathematics anchored in the certainties of 'number' within arithmetic rather than in subtleties (not yet deductively proven) about 'dimension' and the 'mathematical continuum' within analysis. In theoretical physics, experimental discoveries and electrodynamic theories challenged Newtonian notions of space, time and matter. Philosophical discourse about the meaning, methods and purpose of science stumbled on questions about space, absolute motion and the nature of the ether.

This chapter assesses the following question: how did Poincaré's vast corpus of mathematical innovation engage the rationale, and impact the fate, of the notion of the ether in physics? It asks what *Poincaré* was thinking, and it seeks understanding through his voice (speaking mathematics and philosophy), not through the arguments of contemporaries and later interpreters who did not grasp his full meaning (often completely distorting it), nor through categories foreign to his unique way of thinking. It finds that Poincaré had no ownership of the physicists' ether concept, and that he viewed the ether as neither necessary nor necessarily a hindrance for further advance. Rather, Poincaré attended to the profound and subtle needs within physics by creating profound and subtle mathematics—utterly new theoretical and interpretive concepts, tools and structures—to capture the 'true relations of things', rendering the physicists' ether superfluous to that goal while also creating mathematical structures for gravitational and quantum phenomena.

In his scientific practice and philosophy of science, Poincaré sought the 'true relations' that adhere in the phenomena—relations that persist irrespective of the choice of a metric geometry and a change in physical theory. This chapter is structured to aid understanding of how Poincaré's lifeworks 'hang together'[4]—how they cohere within Poincaré's way of thinking—which enables us to assess how his work instantiates what he means by the 'true relations of things' that unify physics.

Taking a historicist perspective anchored in detailed assessments of Poincaré's corpus and legacy by mathematicians and historians of mathematics, the chapter traces key strands in Poincaré's 1880s engagement with the subtleties of space and time and the structure we know as spacetime, and, concurrently, with the conceptual possibilities that opened to him as he sought to master and exploit topological intuitions in creating *analysis situs* (algebraic topology). It traces how Poincaré embedded these utterly new geometric and topological ways of thinking at the heart of pure mathematics, mathematical physics and philosophy.

Section 4.1 explains how Poincaré emphasised the need to break free from the geometry habituated by our senses, altered the discourse about the geometry of physical space and set out to create a new mathematics for relations 'so delicate, so rich, and so precise'. Section 4.2 examines Poincaré's 1891 essay, detailing how he maps the path from his 1880 hyperboloid model to his 1887 'fourth geometry'—offering different lenses into the geometry that would become Minkowski spacetime. Section 4.3 traces through Poincaré's philosophical writings (from 1901 to 1905) and his 1904 St. Louis address on the dynamics of the electron, documenting his view of the physicists' ether as a disposable 'garment' in the search for the 'true relations' that persist within natural phenomena. Section 4.4 examines the reaction to Poincaré's work at Göttingen, where Poincaré's corpus was assiduously studied and built upon, illustrating how Poincaré's geometric creations from 1880 to his four-dimensional geometric interpretation of the Lorentz transformation was a ready resource for Hermann Minkowski as he developed his spacetime geometry based on Einstein's theory of special relativity. Section 4.5 offers concluding comments about Poincaré's last year of life and his legacy, witnessing a juxtaposition of his works on space and time, analysis situs and dimension, quantum theory and statistical mechanics, and documenting the profundity that Poincaré and Einstein recognised in each other's works.

4.2 SEEKING A MATHEMATICS TO EXPRESS 'RELATIONS SO DELICATE, SO RICH, AND SO PRECISE'

The question of the geometry of physical space was rooted in early nineteenth-century discoveries regarding non-Euclidean geometries, intrinsic curvature and topological shape. Mathematicians mid-century were intrigued by topological possibilities for our space, but philosophers and physicists focused more on the question whether it might exhibit a positive or negative curvature at an astronomically large scale—whether ours might be a spherical world (with positive curvature) or a hyperbolic world (with negative curvature) rather than a Euclidean one (with zero curvature). The notion that one might empirically measure the curvature of our space by means such as stellar parallax became a tantalising possibility, indeed, an increasingly confident and explicit goal within informed scientific communities in the 1890s.

In an 1891 essay entitled *'Les géométries non euclidiennes'*, Poincaré radically altered the possibilities and stakes regarding the question of the 'true' geometry and topology of our space and how we can come to know it.[5] In a highly provocative thought experiment, Poincaré explained how intelligent creatures from a hyperbolic world, whose geometry is based on its freely chosen non-Euclidean conventions, would, if transported to our world, observe the same phenomena we do but express the physical laws differently. We, likewise, would easily enunciate the laws of their world using our Euclidean conventions. Yet, in each such world, while understanding of the 'true relations' of its physical phenomena would be secure, no experiment would be able to determine *by metric means alone* the actual geometry of that world.

Poincaré's 'fiction' of the hyperbolic world (introduced in his 'Letter to Mouret')[6] challenged the validity of the assumptions of *geometric empiricism*—the claim that experiment (measurement) was sufficient to determine the geometry of cosmological space—while at the same time offering an alternative epistemology (*geometric conventionalism*) and promising a new mathematics, a new type of geometric reasoning (which he called 'analysis situs') that would help us 'find a way' to secure knowledge of the geometry and topology of spaces of higher dimension.[7]

As Poincaré explained in 1901, deep challenges throughout his varied research pushed him to engage in this ongoing mathematical quest and encouraged his trust in profound rewards for mathematics and physics: 'As for me, all the diverse paths on which I was successively engaged [1879–91] have led me to *Analysis Situs*.'[8] Poincaré recognised that mathematics, physics and celestial mechanics needed the certainty of theorems accessible only with this new type of reasoning.

Poincaré had entered into a philosophical discourse, couched in the language of neo-Kantian philosophy, which sought foundations for the geometry of our space. The discourse first centred on Hermann von Helmholtz's notion of freely moving rigid (measuring) bodies, then on infinitesimal transformation groups, becoming known as the Helmholtz–Lie (classical) space problem.[9] Poincaré saw the need to escape its metrical (measurement) requirements, and his philosophical stance was inextricably tied to his 1887 establishment of the 'fourth geometry' (explained in Section 4.3) and to the new mathematics he was creating.[10]

Poincaré's 'Analysis Situs' requires that our imagination break free from the geometry habituated by our senses, ignore metrical properties of geometric objects (properties involving measurement of distance and angles) and pursue the difficult art of 'reasoning well based on badly drawn figures' by focusing on relations unchanged by any continuous deformation.[11] The years 1887–91 mark the momentous interval during which Poincaré set himself the task to establish this new field of mathematics for the study of space. By 1895, Poincaré had single-handedly created the field called 'algebraic topology', developing over the next decade entirely new concepts, tools and intuitions to assess topological spaces of three and higher dimensions.[12]

The mathematical concept of the *amorphous continuum* provides the basis (realm of action) for mathematicians to rigorously conceptualise in higher dimensions, learning how we can 'supplement' our senses so as to reason in 'hyperspace'. The distinction between the mathematical (amorphous) continuum and the physical continuum of experience is crucial, as is Poincaré's emphasis that we do not have intuition about space itself. We hear Poincaré's philosophical voice, speaking mathematics, in the section entitled

'Qualitative Geometry' in his 1903 article 'Space and Its Three Dimensions',[13] which was reprinted as Chapter 3, 'The Notion of Space', in his 1907 collection of essays, *The Value of Science*:

> Euclidean space is not a form imposed upon our sensibility, since we can imagine non-Euclidean space; but the two spaces, Euclidean and non-Euclidean, have a common basis, that amorphous continuum of which I spoke in the beginning. From this continuum we can get either Euclidean space or Lobachevskian space... This continuum has a certain number of properties, *exempt from all idea of measurement*... The theorems of analysis situs have, therefore, this peculiarity that *they would remain true if the figures were copied by an inexpert draftsman* who should grossly change all the proportions and replace the straights by lines more or less sinuous. In mathematical terms, they are not altered by any 'point-transformation' whatsoever... Of all the theorems of analysis situs, the most important is that which is expressed in saying that space has three dimensions.[14]

Key to understanding subtleties of this need to pass from the world of the amorphous continuum to the world of space and geometry is Poincaré's claim that experience alone cannot 'engender mathematical notions', in particular the notion of mathematical continuity itself. Adding a *metric structure* to the amorphous continuum yields a *space* with attendant cosmological and epistemological implications.

In his works in analysis situs, Poincaré was after a precise mathematical notion of *dimensionality* anchored in the subtle conceptual tools and theorems he was creating. As he explained in 1908 to the fifth ICM in Rome, such deep penetration into unexplored terrains of thought can 'enable us really to see into hyperspace and to supplement our senses'.[15] Poincaré insisted that analysis situs is 'the only true domain of geometric intuition' and, once accessed, promised entry into vast new realms of mathematical activity.[16]

4.3 POINCARÉ'S 1891 ESSAY, FROM HIS 1880 HYPERBOLOID MODEL TO HIS 1887 FOURTH GEOMETRY: GEOMETRIC REASONING UNTETHERED, SPACE 'STRUCTURES' ENGAGED

Poincaré's essay '*Les géométries non euclidiennes*' appeared 15 December 1891 in a recently launched French journal with a diverse scientific audience for whom Poincaré sought to capture the *mathematical* and *epistemological* challenges confronting a geometric (*metrically empirical*) understanding of space.[17] Many scholars who assess the geometric conventionalism Poincaré introduced there focus on its conclusion, namely that, since hyperbolic and Euclidean metrics are inter-translatable, one can choose the simplest, Euclidean geometry. But Poincaré takes us to that conclusion by making a much deeper argument about geometry and what it yet lacks.

In a breathless litany, he takes us through Riemann's spherical world (finite without boundary) and the accelerating mathematical innovations that had radically altered the nature and role of geometry throughout mathematics and in the formulation of physical laws. Poincaré repeatedly asserts that we must release geometry from experiences that restrict our thinking—by contemplating how beings in a hyperbolic three-dimensional world would create their own geometry, and by recognising what mathematicians have already done by means of unrestricted geometric reasoning. Immediately after establishing a dictionary between the three-dimensional hyperbolic and Euclidean spaces, he states, 'But this is not all' and starts recounting the fruits of these striking shifts in geometric reasoning.[18]

'Consider what Klein and myself have done by using them in the integration of linear equations.' A decade earlier Poincaré had changed the essence of hyperbolic geometry from a mere curiosity of the geometer and an intriguing 'possibility' to a profound necessity for the analyst that lay hidden within much of mathematics. In 1880, examining issues regarding integrals of linear differential equations with algebraic coefficients, Poincaré had created the theory of automorphic functions, exploiting a series of deep and subtle insights all the way to the concept of the universal cover and to establishing and proving the uniformisation theorem that classified solutions of all analytic functions (as rational, elliptic or Fuchsian functions). As the historian of mathematics Jeremy Gray argues, Poincaré had devised a unique way of 'deriving' the essence of Riemann surfaces; he constructed them naturally from discontinuous groups, obtaining them as quotient spaces of the unit disc (rather than as branched coverings of the Riemann sphere), once he had the wonderful insight that the Möbius transformations he had used to define the class of Fuchsian functions were identical to the groups of motions of hyperbolic geometry. Poincaré's 'almost effortless introduction of Riemann surfaces' into his analysis was viewed as a 'dramatic novelty... especially since Riemann's ideas were still generally considered obscure and lacking in rigor'.[19]

A key move in his 1880 epochal advance occurred when Poincaré, while engaging research on indefinite ternary quadratic forms within number theory, conceptualised a new model of the hyperbolic plane, a hyperboloid model that projected to the open unit disc—establishing the existence of his Fuchsian groups and hyperbolic geometry at the heart of pure mathematics.[20] Poincaré's conceptualisation here is of a completely different nature from the establishment of Weierstrass coordinates, the Helmholtz hyperboloid model and other such models of much earlier date.[21] Poincaré details how to picture and construct this profound relationship in a self-analysis of his work (1884, 1886):

> One of the most important problems in the subject of indefinite ternary quadratic forms is the study of the discontinuous groups formed by the similarity substitutions, that is, linear substitutions which preserve the form. Let $F(x, y, z)$ be an indefinite quadratic form.
>
> We can choose the constant K so that $F(x, y, z) = K$ represents a hyperboloid of two sheets. The similarity substitutions then map a point on the hyperboloid to another point on the same sheet and, since the group is discontinuous, the hyperboloid becomes partitioned into infinitely many curvilinear polygons whose sides are diametric sections of the surface. A similarity substitution changes each polygon into another. We now take a perspective view by placing the eye at an umbilic of the surface and taking the plane of projection to be a

> circular section. One sheet of the hyperboloid is projected inside a circle, and the polygons drawn on this sheet project to curvilinear polygons bounded by circular arcs of the kind we have discussed in the theory of Fuchsian groups. Thus the study of similarity substitutions of quadratic forms reduces to that of Fuchsian groups, which is an unexpected rapprochement between two very different theories, and a new application of non-Euclidean geometry.[22]

Poincaré's surprising linkage of his hyperbolic disc and the two-sheeted hyperboloid created a new world of action within mathematics of profound power and import. The indefinite quadratic form producing the figure Poincaré instructs the reader to visualise (and construct) is equivalent to the indefinite quadratic form of flat two-dimensional spacetime; we will see that Minkowski carefully studied Poincaré's reasoning from these 1880 insights to Poincaré's 1905–6 creation of the four-dimensional metric (indefinite quadratic form) linking light, space and time.

Upon discovering Poincaré's theory of Fuchsian (automorphic) functions, Felix Klein initiated a correspondence with Poincaré the next day (12 June 1881), claiming in his second letter that Poincaré's 'analogy with non-Euclidean geometry does not hold' for more complicated cases without a limit circle. Poincaré immediately demonstrated that his analogy with non-Euclidean geometry *does* hold by creating his three-dimensional 'hyperbolic ball' model (the hyperbolic world), and fully generalising his theory of automorphic functions to all cases (naming cases without a limit circle Kleinian groups and functions).[23]

The litany continues. There is also a 'fourth geometry' with far more surprising properties than Riemann's or Lobachevsky's non-Euclidean geometries. Poincaré had established this fourth geometry and its astonishing properties in an 1887 article on the 'fundamental hypotheses of geometry', a geometry that implied three propositions 'so contrary to our habits of thought that the founders of geometry have denied them'.[24] In his 1891 litany, Poincaré mentions only one of its theorems and 'not the most surprising: a real line can be perpendicular to itself'.[25] This fourth geometry is the single-sheeted hyperboloid, which is precisely the two-dimensional hypersurface of de Sitter space, and one of its degeneracies is the geometry of two-dimensional Minkowski space.[26]

Poincaré continues to higher dimensions: from a group-theoretic theorem of Sophus Lie, we know there are a finite number of constant curvature n-dimensional geometries; but there are an infinite number of variable curvature Riemannian geometries, all depending on how the length of a curve is defined. The fundamental hypotheses of geometry take us much deeper and far beyond the need for a consistent and convenient metric geometry which can readily be translated into a different metric geometry if the need arises. At those depths lie the topological issues that Poincaré confronted in all of his mathematical creations—and that would shortly lead him to the topological link of the universal covering space and the fundamental group.

Poincaré then poses the following question: if several geometries are possible, which is true? He goes on to argue to his geometrical conventionalism and the puzzles regarding the hyperbolic world. He would reintroduce this 'hyperbolic world' in his 1895 article '*L'espace et la géométrie*',[27] coinciding with publication of his first innovations in 'Analysis Situs'. There, as in his 1903 fictions featuring the equivalence and hence 'indistinguishability' of the Hyperbolic and Euclidean worlds, Poincaré was getting at the notion of what space 'is'—not geometrically but topologically. No experiment can ever tell us which geometry (metric) is 'true' (natural) for our space, he argued, because we can only observe results of measurements involving objects—not of space itself nor of relations between objects and space. Poincaré established the 'amorphous continuum' as the underlying 'common basis' of the Euclidean and non-Euclidean metrics we can impose on the space.

As Poincaré emphasised in 1903, the 'worlds' so constructed are indistinguishable as different spaces 'if we can pass from one to the other by any point transformation whatever', adding that it is in this sense that it would be proper to understand the relativity of space.[28] Understanding the structure of our space, and our physics, requires mathematical constructions and the theorems of 'Analysis Situs' by which we delve beyond the impressions of experience.

4.4 THE 'TRUE RELATIONS' THAT PERSIST (A 'NATURAL KINSHIP') VERSUS THE ETHER (A 'GARMENT')

How did Poincaré engage the rationale, and impact the fate, of the notion of the ether in physics? This section seeks understanding through Poincaré's voice by tracing through his philosophical writings from 1901 to 1905—his 1901 updated self-analysis of his work and his first two collections of reprinted addresses and articles.

4.4.1 1886–1901: The 1900 Congresses, Poincaré's 1901 update, *Science and Hypothesis*

In 1901, Gösta Mittag-Leffler, the Swedish mathematician and editor of *Acta Mathematica*, asked Poincaré to prepare an update of the seventy-five-page self-analysis of his works that Poincaré had composed early in his career (1884, 1886) for induction into the Académie des Sciences. Poincaré's one-hundred-page update covers the fifteen-year period 1886–1901 and is organised into seven parts.[29] It was not published until 1921; Poincaré never wrote an update for 1901–12. Part 3 of the update, 'Diverse Questions of Pure Mathematics', explains how his need to create analysis situs deepened amid advances throughout his mathematical corpus. Part 4 of the update, 'Celestial Mechanics', features the concepts and methods (confronting analytic issues with topological and geometric reasoning) by which he revolutionised celestial mechanics and which became widely fruitful in astronomy and physics.

Part 5 of the update, 'Mathematical Physics', traces his contributions to the theory of partial differential equations and his critiques of physical theories, from early theoretical doubts about Fresnel's optical theory to Maxwell's electromagnetic theory of light and Lorentz's theory of the electron—loci of the physicists' ether. The long-standing concerns related to stellar aberration and Fresnel's dependence on the index of refraction, a shift in focus to time intervals involved in the transmission of light, and

Poincaré's 1898 considerations of the notion of simultaneity in astronomy and in measuring longitude—all this coalesced in Poincaré's 1900 metrological interpretation of 'local time' in Lorentz's theory.[30]

Part 6 of the update, 'Philosophy of Science', features, in a subsection entitled 'Physics', both Poincaré's 1897 ICM address and his 1900 ICP lecture assessing the status of the ether in mathematical, experimental and theoretical physics.[31] After stipulating that 'belief in the unity and simplicity of nature...is necessary for science', Poincaré insists on the need to distinguish between 'the foundation [*le fond*] and the form [*la forme*]' in physical theories—a terminology of his own invention that is key to how he views and contends with the physicists' ether within his mathematical physics:[32]

> The foundation is the existence of certain relationships between inaccessible objects. These relationships are the only reality we can achieve, and *all we can ask for is that there be the same relationships between these unknown real objects and the images we put in their place.*
>
> The form [image, analogy] is only a sort of garment with which we dress this skeleton [foundation, system of relationships]; we frequently change this garment, to the astonishment of the people of the world...But if the form changes often, the foundation remains.
>
> The hypotheses regarding what I have just called the form cannot be true or false, they can only be convenient or inconvenient. For example, the existence of the ether, [and] the existence of external objects, are only convenient hypotheses.[33]

For Poincaré, there is neither truth nor falsity regarding the existence of the ether; it is merely a convenient hypothesis, a garment freely chosen and discarded.

Most significant are two conclusions, and a warning, that Poincaré draws from the fact that the 'form' of the physical theory is merely a garment. First is his historical observation about the strength of science amid theory change: 'It is for this reason that certain theories which were believed to be permanently abandoned are reborn from their ashes.' Second is the conundrum of underdetermination: 'There are certain categories of facts which are equally well explained in two or more different theories, without any experience ever being able to decide between them.' Indeed, this underdetermination is 'particularly true for mechanistic theories', for 'it can be shown that if a phenomenon includes one mechanical explanation, it will have an infinity'.[34] Poincaré continues with a caution about the particularly dangerous allure of 'mechanism' in explaining physical phenomena:

> In any case, Mechanism is only one of the garments with which truth can be dressed, and if it satisfies our mind, we must not attach more importance to it than it deserves. It obliges us to introduce the hypothesis of auxiliary fluids such as ether; I present some views on the greater or lesser reality of this fluid.[35]

An infinite number of possible mechanical ether theories could be constructed and, indeed, a proliferating number appeared during the previous decade.

In contrast, for Poincaré, the ether is merely a term attached to relations that adhere in the physical phenomena of electromagnetism. His 1900 ICP address concludes: 'We must not forget that the goal of science is not Mechanism but unity [of explanation and understanding].' Poincaré refutes the so-called *bankruptcy of science* (the layman's view that the 'ephemeral nature of scientific theories' implies that 'they are absolutely in vain') with the example of Fresnel's theory of light as 'movements of the ether' in contrast to Maxwell's electromagnetic theory of light. The reality of the ether, and the particular form it took in the different theories, was not the relevant question: 'Fresnel's object was not to know whether there really is an ether... his object was to predict optical phenomena.' Fresnel's theory enables such prediction today 'as well as it did before Maxwell's time', not because of the reality of the theoretical objects or mechanical structure of Fresnel's theory, but because Fresnel's 'differential equations are always true'. These differential equations 'express relations, and if the equations remain true, it is because the relations preserve their reality'. The 'true relations' among these 'real objects which Nature will hide forever from our eyes', Poincaré insists, 'are the only reality we can attain'. What matters is that 'the same relations shall exist between these objects as between the images we are forced to put in their place'. We might find one image more convenient than another, but that choice does not change the underlying 'relations [that] are known to us' and that persist.[36]

Also in Part 6, in a subsection entitled 'Mechanics', Poincaré explains how concerns raised in his 1898 paper on the 'measure of time' reverberated throughout his 1900 ICPHIL critique of the 'principles of mechanics' (shortly reprinted in *Science and Hypothesis*), where he asks:

> May we not someday be compelled by new experiments to modify or even to abandon [the principles of mechanics]? These are questions that naturally arise, and the difficulty of solution is largely due to the fact that treatises on mechanics do not clearly distinguish between what is experiment, what is mathematical reasoning, what is convention, and what is hypothesis. That is not all.[37]

Poincaré argues that these distinct categories of practice enter into the conceptual difficulties within mechanics that he proceeds to itemise:

1. There is no absolute space, and we only conceive of relative motion; and yet in most cases mechanical facts are enunciated as if there were an absolute space to which they can be referred.
2. There is no absolute time. When we say that two periods are equal, the statement has no meaning, and can only acquire a meaning by convention.

It is a conceptually flawed and misleading practice, Poincaré warns, to enunciate mechanical facts with reference to an absolute space *as if it exists*. And, since there is no 'absolute time' (and since we cannot directly intuit equality for intervals), we can only establish a 'meaning' regarding the 'equality' of two time intervals *by convention*. Moreover, he cautions:

3. Not only have we no direct intuition of the equality of two periods, but we have not even direct intuition of the simultaneity of two events occurring in two different places. I have explained this in an article entitled '*Mesure du Temps*'.

Poincaré references his 1898 article explaining how the *problem of simultaneity of two distant events* is linked to the *problem of measuring time*, and how astronomers measure the velocity of light by supposing that it is constant and the same in all directions; thus, any statement about the simultaneity of spatially separated events is based on *a freely chosen convention*.[38] He concludes:

> 4. Finally, is not our Euclidean geometry in itself only a kind of convention of language? Mechanical facts might be enunciated with reference to a non-Euclidean space which would be less convenient but quite as legitimate as our ordinary space; the enunciation would become more complicated, but it still would be possible.[39]

Here is the challenge and the promise of Poincaré's geometric conventionalism—the invitation to consider an alternative non-Euclidean geometry for enunciating the facts of mechanics. Notwithstanding their experiential character, the principles of mechanics might just as legitimately (and perhaps more fruitfully) be 'enunciated' with reference to a 'space' that does not have the Euclidean metric.

Poincaré soon wrote the preface to *Science and Hypothesis*, there distinguishing variants of hypotheses for geometry, mechanics and the physical sciences:

> We therefore conclude that the *principles of geometry* are only conventions; but these conventions are not arbitrary, and if transported into another world (which I shall call the non-Euclidean world, and which I shall endeavour to describe), we shall find ourselves compelled to adopt more of them.
>
> In *mechanics* we shall be led to analogous conclusions, and we shall see that the principles of this science, although more directly based on experience, still share the conventional character of the geometrical postulates...
>
> But we now come to the *physical sciences*, properly so called, and here the scene changes. We meet with hypotheses of another kind, and we fully grasp how fruitful they are. No doubt at the outset theories seem unsound, and the history of science shows us how ephemeral they are; but they do not entirely perish, and of each of them some traces still remain. It is these traces which we must try to discover, because in them and in them alone is the true reality.[40]

The 'traces' that remain from now-discarded physical hypotheses—those relations which remain *despite theory change*—are the 'true relations of things' that are our sole access to 'true reality'.

Poincaré differentiates between types of hypotheses: (1) 'Some are *verifiable*, and when once confirmed by experiment become *truths of great fertility*', for empirically verifiable hypotheses propel science's advance. (2) Others 'may be useful to us in fixing our ideas' and guide our path to hypotheses of the first kind. (3) Still others 'are hypotheses only in appearance, and reduce to definitions or to conventions in disguise':

> The latter are to be met with especially in mathematics and in the sciences to which it is applied. From them, indeed, the sciences derive their rigor; such conventions are the result of the unrestricted activity of the mind, which in this domain recognizes no obstacle. For here the mind may affirm because it lays down its own laws; but let us clearly understand that while these laws are imposed on *our* science, which otherwise could not exist, they are not imposed on nature. Are they arbitrary? No; for if they were, they would not be fertile. Experience leaves us our freedom of choice, but it guides us by helping us to discern the most convenient path to follow.[41]

He distinguishes freely chosen conventions guided by experience from two antithetical extremes—conventionalism is *not* nominalism (mere names with no connection to physical reality), and conventionalism is *not* positivism (accessing 'things themselves'). Poincaré's argument is unambiguous: the aim of science is 'the relations between things; outside those relations there is no reality knowable' to science.[42]

4.4.2 1901–5: The crisis of theory, the 1904 St. Louis Congress, *The Value of Science*

Meanwhile, heated debates about the bankruptcy of science had fuelled distortions of Poincaré's 1900 ICP and ICPHIL lectures, issues he confronts in Part III, 'The Objective Value of Science', of his 1905 collection *La valeur de la science*. Chapters entitled 'Is Science Artificial' and 'Science and Reality' ask: 'Can science teach us the true relations of things?' Yes, Poincaré answers, distinguishing the *'form' taken by a physical theory* from the *underlying 'foundation' of 'true relations'* that remain true across different physical theories. While the form may change—as between Fresnel's and Maxwell's theories of light—the relations in the background foundation remain. Despite changing scientific hypotheses, the true relations 'will be found again under a new disguise in the other theories which will successively come to reign in place of the old'. Between 'the hypothetical currents which Maxwell supposes there are the same relations as between the hypothetical motions that Fresnel supposed'. Relations that remain true are our link to objective reality. So, 'what is the measure of their objectivity?'

> Well, it is precisely the same as for our belief in external objects... It may be said, for instance, that the ether is no less real than any external body; to say this body exists is to say there is between the color of this body, its taste, its smell, an intimate bond, solid and persistent; to say the ether exists is to say there is a *natural kinship between all the optical phenomena*, and neither of the two propositions has less value than the other... In sum, *the sole objective reality* consists in the relations of things whence results the universal harmony... [They] are objective because they are, will become, or will remain, common to all thinking beings.

The 'ether' for Poincaré is a name attached to the '*natural kinship* [*that exists*] *between all the optical phenomena*'—the objectively real relations that adhere in the phenomena and *persist* even with changes in the 'garment' of electromagnetic theory.[43]

Poincaré wrote these essays for *The Value of Science* to correct the 'strangest interpretations' (as if it were a defence of the Church against Galileo) of a comment Poincaré made at the 1900 ICPHIL regarding the truth status of propositions.[44] His address asserted that there is no absolute space and that we can only conceive of relative motion. In these essays, Poincaré explains the contested philosophical issue:

> No, there is no absolute space; these two contradictory propositions: 'The earth turns round' and 'The earth does not turn round' are, therefore, neither of them more true than the other. To affirm one while denying the other, *in the kinematic sense*, would be to admit the existence of absolute space.[45]

Poincaré relates the *kinematical impossibility* of determining absolute motion to the overriding question, can science teach us the true relations of things? At issue were profound challenges to our notions of space and time, high-stakes concerns he assessed in Part I, 'The Mathematical Sciences', in three reproduced articles: 'The Measure of Time' (1898), 'The Notion of Space' (1903), and 'Space and Its Three Dimensions' (1903)—directly linking 1898, when Poincaré asserted the metrological function of light and the conventionality of simultaneity, with 1903, when he completed his series 'Analysis Situs' regarding space of any dimension.

Poincaré's treatment of the crisis in theoretical physics putting these concepts at risk comprised Part II, 'The Physical Sciences', which links his 1897 ICM address 'Analysis and Physics' with his 1904 address to the St. Louis Congress (reproduced as three chapters assessing the history, present crisis and future of mathematical physics).[46] The St. Louis address encapsulates Poincaré's assessment of this crisis and contribution to its resolution during the intervening seven years, notably his statement of the principle of relativity and his oft-quoted concluding paragraph announcing the need for 'an entirely new mechanics' in which no apparent velocity can exceed that of light and in which observers in motion use a watch giving the 'local time'.[47]

A seminar analysing Poincaré's St. Louis address was held 31 January 1905, opening the Minkowski–Hilbert seminars on electron theory at Göttingen (held weekly until 31 July 1905); attendees of these seminars included Max Born, Max von Laue and Jakob Laub. While Poincaré's (5 June) and Einstein's (30 June) papers were not treated in this seminar, the seminar enabled the participants later to recognise the unique origin and substance of Einstein's purely kinematical theory of special relativity, in which ether had no role.[48] Our concern here is the unique substance of Poincaré's paper and how it established the mathematics of spacetime.

4.5 GÖTTINGEN, THE EXPANDING POINCARÉ CORPUS, AND MINKOWSKI'S PATH TO SPACETIME

This section discusses Minkowski's path to spacetime and sheds light on how Poincaré's contributions—his profound geometric and topological creations, his geometric conventionalism, his notions of space and time and his creation of spacetime geometry to establish 'true relations' that endure—became lost or distorted.

4.5.1 1890–1905: Pursuit of Poincaré's mathematics in the Klein–Hilbert–Minkowski seminars

For seventeen years (from 1890 to 1907), Klein ran a series of seminars (whose participants included Luigi Bianchi, William Osgood, Arthur Sommerfeld, Karl Schwarzschild and Hermann Weyl) that maintained a focused attention on Poincaré's works—reading and lecturing on them, dissecting and critiquing them and absorbing and incorporating their methods and results into the practices and sensibilities of the seminar participants. Poincaré's continuing creative flow kept Göttingen on high alert for his latest works.

David Hilbert's 1900 lecture to the second ICM identified twenty-three unsolved problems of great import, energising the goal to axiomatise and rigorise (thus render 'obsolete') what came before; but such work 'drew much of its strength from already flowing currents in 19th-century mathematical research'.[49] We see this strikingly in how Göttingen mathematicians intensively pursued and absorbed subtleties of Poincaré's reasoning.

In 1902, Poincaré returned a third time to his fourth geometry, highlighting its strange properties in his review of Hilbert's *Grundlagen der Geometrie*, prodding mathematicians to puzzle over why Poincaré finds it so important:

> Where would this new geometry rank in Hilbert's classification? We are glad to see that, as for the geometry of Riemann, all the axioms hold, except those of order and the axiom of Euclid; but whereas in the geometry of Riemann the axioms of order are false on all lines, in contrast, in the new geometry, *lines fall into two classes*, those on which the *axioms of order* are *true*, and those on which they are *false*.[50]

Poincaré would call upon these ('spacetime') relationships extensively in his 'Fifth Complement' to 'Analysis Situs'. Hilbert and his students and colleagues took note, as would Minkowski, for whom a third chair in mathematics was created at Göttingen in 1902. Hilbert and Minkowski held seminars featuring Poincaré's physics, focusing in 1903 on Poincaré's *New Methods of Celestial Mechanics* and many papers on 'figures of equilibrium of fluid masses', followed by mechanics (in 1904), and Lorentz's electron theory (in 1905, beginning with Poincaré's St. Louis lecture as noted in Section 4.4.2).[51]

In contemporaneous mathematics seminars, Minkowski lectured on Poincaré's 'Fifth Complement' (in late 1904), arithmetic and hyperbolic geometry (11 July 1905) and Fuchsian and Kleinian groups (25 July 1905). The claim that Minkowski's 11 July 1905 lecture presented a new model of hyperbolic geometry[52] is based on an article by Hans Jansen (from 1909)[53] that gave a detailed account of the hyperboloid model. But Jansen opens with a reference to Poincaré's (April 1881) Algiers paper[54] (where Poincaré derived the hyperboloid model within arithmetic in relation to his Fuchsian groups and functions), and then references Minkowski's lecture[55] as reporting arithmetic results he obtained based on the hyperboloid model. Minkowski's lecture on Poincaré's Fuchsian and Kleinian groups two weeks later would engage Poincaré's (June 1881) hyperbolic ball model. The focus on Poincaré's geometric moves was intensifying.

4.5.2 June 1905 through 1908: Poincaré's four-dimensional geometry, Minkowski's *Raum und Zeit*

Poincaré's 'On the Dynamics of the Electron' (a summary form) was read to the French academy on 5 June 1905; the paper appeared in January 1906.[56] As Darrigol explains, much in it was 'novel and important': a fully covariant formulation of 'Lorentz's transformations' (so named by Poincaré), 'its relativistic interpretation, its group-theoretic formulation, and its application to non-electromagnetic forces of cohesion or gravitation'.[57] Theoretical physicist Thibault Damour explains that most of Poincaré's 'key new results' are in the final section 'in a rather untransparent and unpedagogical form' but are mathematically complete: Poincaré had pioneered an 'elegant 4-dimensional geometrical formulation of Special Relativity' which Minkowski would expand upon in 1907–8 using more 'transparent notations'.[58]

We turn to Klein's opening lecture (1 November 1905) of the 1905–6 Klein–Hilbert–Minkowski mathematics seminar.[59] Stimulated by newly discovered notes from an 1859 course by Riemann on hypergeometric functions, the seminar would review related German works in hopes of attaining a proof (based on Riemannian principles) of Poincaré's 1883 uniformisation theorem (Hilbert's Problem #22). They would examine 'the work of Poincaré, beginning in 1881, published in the first five volumes of *Acta Mathematica*' and, in a few weeks, Klein would report 'to the [Mathematical Society] on the advances obtained by Poincaré and their relationship to my own investigations'.[60]

At the second session, Minkowski lectured again on Poincaré's 'Analysis Situs', featuring its results on multidimensional manifolds, and H. Mueller reported on Poincaré's work on Weierstrassian function theory. During the next five weeks, Gustav Herglotz gave lectures on Poincaré's uniformisation and Poincaré's '*Sur les résidus integrals doubles*' and Erhard Schmidt lectured on Poincaré's '*Sur les fonctions de deux variables*' while, at the Göttingen Mathematical Society, Ernst Zermelo lectured on Poincaré's work on boundary-value problems, and Schmidt on Poincaré's theory of differential equations—all this and more before the end of 1905.

At the eleventh session (31 January 1906), Klein lectured on Poincaré's development of Fuchsian functions, featuring Poincaré's stance on viewing the substitution group as a group of non-Euclidean rotations. Klein took the seminar participants through all of Poincaré's geometric moves, including a detailed account ('considered advantageous' by Minkowski) of Poincaré's description (quoted in Section 4.3) of how in 1880 he mapped from a calotte of one sheet of a two-sheeted hyperboloid to the unit disc.[61] The next five sessions continued reviewing Poincaré's theory of Fuchsian and Kleinian functions and groups, featuring Poincaré's general theory of automorphic forms that Klein (after Friedrich Schottky) had incorporated into his own programme with collaborators Ernst Ritter and Robert Fricke.[62]

On 13 June 1906, Poincaré submitted his proof of general uniformisation to *Acta Mathematica*.[63] Unaware of Poincaré's proof, Klein's seminars continued *almost a year* before Klein presented to the Göttingen Mathematical Society a proof of general uniformisation proposed by Paul Koebe (a seminar attendee). Koebe would provide a second proof on 19 November 1907 *that built upon (simplified, axiomatised) subtle novelties in Poincaré's proof* (which *Acta* had released earlier that month), novelties immediately absorbed into mainstream mathematical practice at Göttingen.

Klein's presentation of Koebe's first proof occurred on *the very same day* (11 May 1907) as Minkowski's first lecture on the equations of electrodynamics. Six months later (5 November 1907, *two weeks before Koebe's celebrated second proof*), Minkowski presented a second lecture, 'The Principle of Relativity',[64] in which he acknowledged much debt to Poincaré's January 1906 (Palermo) paper, highlighting Poincaré's four-dimensional treatment of gravitation. Minkowski's 1907 paper is evidence that 'it was Poincaré who most directly influenced the mathematics of Minkowski's space-time'.[65] And, as a founder of the Minkowski Institute for Foundational Studies states, it was 'Poincaré who first realized (before July 1905) that the Lorentz transformations have a natural geometric interpretation as rotations in a four-dimensional space whose fourth dimension is time'.[66]

Minkowski brought together Poincaré's 'true geometrical relations' of the four-dimensional space with the 'true physical relations' established by Einstein: as Minkowski wrote in his Cologne lecture '*Raum und Zeit*' on 21 September 1908, 'the credit of first recognising clearly that the time of one of the electrons is just as good as that of the other, that is to say, that t and t' are to be treated identically, belongs to A. Einstein'.[67] But Minkowski did not acknowledge Poincaré's crucial contribution in his Cologne lecture.[68] In the context of the Göttingen milieu, Minkowski's omission is not surprising, for both Klein and Hilbert, in different ways, had made it standard practice to absorb Poincaré's and others' contributions as elements of their own programmes, which they saw as superseding what came before.[69] The coincidence, in terms of both place and timing, of Koebe's two proofs of uniformisation with Minkowski's first two 1907 lectures on the principle of relativity may have played a role.

4.6 CONCLUDING COMMENTS: 1908–12 AND BEYOND

Minkowski's November 1907 lecture drawing upon Poincaré's Palermo paper remained unpublished until 1915. Poincaré never entered into priority disputes, but in 1912 he was asked to clarify his stance on the new mechanics. While many interpret Poincaré's 'Space and Time' paper as rejecting spacetime, mathematician Shlomo Sternberg argues that Poincaré's last four paragraphs establish his long-standing ownership of the geometry of the spacetime manifold. Poincaré died while this paper was in press, as was his paper 'Why Space Has Three Dimensions', where he developed a topological concept of dimension and emphasised how the amorphous continuum becomes space when a metric and other structures are imposed.[70] As these works unambiguously show, Poincaré had broken free from limitations of the physicist's ether concept, establishing 'true relations' that endure—a foundation for the electromagnetic and gravitational fields of relativistic and quantum phenomena.

Within a month after attending the 1911 first Solvay Conference as 'a newcomer to quantum ideas', Poincaré published three proofs at different levels of generality of the necessity of the quantum hypothesis. His proof made some prominent sceptics feel 'logically compelled to accept the quantum hypothesis in its entirety' and, in the 1920s, quantum physicists cemented Poincaré's newly invented 'integral-over-states' method at the heart of statistical mechanics, showing it to be 'completely justified'.[71]

Poincaré also wrote a recommendation for Einstein for a position at ETH Zürich, praising Einstein's unrivalled physical intuition and creativity:

> Einstein is one of the most original minds I have known...[The] facility with which he has adapted to new conceptions and from which he knows how to draw the consequences... translates immediately in his mind into the prediction of new phenomena, susceptible of being one day verified by experiment...The future will show more and more the value of Mr. Einstein, and the university that finds a way to secure this young master is assured of drawing from it great honor.[72]

And, despite legends to the contrary, Einstein respected Poincaré's comments on quantum theory and dynamical systems at the conference, drawing upon them in his highly creative 1917 paper on quantum chaos.[73] Einstein was also well aware that many philosophers misinterpreted Poincaré's ideas, noting in 1919 that the German mathematician and philosopher Eduard Study (an early Klein protégé),[74]

> has treated him quite badly by pinning him down on a *truly superficial* comment about the practical significance of Euclidean geometry. Poincaré's expositions on the place of geometry within the whole scientific system appear to me to be considerably more profound.[75]

Einstein here picks out *precisely* the point which others—missing Poincaré's 'considerably more profound' mathematics and philosophy (and associated cosmological considerations)—distort into a dogma or premise.

Even physicists and mathematicians unaware of Poincaré's 'profound' contributions were immersed within the mathematical world he helped create, and, however unknowingly, used his tools, methods, reasoning and language. As the mathematician and physicist Hermann Weyl acknowledged in 1931:

> We differentiate now between the amorphous continuum and its metrical structures. The first has retained its *a priori* character... whereas the structural field is completely subjected to the power-play of the world; being a real entity, Einstein prefers to call it the ether.[76]

The amorphous continuum, the mathematics of analysis situs, enables creation of the various 'structures' required in physics. Here we find Poincaré's legacy: the mathematical realm and methods of action needed to express physical relations within our space—however one choses to label them—relations so delicate, rich and precise.

NOTES

1. Henri Poincaré, 'Analysis and Physics', in *The Value of Science (VOS)*, trans. George Bruce Halsted (New York: The Science Press, 1907), 75–83, p. 76; orignally published as Henri Poincaré, '*Sur les rapports de l'analyse pure et de la physique mathématique*', *Acta Mathematica* 21 (1897): 331–41; republished (with the mathematical formulae removed) as Henri Poincaré, '*L'analyse et la physique*', in *La valeur de la science* (Paris: Flammarion, 1905).
2. Poincaré, 'Analysis and Physics', pp. 75–6.

3. Poincaré, 'Analysis and Physics', p. 76.
4. Albert Einstein, *Autobiographical Notes*, trans. and ed. Paul Arthur Schilpp (La Salle: Open Court, 1949), pp. 88–9. Einstein's final sentence reads: 'This exposition has fulfilled its purpose if it shows the reader how the efforts of a life hang together and why they have led to expectations of a certain kind.'
5. Henri Poincaré, '*Les géométries non euclidiennes*', *Revue générale des sciences pures et appliquées* 2 (1891): 769–74; trans. as 'Non-Euclidean Geometries' *Nature* 45 (1892): 404–7; reprinted as Chapters 3, 4 and 5 ('*Les géométries non euclidiennes*', '*L'espace et la géométrie*' and '*L'expérience et la géométrie*') in Part II: *L'Espace*' in Henri Poincaré, *La science et l'hypothèse* (Paris: Flammarion, 1902), pp. 49–67, 68–91, 92-109; this book was later translated into English as Henri Poincaré, *Science and Hypothesis* (*S&H*), trans. William Scott Greenstreet (London: Scott, 1905).
6. Henri Poincaré, '*Correspondance sur les géométries non euclidiennes* (*lettre à* M. Mouret),' ***Revue générale*** 3 (1892): 74–5. English translation by Peter Pesic, *Beyond Geometry: Classic Papers from Riemann to Einstein* (Mineola, NY: Dover, 2007): 105–7, p. 105.
7. Poincaré's conventionalism dominated much of twentieth-century philosophy of science, later joined by structural realist interpretations. See Michael Friedman, 'Einstein, Kant, and the Relativized *A Priori*', in Michel Bitbol, Pierre Kerszberg and Jean Petitot, eds., *Constituting Objectivity* (Dordrecht: Springer, 2009): 253–67; Yemima Ben-Menahem, *Conventionalism* (Cambridge: Cambridge University Press, 2006); 'Poincaré's Impact on Twentieth-Century Philosophy of Science', *HOPOS* 6 (2016): 257–73; Janet Folina, 'Poincaré and the Invention of Convention', in Maria de Paz and Robert DiSalle, eds., *Poincaré, Philosopher of Science* (Dordrecht: Springer, 2014), 25–45; Gerhard Heinzmann, 'Henri Poincaré and His Thoughts on the Philosophy of Science', in Éric Charpentier et al., eds., *The Scientific Legacy of Poincaré* (Providence: American Mathematical Society, 2010), 373–91, p. 374; Katherine Brading and Elise Crull, 'Epistemic Structural Realism and Poincaré's Philosophy of Science', *HOPOS* 7 (2017): 108–29.
8. Henri Poincaré, '*Analyse des travaux scientifiques de Henri Poincaré faite par lui-même*' (1901 update), *Acta Mathematica* 38 (1921; section printed 28 March 1913): 3–135, p. 101; my translation.
9. See Michael Friedman, 'Kant–*Naturphilosophie*–Electromagnetism', in Michael Friedman and Alfred Nordmann, eds., *The Kantian Legacy in Nineteenth-Century Science* (Cambridge: MIT Press, 2006), 51–79.
10. Several poets and artists (symbolists, cubists) seized upon Poincaré's mathematical conceptions: Linda Dalrymple Henderson, *The Fourth Dimension and Non-Euclidean Geometry in Modern Art, Revised Edition* (Cambridge: MIT Press, 2013); Romano Nanni, '*Voir par l'intellect, voir par les yeux*', in Christina Vogel, ed., *Valéry et Léonard* (Frankfurt: Peter Lang, 2007), 141–52; Steven Cassedy, *Flight from Eden* (Berkeley: University of California Press, 1990).
11. Henri Poincaré, 'Analysis Situs', *Journal de l'Ecole Polytechnique* 1 (1895): 1–121; in Henri Poincaré, *Papers on Topology: Analysis Situs and Its Five Supplements*, trans. John Stillwell (Providence: American Mathematical Society, 2010), 5–74, p. 5; emphasis added.
12. Connemara Doran, *Seeking the Shape of the Universe: Confronting the Hyperbolic World, from Henri Poincaré to the Cosmic Microwave Background* (PhD dissertation, Harvard University, 2017).
13. Henri Poincaré, '*L'espace et ses trois dimensions*', *Revue de métaphysique et de morale* 11 (1903): 281–301, 407–29.
14. Poincaré, *VOS*, pp. 40–41; emphasis added.
15. Henri Poincaré, '*L'avenir des mathématiques*', Plenary Lecture at the ICM in Rome (April 1908). Translation by Jeremy Gray, 'Poincaré Replies to Hilbert: On the Future of Mathematics ca. 1908', *The Mathematical Intelligencer* 34 (2012): 15–29, p. 23.

16. Henri Poincaré, 'Why Space Has Three Dimensions', *Mathematics and Science: Last Essays*, trans. John W. Bolduc (New York: Dover Publications, 1963): 25–44, p. 42; originally published as Henri Poincaré, *'Porquoi l'espace a trois dimensions'*, *Revue de métaphysique et de morale* 20 (1912): 483–504; *'Porquoi l'espace a trois dimensions'* in Henri Poincaré, *Dernières pensées* (Paris: Flammarion, 1913), 55–98.
17. Poincaré, *'Les géométries non euclidiennes'*.
18. Poincaré, *S&H*, p. 43.
19. Poincaré, *S&H*, p. 43; Jeremy Gray, 'The Three Supplements to Poincaré's Prize Essay of 1880 on Fuchsian Functions and Differential Equations', *Archives internationals d'histoire des sciences* 32 (1982): 221–35, p. 221. Henri Poincaré, *Three Supplements on Fuchsian Functions*, ed. Jeremy Gray and Scott Walter (Berlin: Akademie-Verlag, 1997). See assessments in Jeremy Gray, *Linear Differential Equations and Group Theory from Riemann to Poincaré* (Boston: Birkhäuser, 1986; second edition 2000); *Henri Poincaré: A Scientific Biography* (Princeton: Princeton University Press, 2013); '"The Soul of the Fact": Poincaré and Proof', *Studies in History and Philosophy of Modern Physics* 47 (2014): 142–50.
20. Poincaré reports this astonishing advance in the second supplement, submitted 6 September 1880 to the Académie des Sciences in Paris; and in one of two papers he presented at the Algiers meeting of the Académie des Sciences on 16 April 1881 (*Comptes rendus des sessions de l'Association Française pour l'Avancement des Sciences*, 10th Session, Algiers, 1881).
21. William Reynolds, 'Hyperbolic Geometry on a Hyperboloid', *The American Mathematical Monthly* 100, 5 (May 1993): 442–55, p. 453. Thomas Hawkins, 'Non-Euclidean geometry and Weierstrassian mathematics', *Historia Mathematica* 7 (1980): 289–342.
22. Henri Poincaré, *Sources of Hyperbolic Geometry*, trans. John Stillwell (Providence: American Mathematical Society, 1996), p. 121 (originally published as Henri Poincaré, *Notice sur les travaux scientifiques de M. Poincaré (rédigée par lui-même)* (Paris, Gauthier-Villars, 1884; second edition 1886), explaining his 1880–1 work); a figure by Konrad Polthier, Freie Universität Berlin, is reproduced at p. 120 and as the cover image.
23. Felix Klein to Henri Poincaré, 19 June 1881. Henri Poincaré to Felix Klein, 22 June 1881. *'Correspondance d'Henri Poincaré et de Felix Klein'*, *Acta Mathematica* 39 (1923): 94–132. Poincaré delivered his generalisation to the Académie des Sciences in publications dated 27 June and 4 July 1881. Connemara Doran, 'Poincaré's Path to Uniformization', in Lizhen Ji and Shing-Tung Yau, eds., *Uniformization, Riemann–Hilbert Correspondence, Calabi–Yau Manifolds, and Picard–Fuchs Equations*, Advanced Lectures in Mathematics (Boston, International Press, 2018), 55–79.
24. Henri Poincaré, *'Sur les hypothèses fondamentales de la géométrie'*, *Bulletin de la Société mathématique de France* 15 (1887): 203–16. English translation by Shlomo Sternberg, 'Review of *Imagery in Scientific Thought* by Arthur I. Miller', *The Mathematical Intelligencer* 8, 2 (1986): 65–74, p. 68. Poincaré's title, like Riemann's, featured 'the fundamental hypotheses' (not 'the facts') of geometry.
25. Poincaré, *'Les géométries non euclidiennes'*, p. 772; my translation.
26. Sternberg, 'Review of *Imagery in Scientific Thought*', p. 68.
27. Henri Poincaré, *'L'espace et la géométrie'*, *Revue de métaphysique et de morale* 3 (1895): 631–46.
28. Poincaré, *VOS*, p. 39; Poincaré, *'L'espace et ses trois dimensions'*.
29. Poincaré, *'Analyse des travaux scientifiques'*, pp. 129–31; my translation.
30. John Stachel, 'Fresnel's (Dragging) Coefficient as a Challenge to 19th Century Optics of Moving Bodies', in Anne Kox and Jean Eisenstaedt, eds., *The Universe of General Relativity* (Boston: Birkhäuser, 2005): 1–13. Peter Galison, *Einstein's Clocks, Poincaré's Maps: Empires of*

Time (New York: Norton, 2003). Max Jammer, *Concepts of Simultaneity* (Baltimore: Johns Hopkins University Press, 2006). Olivier Darrigol, 'Poincaré's Light', in *Poincaré, 1912–2012*, ed. Bertrand Duplantier and Vincent Rivasseau (Basel: Springer Birkhäuser, 2012): 1–50.

31. The update cites Henri Poincaré, *'Préface'*, in *Théorie mathématique de la lumière* (Paris: Publications chez G. Carré et Naud, 1889), pp. i–iv, and the 1900 ICP address *'Sur les rapports de la physique mathématique et de la physique expérimentale'*, which was immediately published as Henri Poincaré, *'Les relations entre la physique expérimentale et la physique mathématique'*, in C.-E. Guillaume and L. Poincaré, eds., *Rapports présentés au Congrès international de physique réuni à Paris en 1900 sous les auspices de la Société française de physique, rassemblés et publiés*, vol. 1 (Paris: Gauthier-Villars, 1900), 1–29, and Henri Poincaré, *'Les relations entre la physique expérimentale et la physique mathématique'*, *Revue générale des sciences pures et appliquées* 11 (1900): 1163–75; it was then republished in 1902 as Henri Poincaré, *'Les hypothèses en physique'*, in Poincaré, *La science et l'hypothèse*, 167–88, and Poincaré, *'Les théories de la physique moderne'*, in Poincaré, *La science et l'hypothèse*, 189–212.
32. Poincaré, *'Analyse des travaux scientifiques'*, p. 130.
33. Poincaré, *'Analyse des travaux scientifiques'*, p. 130; emphasis added.
34. Poincaré, *'Analyse des travaux scientifiques'*, p. 130.
35. Poincaré, *'Analyse des travaux scientifiques'*, p. 131.
36. Poincaré, *S&H*, pp. 160–1.
37. Henri Poincaré, *'Sur les principes de la mécanique'*, *Bibliothèque du Congrès international de philosophie*, vol. 3 (Paris, 1901): 457–94; p. 457; Poincaré, 'The Classical Mechanics', in *S&H*, 89–110, p. 89, and Poincaré, 'Relative and Absolute Motion', in *S&H*, pp. 111–22.
38. Henri Poincaré, *'La mesure du temps'*, *Revue de métaphysique et de morale* 6 (1898): 1–13; Henri Poincaré, The 'Measure of Time', in Poincaré, *VOS*, 26–36.
39. Poincaré, Chapter 6 'The Classical Mechanics', *S&H*, p. 90.
40. Poincaré, *S&H*, pp. xxv–xxvi.
41. Poincaré, *S&H*, pp. xxii–xxiii.
42. Poincaré, *S&H*, p. xxiv.
43. Poincaré, *VOS*, pp. 138–40; emphasis added.
44. Poincaré, *VOS*, p. 140.
45. Poincaré, *VOS*, p. 141; Poincaré's emphasis.
46. Henri Poincaré, *'L'état actuel et l'avenir de la physique mathématique'*, *Bulletin des sciences mathématiques* 28, 2nd series (1904): 302–24; trans. George Bruce Halsted, *The Monist* 15 (1905): 1–24.
47. Poincaré, *VOS*, p. 111.
48. Richard Staley, *Einstein's Generation: The Origins of the Relativity Revolution* (Chicago: University of Chicago Press, 2008), pp. 369–75.
49. David Rowe, 'Klein, Hilbert, and the Göttingen Mathematical Tradition', *Osiris*, 2nd Series, 5 (1989): 186–213, p. 199; 'The Calm before the Storm', in Vincent Hendricks, Stig Andur Pedersen and Klaus Frovin Jørgensen, *Proof Theory* (Dordrecht: Springer, 2000), p. 71.
50. Henri Poincaré, *'Les fondements de la géométrie: Grundlagen der Geometrie par* M. Hilbert, *professeur à l'Université de Göttingen'*, *Bulletin des sciences mathématiques* 26 (1902): 249–72, reprinted in Henri Poincaré, *Scientific Opportunism, L'Opportunisme scientifique: An Anthology*, comp. Louis Rougier, ed. Laurent Rollet (Basel: Birkhäuser, 2002), 33-46, p. 44; my translation; emphasis added.
51. Scott Walter, 'Breaking in the 4-Vectors', in Jürgen Renn and Matthias Schemmel, eds., *The Genesis of General Relativity*, vol. 3 (Berlin: Springer, 2007): 193–252, p. 214; 'Minkowski, Mathematicians, and the Mathematical Theory of Relativity', in Hubert Goenner, Jürgen Renn, Jim Ritter and Tilman Sauer, eds., *The Expanding Worlds of General Relativity* (Boston:

Birkhäuser, 1999): 45–86; 'Minkowski's Modern World', in Vesselin Petkov, ed., *Minkowski Spacetime: A Hundred Years Later* (Dordrecht: Springer, 2010), 43–61. Leo Corry, 'Hermann Minkowski, Relativity and the Axiomatic Approach to Physics', in Petkov, *Minkowski Spacetime*, 3–41; Peter Galison, 'Minkowski's Space-Time: From Visual Thinking to the Absolute World', *Historical Studies in the Physical Sciences* 10 (1979): 85–121.

52. Apparently misreading Reynolds, 'Hyperbolic Geometry on a Hyperboloid', which correctly situates Poincaré's 1880–1 papers (see notes 19 and 20), Poincaré's 1887 paper (Poincaré, '*Sur les hypothèses fondamentales de la géométrie*'), Minkowski's 1907 article (published posthumously as Hermann Minkowski, '*Das Relativitätsprinzip*', *Annalen der Physik* 352 (1915): 927–38) and Jansen's 1909 article, published as Hans Jansen, '*Abbildung der hyperbolischen Geometrie auf ein zweischaliges Hyperboloid*', *Mitteilungen der Mathematische Gesellschaft in Hamburg* 4 (1909): 409–40 (not noticing Poincaré's 1905–6 articles (see note 56)).
53. Jansen, '*Abbildung der hyperbolischen Geometrie auf ein zweischaliges Hyperboloid*', p. 409.
54. Jansen references H. Poincaré, '*Sur les applications de la géométrie non euclidienne à la théorie des formes quadratiques*', in *Association française pour l'avancement des sciences, Compte Rendu de la 10E Session. Alger. 1881* (Paris: *Association française pour l'avancement des sciences*, 1881): 132–8.
55. Jansen references: Hermann Minkowski, '*Jahresbericht der Deutschen Mathematiker-Vereinigung*' ed. by August Gutzmer (Leipzig: Teubner, 1905).
56. Henri Poincaré, '*Sur la dynamique de l'électron*', *Comptes rendus* 140 (1905): 1504–8, and '*Sur la dynamique de l'électron*', *Rendiconti del circolo matematico di Palermo* 21 (1906): 129–76.
57. Darrigol, 'Poincaré's Light', p. 39.
58. Thibault Damour, 'What Is Missing from Minkowski's "*Raum und Zeit*" lecture', *Annalen der Physik* 17 (2008): 619–30, p. 625.
59. Felix Klein, 'Mathematical Seminar at Göttingen, Winter Semester 1905/1906 under the Direction of Professors Klein, Hilbert, Minkowski. Talks by F. Klein. Notes by Dr. Phil. Otto Toeplitz', http://www.claymath.org/sites/default/files/klein1math.sem_._ws1905-06.pdf, accessed 29 March 2018.
60. Klein, 'Mathematical Seminar', p. 3.
61. Felix Klein, '*Elfte Sitzung*', in *Klein Protokolle, Band* 23, *Seite* 85, http://page.mi.fu-berlin.de/moritz/klein/#id-988, accessed 29 March 2018.
62. Umberto Bottazzini and Jeremy Gray, *Hidden Harmony: Geometric Fantasies: The Rise of Complex Function Theory* (New York: Springer, 2013), p. 611.
63. Doran, 'Poincaré's Path to Uniformization'.
64. Minkowski, '*Das Relativitätsprinzip*'.
65. Galison, 'Minkowski space-time', 94–5, noting that Minkowski and Poincaré differed in their ontological commitments.
66. Petkov, 'Preface', *Minkowski Spacetime*, p. vi.
67. Hermann Minkowski, '*Raum und Zeit*', *Jahresberichte der Deutschen Mathematiker-Vereinigung* (Leipzig: Teubner, 1909); 'Space and Time', trans. Dennis Lehmkuhl in Petkov, *Minkowski Spacetime*, p. xxv.
68. See Damour, 'What Is Missing'.
69. Doran, 'Seeking the Shape of the Universe'.
70. Henri Poincaré, '*L'espace et le temps*', in Poincaré, *Dernières pensées*, 33–54; Henri Poincaré, '*Pourquoi l'espace a trois dimensions*', in Poincaré, *Dernières pensées*, 55–98; Sternberg, 'Review of *Imagery in Scientific Thought*'.
71. Jeffrey Prentis, 'Poincaré's Proof of the Quantum Discontinuity of Nature', *American Journal of Physics* 63 (1995): 339–50, pp. 340, 348; Henri Poincaré, '*Sur la théorie des quanta*', *Comptes*

rendus de l'Académie des sciences 153 (1911): 1103–8; Henri Poincaré, '*Sur la théorie des quanta*', *Journal de physique théorique et appliquée*, 5th series, 2 (1912): 5–34; Henri Poincaré, '*L'hypothèse des quanta*', *Revue scientifique*, 4th series, 17 (1912): 225–32.

72. Henri Poincaré to Pierre Weiss, *c.* November 1911. English translation in Galison, *Einstein's Clocks, Poincaré's Maps*, p. 300.
73. A. Douglas Stone, 'Einstein's Unknown Insight and the Problem of Quantizing Chaos', *Physics Today* (2005): 1–7, pp. 1–3; Albert Einstein, '*Zum Quantensatz von Sommerfeld und Epstein*', *Deutsche Physikalische Gesellschaft. Verhandlungen* 19 (1917): 82–92.
74. Eduard Study, *Die Realistische Weltansicht und die Lehre vom Raume*, 1914.
75. Albert Einstein to Hans Vaihinger, 3 May 1919, in *The Collected Papers of Albert Einstein, Volume 9, The Berlin Years: Correspondence, January 1919–April 1920 (English translation supplement)*, ed. Diana Kormos Buchwald, Robert Schulmann, Jozsef Illy, Daniel J. Kennefick and Tilman Sauer and trans. Ann Hentschel (Princeton: Princeton University Press, 1989), 29; emphasis added.
76. Hermann Weyl, '*Geometrie und Physik*', *Die Naturwissenschaften* 19 (1931): 49–58, p. 51; translation from John L. Bell and Herbert Korté, 'Hermann Weyl', in Edward N. Zalta, ed., *The Stanford Encyclopedia of Philosophy* (Winter 2016 Edition), http://plato.stanford.edu/archives/win2016/entries/weyl/, accessed 29 March 2018.

5

Ether and Electrons in Relativity Theory (1900–11)

Scott A. Walter

5.1 INTRODUCTION

The history of twentieth-century physics is scanned by the 'classical', or non-quantum theory of the electron, introduced by H. A. Lorentz, Joseph Larmor and others from the early 1890s. As a research programme in physics, electron theory became more promising at the end of the decade, with the measurement of the electron mass-to-charge ratio in cathode-ray deflection experiments by J. J. Thomson and Emil Wiechert. It was successful in explaining magneto-optical phenomena, beginning with the Zeeman effect, and soon appeared to point the way to unifying the explanation of all forces—including the force of gravitation—by means of interactions between electrons and the ether. During the first two decades of the twentieth century, the physicists Lorentz, Henri Poincaré, Albert Einstein, Ebenezer Cunningham and Max Laue, all of whom were familiar with the idea of electromagnetic waves and electrons propagating in the ether, were led to engage with new technologies, mathematical formalisms and ideas about relative time and space. This chapter sketches some of the ways in which leading physicists sought to adapt pre-relativistic ether-based electron theories to the physics of relativity, from the emergence of the electromagnetic world picture in 1900, to the publication of Emil Wiechert's post-Minkowskian defence of the ether in 1911.

The history of the electron and that of the ether are tightly knit in the early twentieth century, although historians have paid much more attention to the former than the latter, leaving gaps in the literature concerning electron theory.[1] Since E. T. Whittaker's history of theories of 'æther' and electricity (1953),[2] historians of physics have considered the emergence of relativity theory in the context of ether and electron theory. A one-time participant in this history, Whittaker created a stir late in his life by attributing the theory of relativity not to Albert Einstein but to Lorentz and Poincaré. Much effort was subsequently expended to discredit Whittaker's view and reinforce Einstein's priority of discovery. The historian Stanley Goldberg, for example, argued that the theories of

Walter, S.A., 'Ether and Electrons in Relativity Theory (1900–11)' in *Ether and Modernity*, edited by J. Navarro.
 DOI: 10.1093/oso/9780198797258.003.0005

Poincaré and Einstein were distinct, except for the formalism.[3] Many commentators since Goldberg have focused on the contrast between, on the one hand, Einstein's provocative claim that the ether is a superfluous concept in electrodynamics and, on the other hand, Poincaré's advocacy of an abstract ether.[4]

Most recently, Peter Galison showed that Poincaré and Einstein shared an interest in the topics of microphysics and clock synchronisation.[5] For Galison, the emergence of relativity theory was the result of a chance confluence of three separate conceptual streams: physics, philosophy and technology. To these three streams I will add a fourth: applied mathematics, which Poincaré and Einstein brought into play in 1905, with profound and far-reaching consequences for conceptions of ether, electrons and relativity.

The first three streams came together in Leiden, in December 1900. The occasion for this meeting was the jubilee celebration of Lorentz's defence of his doctoral thesis; Poincaré had been invited to this celebration, along with one of his students and most of Europe's electrodynamicists. Poincaré declined the invitation but contributed a paper in which he analysed the status of Newton's third law in Lorentz's electron theory.[6]

5.2 LORENTZ ELECTRONS AND LOCAL TIME IN LEIDEN

For purely mathematical reasons, Lorentz had introduced a fictive time which he called '*Ortszeit*' and which Poincaré called '*temps local*', or 'local time', following the translation of his former student, Alfred Liénard.[7] Poincaré realised that local time could be construed as a measurable quantity: local time is the time measured at relative rest by observers in uniform motion with respect to the ether, provided that they synchronise their watches in a certain way. The watches of the co-moving observers are synchronised, Poincaré explained, via an exchange of light signals, whereby light time-of-flight between observers is taken into account when zeroing the watches, but their common motion with respect to the ether is ignored.[8]

Poincaré knew quite well that he was breaking new ground, as he described his time-synchronisation protocol as a *definition* of local time. In providing this definition, Poincaré turned Lorentz's electron theory into a regular theory of physics, that is, one having measurable consequences, whereas, before, it had been an abstract model. Lorentz seems not to have appreciated this point, although he thanked Poincaré for his contribution.[9]

In addition to being a contribution to physics, Poincaré's paper was a result of philosophical reflection on the nature of distant simultaneity, in relation to technical advances in astronomy and geodesy. As a member of the French Bureau of Longitudes, Poincaré was charged with the scientific oversight of national and colonial mapping projects, and French participation in international collaborations in astronomy and geodesy.

One of the latter projects sought to measure an arc of meridian passing through Peru, Ecuador and Colombia. Longitude measurement at the end of the nineteenth century involved several techniques, including that of fixing the position of celestial objects, and noting the temporal instant of astronomical events such as occultations and zenith passages. To determine the exact time of a given event, geodesists relied on time signals relayed by telegraph, as Poincaré observed in an original essay entitled 'The Measurement of Time' (1898).[10]

Poincaré's essay on time measurement and his definition of local time both deeply influenced the later development of the principle of relativity, while Poincaré's own conceptions evolved in the first years of the twentieth century, partly as a result of new ideas in electron theory, including those of Paul Langevin.

5.3 LANGEVIN'S ELECTRON WAVES IN ETHER

Paul Langevin's theory of the electron drew on several sources, especially Joseph Larmor's *Æther and Matter,*[11] J. J. Thomson's *Notes on Recent Researches*[12] and George Searle's calculation of the energy of a Heaviside ellipsoid.[13] He introduced a distinction between the velocity fields and acceleration fields of the electron, and published a graphical depiction of the velocity waves of a spherical electron in motion. This led in turn to a geometrical derivation of the field of a Heaviside ellipsoid, introduced by Heaviside in 1889 and glossed by J. J. Thomson in 1893.[14]

Langevin was concerned, as the title of his paper implies, with the source of electromagnetic radiation and the inertia of the electron.[15] He supposed (following Larmor) that electron radiation was due entirely to acceleration, and he calculated thereby the energy of an electron in uniform motion. The 'electromagnetic mass' of such an electron was given to be a function of the 'wake' of the electron in motion, represented by Langevin's velocity waves. The electron's wake, in turn, depended on the charge distribution of the electron; Langevin considered uniform distributions by surface and volume.

Langevin's distinction between velocity and acceleration waves recalled the retarded potentials introduced to electrodynamics by Langevin's former teacher Henri Poincaré in the early 1890s, as well as Alfred Liénard and Emil Wiechert's formulation of the potentials for a moving point charge.[16] Langevin's theory, like all contemporary electron theories, assumed that his waves propagated in an ether at absolute rest. According to Langevin's theory, electrons travelled through the ether at sub-light velocity, generating velocity waves and, in the case of non-inertial motion, acceleration waves. These waves propagated in the ether with the speed of light; velocity waves dissipated rapidly, such that only acceleration waves could be detected far from the electron.[17]

5.4 POINCARÉ ON APPARENT TIME AND THE RELATIVISTIC ELECTRON

As a student in the mid-1890s, Langevin had followed Poincaré's lectures on Sommerfeld's theory of diffraction, but he did not engage personally with Poincaré until September, 1904, when they were both members of the French delegation to the Congress of Arts and Sciences, held at the World's Fair in St. Louis. The younger man was flattered by the attention of his former teacher, as he recounted the meeting by letter to his wife back in Paris.[18] By that time, the two men had a mutual interest in the theory of electrons, which was the topic of Langevin's lecture in St. Louis.[19]

We do not know if Langevin discussed with Poincaré his forthcoming paper on the inertia of the electron, but we do know that Poincaré found inspiration from the latter paper for his discovery of the Lorentz group. Under the coordinate transformations of the Lorentz group, Poincaré demonstrated in 1905, the laws of electrodynamics retain their form. What impressed Poincaré most was not Langevin's constant-volume model of the electron but his explanation of the velocity and acceleration waves produced by an electron; according to Langevin, these waves propagated in free ether at the speed of light. Instead of Langevin's model, Poincaré preferred the deformable electron model proposed by Lorentz, as this model had the advantage, as Poincaré proved, of preserving the principle of relativity. Poincaré noticed further that, by applying the Lorentz transformations to Langevin's acceleration waves, he could recover Hertz's solution of Maxwell's equations for an oscillator at rest in the absolute ether.[20]

In June 1905, Poincaré supposed that all laws of physics were, likewise, form invariant with respect to the transformations of the Lorentz group, including the law of gravitation.[21] In a letter to Lorentz announcing his discovery, Poincaré observed that the requirement of Lorentzian form invariance spelled the end of what he called the 'unity of time'.[22] Yet, Poincaré was not ready to abandon the traditional definition of time and space in this new theoretical context. He deftly elided the question of time and space deformation in his memoir on the dynamics of the electron by focusing on active transformations alone.[23]

In his university lectures of 1906–7, Poincaré explained how, in principle, one could measure Langevin waves and thereby determine the shape of an electromagnetic pulse generated by a source in motion with respect to the ether. According to lecture notes recorded by a student note-taker, Poincaré recalled Langevin's paper and reproduced the latter's schematic of an electron in motion (Fig. 5.1) along with a diagram of his own creation (Fig. 5.2), which showed how the electromagnetic pulse was related to the Lorentz transformations. The pulse created by the point source had the form of an ellipsoid, which was elongated in the direction of motion of the source and had a focus located at the source. A section through a meridian of the ellipsoid produces the ellipse shown in Fig. 5.2.

Poincaré's diagram illustrates the Lorentz contraction, whereby all material objects contract by a Lorentz factor, but only in the direction of their motion with respect to the ether. Electromagnetic waves are immaterial, and are thereby unaffected by this contraction. If we measure the form of an electromagnetic pulse with a material rod at any instant of time, the result will depend on the speed of the rod with respect to the ether. For any non-zero rod velocity, the 'true' form of the pulse is then an ellipsoid with an elongation proportional to the velocity of the measuring rod, assumed to be at relative rest with respect to the moving point source.

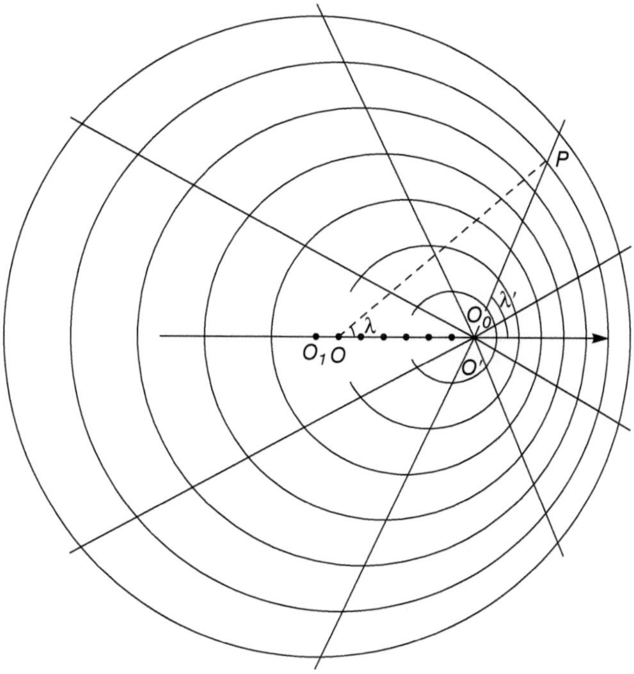

Fig. 5.1 Schematic diagram of an electron moving uniformly from left to right, with its velocity waves. Redrawn from Langevin's '*Sur l'origine des radiations*'.

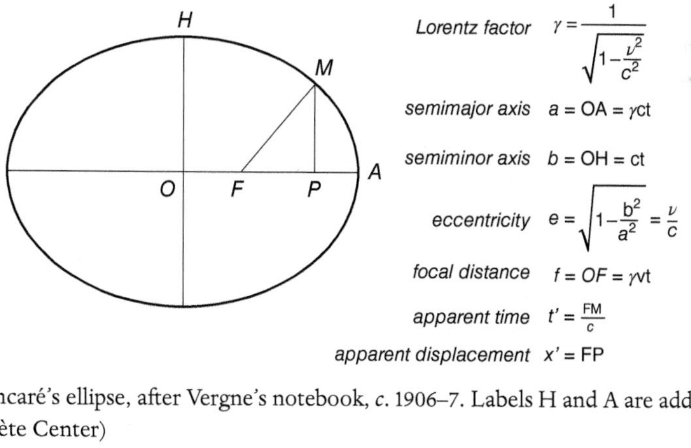

Fig. 5.2 Poincaré's ellipse, after Vergne's notebook, *c*. 1906–7. Labels H and A are added for legibility. (Viète Center)

Poincaré's ellipse was designed to explain the meaning of the primed variables appearing in the Lorentz transformations, which is to say, in Poincaré's terminology, the 'apparent' time and space coordinates of a reference frame moving uniformly with respect to the ether. While Poincaré affirmed the reality of Lorentz contraction of material bodies, he recoiled from affirming the corresponding reality of temporal deformation. According to student notes, he explained the situation as follows: 'So Lorentz assumes that all bodies undergo a contraction in the direction of motion proportional to the square of velocity. Lengths are then altered, and durations are altered by the impossibility of setting watches truly, such that the apparent velocity of light is constant.'[24]

Why did Poincaré feel it was impossible to 'truly' synchronise mobile timekeepers in a state of relative rest? He did not explain this statement, but his intention is not hard to discern. Poincaré always referred to primed time and space coordinates of a reference frame in motion with respect to the ether as 'apparent' coordinates, while the unprimed, 'true' coordinates belonged to the frame of the absolute ether. In a manner consistent with this view, Poincaré's discussion of the electromagnetic ellipse referred only to measurements made with rods and timekeepers at rest with respect to the ether, or to rods in motion, but not to timekeepers in motion. Thus, Poincaré had greater faith in mobile rods than in mobile clocks, even though the rods and clocks in physics laboratories and astronomical observatories around the world were all understood to be in motion with respect to extraterrestrial objects, for example the centre of the Sun. Poincaré was not alone in his distrust of mobile clocks in 1906–7, although he later had a change of heart, to be explained in Section 5.7.

5.5 EINSTEIN EXILES THE 'LIGHT-ETHER'

Even after relativity theory became a part of mainstream physics, Poincaré remained attached to the notion that the ether was a useful construct for the theorist. When he first affirmed that the laws of physics exhibit Lorentz covariance, electron theorists were agreed that the ether was an essential element of the physical world, with one exception: Albert Einstein. Shortly following Poincaré's discovery of the Lorentz group, Einstein derived the Lorentz transformations from kinematic arguments, based on two postulates: the relativity of phenomena for inertial frames of motion, and the universal invariance of the speed of light in empty space. One consequence of Einstein's theory, announced in the second paragraph of his essay 'On the Electrodynamics of Moving Bodies', was that that the ether (or 'light-ether', in Einstein's terms) was not needed in electrodynamics: 'The introduction of a "light-ether" will prove to be superfluous, inasmuch as the view to be developed here will neither introduce an "absolutely resting space" endowed with special properties, nor associate a velocity-vector with a point of empty space at which electromagnetic processes occur.'[25] Einstein's remark was certainly designed to capture physicists' attention, and so it did.

For a theory of the electrodynamics of moving bodies, refusing to assign a velocity to points of empty space where electromagnetic processes occur would have been an obvious non-starter. However, Einstein had figured out a way around this problem. The physicist needed only to describe electromagnetic processes in a frame with constant rectilinear velocity. Knowledge of the processes in all such frames was then obtained via the Lorentz transformations, which depend on the inter-frame velocity. All measurements are performed in inertial frames, and there is no recourse to an absolute frame of reference, or to a luminiferous ether.

Einstein's decision to exile the ether from electrodynamics (and from physics in general) came at a high cost to what Poincaré referred to (see Section 5.4) as the unity of time. While physicists at the turn of the twentieth century were familiar with the concept of the luminiferous ether, they were unfamiliar in general with the idea that time and space were relative notions.[26] Naturally, Einstein sought to render such ideas more plausible. The details of his arguments in favour of the relativity of time and space have been discussed by many authors, and need not be rehearsed here.[27] I will focus instead on how Einstein tried to persuade his readers of the logical compatibility of his two postulates.

The compatibility of Einstein's postulates of relativity and light-speed invariance followed for Einstein from an argument which may be summarised as follows.[28] Let a spherical light wave be transmitted from the coordinate origin of two inertial frames designated S and S' at time $t = \tau = 0$. In frame S, the light wave spreads with velocity c such that the wavefront is expressed as

$$x^2+y^2+z^2=c^2t^2. \tag{5.1}$$

To obtain the equation of the wavefront in frame S' moving with velocity v with respect to S, we apply a certain linear transformation of coordinates from S to S', depending on a factor that depends on velocity.[29] The result of the transformations, Einstein found, was just

$$\xi^2+\eta^2+\zeta^2=c^2\tau^2. \tag{5.2}$$

Since Equation 5.1 goes over to Equation 5.2 via these transformations, Einstein observed, the light wave that is spherical in S is also spherical in S', it propagates with the same velocity c and, consequently, 'our two basic principles are mutually compatible'.[30]

Einstein's compatibility demonstration addressed one of the more immediate objections to be raised against his theory: that the propagation of light implied the existence of a substrate. This substrate, known as the ether, was common to the electron theories of Lorentz, Joseph Larmor, Alfred Bucherer, Paul Langevin and Max Abraham. Some, but not all of these theorists, went on to adopt relativity, but none of them ever saw fit to banish the luminiferous ether from physics.

5.6 CUNNINGHAM'S MULTIPLE ETHERS

Among British theorists, relativity theory had few proponents, if any, until Ebenezer Cunningham (1881–1977) took it up. A Cambridge-trained mathematician, senior wrangler in 1902 and lecturer in applied mathematics at University College London in 1907, Cunningham understood Einstein's theory to be consistent with the existence of multiple ethers, provided that every inertial frame is associated with an ether.[31] Inspired by

Larmor's electron theory, Cunningham's multiple-ether approach to relativity recalls the mechanics proposed by the Leipzig mathematician Carl Neumann. Newton's laws of mechanics, Neumann observed, give one the freedom to consider any inertial frame to be at rest with respect to a fixed set of coordinate axes he called the 'Body Alpha'.[32] Views equivalent to Cunningham's, but stripped of reference to the ether, were subsequently advanced by Minkowski and Laue.[33]

Cunningham's first paper on relativity set out to overturn an objection raised by Max Abraham with respect to Lorentz's electron theory. Abraham believed that energy conservation required a fundamental modification of Lorentz's deformable electron model, in the form of a supplemental internal and non-electromagnetic source of energy. Cunningham challenged Abraham's (frame-dependent) definition of electromagnetic momentum, and found that, under the same quasi-stationary approximation and an alternative momentum definition, the problem vanishes. He concluded that no non-electromagnetic energy was required by Lorentz's electron model, which remained for him a possible foundation for a 'purely electromagnetic theory of matter'.[34]

Along the way, Cunningham assumed that, if Lorentz's deformable electron is spherical when at rest, when put in motion and measured by co-moving observers, it will remain spherical. But, when measured with respect to a frame at rest, the moving electron will have a 'spheroidal shape as suggested by Lorentz'.[35] Cunningham took this suggestion a step further, arguing that a light wave would appear spherical to all inertial observers, in agreement with Einstein on this point (and with reference to Einstein's relativity paper of 1905).

Next, Cunningham took an important step towards the legitimation of the concept of light-sphere covariance, as Goldberg first noticed.[36] Einstein's derivation of the Lorentz transformations could be reduced to a handful of steps, Cunningham realised, by requiring the covariance of the light-sphere equation (Equation 5.1) with respect to these transformations. Cunningham's requirement of covariance of the light-sphere equation entailed the relativity of space and time:

> For it is required, among other things, to explain how a light-wave traveling outwards in all directions with velocity C relative to an observer A, may at the same time be traveling outwards in all directions with the same velocity relative to an observer B moving relative to A with velocity v. This can clearly not be done without some transformation of the space and time variables of the two observers.[37]

Shortly after Cunningham's derivation of the Lorentz transformations appeared in print, Einstein employed the same method of derivation, making Cunningham the first British contributor to what was later known as Einstein's theory of relativity. However, Einstein did not acknowledge Cunningham as the source of the derivation, and he may well have come up with it on his own.[38]

In subsequent years, Cunningham contributed to the generalisation of the principle of relativity beyond frames with uniform motion, and published one of the first English-language textbooks on relativity. The latter work introduced Minkowski's spacetime theory to English readers, and revived Cunningham's earlier view of multiple mobile ethers, in a chapter entitled 'Relativity and an Objective Æther'. The sufficient condition for the objective reality of the mobile ether was announced by Cunningham quite simply to be the conformity of its kinematics to those of the principle of relativity.[39] Cunningham's view of the mobile ether found employment throughout the 1920s, and well beyond Cambridge, thanks to Sommerfeld's celebrated textbook *Atombau und Spektrallinien* and its translations into French and English.[40]

Contributing to Cunningham's confidence in the universal validity of the principle of relativity was Minkowski's spacetime theory, and also Einstein's notion of light quanta, whereby light energy is not distributed uniformly in space but exists in discrete packets.[41] The theory of radiation appeared thereby to conform to relativity, at least by 1914, thanks in part to Einstein's light quanta. Even before light quanta came to be accepted, the theory of radiation was central to discussions of relativity and the ether, as shown in the next section.

5.7 RELATIVE TIME FROM RADIO WAVES

In the fall of 1908, the Göttingen mathematician David Hilbert (1862–1943) extended an invitation to Poincaré, on behalf of the Wolfskehl Foundation, to deliver a series of lectures. Poincaré accepted the honour, and delivered six lectures in Göttingen in April 1909, before an international audience of mathematicians and physicists.[42] One of Hilbert's Göttingen colleagues, and perhaps his closest friend, was Hermann Minkowski (1864–1909), who had recently put forward the theory of spacetime. Minkowski's spacetime theory borrowed key insights from Poincaré's study of the Lorentz-covariant law of gravitation, including the idea of a four-dimensional vector space with one imaginary dimension. Minkowski had planned to pursue the theory but, following an attack of appendicitis, he died in January 1909, just three months before Poincaré's lecture series. There are indications that Minkowski's theory inspired the topic of Poincaré's final Wolfskehl lecture, which was entitled 'The New Mechanics', although the only Göttingen scientist mentioned by name in the published text of the lecture is the electron theorist Max Abraham, whom Poincaré characterised, along with H. A. Lorentz, as one of the 'great demolitionists' of Newtonian mechanics.[43]

One of the novelties of Minkowski's theory was its definition of 'proper time' as the parameter of a four-dimensional trajectory in spacetime, which was given to be the time read by an ideal clock describing this same trajectory. According to this theory, there are an infinite number of temporal axes, and as many corresponding three-dimensional spaces, which may be described with reference to a single spacetime extending infinitely in four dimensions. In Poincaré's original scheme, as noted in Section 5.4, there was only one three-dimensional space—the ether—and one temporal dimension. Only clocks at rest in the ether were reliable timekeepers. But, in Göttingen, Poincaré admitted that ideal clocks in motion with respect to the ether might also tell the right time.

The circumstances of this change are of interest, as they illustrate how technology can shake the foundations of science. Since Heinrich Hertz's discovery of electromagnetic waves in air in the late 1880s, Poincaré had contributed to the theory of Hertzian waves, and engaged with the emerging technology of wireless telegraphy. When the Eiffel Tower was menaced with destruction, Poincaré backed a plan to employ it as the world's tallest wireless transmission antenna, thereby strengthening the tower's chances for preservation.[44]

Poincaré knew that Hertzian waves propagate in the ether with the speed of light, c. In his final Wolfskehl lecture, he imagined an observer B in a vehicle in motion with respect to the ether with speed $2c/3$. Observer B transmits telemetry data to an observer A in a second vehicle moving with equal and opposite speed:

> A and B begin by setting their watches, then B sends telegrams to A indicating his successive positions; putting these signals together, A can give an account of B's motion, and trace its curve. Well, the signals propagate at the speed of light; the watches marking apparent time vary at every instant and it will all go down as if B's watch were fast.[45]

As a result of the deformation of time revealed by the telemetry data, Poincaré explained a few months later, observer B would come to believe his vehicle was advancing not at hyperlight velocity but at sub-light velocity, so that the principle of relativity would not be violated.[46]

Notice that Poincaré's watch in motion runs fast (with respect to a watch at rest in the ether), while Einstein's clocks in motion run slow (with respect to clocks at relative rest). Not only is Einstein's notion of relativistic time dilation readily admitted in physics, it also underpins the accuracy of global positioning systems.[47] What then are we to make of Poincaré's fast watch? His thought experiment supposes that observer B initially judges his speed relative to A to be $4c/3$, in accordance with Newtonian kinematics. Only after exchanging telemetry data at the speed of light does B revise his initial speed estimate downwards to a sub-light value. If we ignore Lorentz contraction, then time contraction is the only plausible means of obtaining a reduction to sub-light vehicle (and watch) velocity. Simply stated, observer B concludes that his watch runs fast, because his speed is slower than Newtonian kinematics had led him to believe.[48]

As for Lorentz, by 1909 he, too, admitted that both rods and clocks in motion with respect to the ether could be used to measure temporal and spatial intervals, respectively. Like Poincaré, he recognised that inertial observers, whether in uniform motion or at rest, would agree on the speed of light, provided that the mobile clocks were optically synchronised in the frame of motion. Lorentz admitted the (Lorentz) contraction of bodies in their direction of motion, and proved that the apparent time of an observer in motion is just that indicated by the Lorentz transformations, such that time is apparently dilated in the observer's frame of motion and, consequently, that his clocks run slower than identical clocks at rest in the ether. In his authoritative monograph *The Theory of Electrons*, Lorentz encouraged his reader to keep in mind that 'in doing all that has been said, the observer would remain entirely unconscious of his system moving (with himself) through the ether, and of the errors of his rod and his clocks'.[49]

The 'errors' of rod and clocks that Lorentz refers to are understood to be relative to the 'true' values indicated by an identical rod and identical clocks at rest in the ether. From the forgoing considerations of Poincaré and Lorentz, we see that their attachment to the ether did not prevent them from admitting that time could be measured in moving frames. It was just not the true time.

5.8 MINKOWSKIAN ELECTRONS IN SPACETIME

Since 1902, Hermann Minkowski had been a professor of mathematics in Göttingen, where he took up several topics in theoretical physics, including heat radiation and the electrodynamics of moving bodies. Before moving to Göttingen, Minkowski had been on the faculty of Zürich Polytechnic (currently ETH Zürich), where Albert Einstein and Mileva Marić were enrolled in his course in analysis.[50] In Göttingen, Minkowski co-led a seminar on electron theory in the summer of 1905, in tandem with David Hilbert. Two years later, in the summer of 1907, Minkowski took notice of Einstein's and Poincaré's foundational contributions to relativity theory, and soon thereafter made a fundamental contribution of his own: the theory of spacetime.[51]

Minkowski presented his theory of spacetime in a lecture entitled 'Space and Time' at the annual meeting of the German Association of Natural Scientists and Physicians in Cologne on 21 September 1908. Central to his semi-popular lecture was a hand-drawn transparency (Fig. 5.3). The transparency depicts two diagrams: on the left is a model of

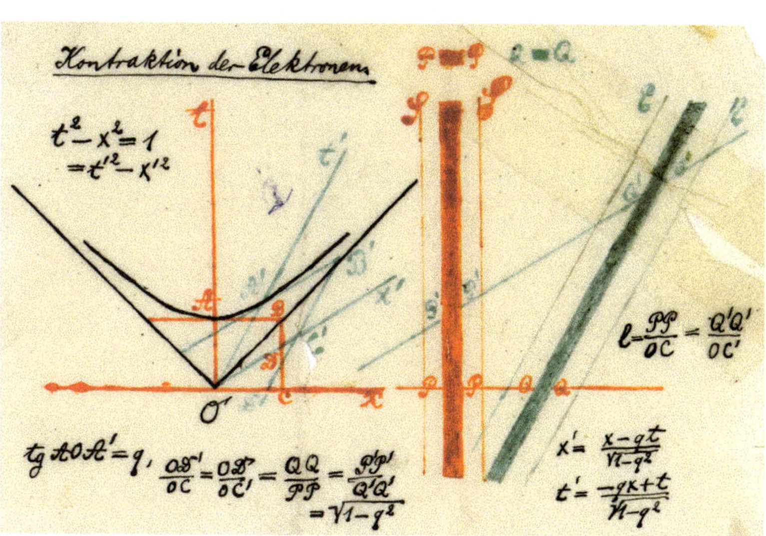

Fig. 5.3 A hand-drawn transparency featuring two of Minkowski's diagrams. (Courtesy of SUB Göttingen, Cod. Ms. Math.-Arch. 60:2)

a two-dimensional Minkowski space and, on the right, a comparison of the trajectories in such a space of two electrons, with one at rest and the other in uniform motion. The transparency features several formulae which illustrate Minkowski's demonstration that the contraction of electrons (and Lorentz contraction in general) is a consequence of the fact that space and time are not separate entities but part of a four-dimensional continuum: spacetime.

The transparency also shows that temporal intervals are dilated for observers in motion, although Minkowski appears not to have explained the details of the interpretation in Cologne. There can be little doubt about Minkowski's recognition of the reality of time dilation, however, as he credited Einstein with 'first clearly recognising that the time of one electron is just as good as that of the other, which is to say, that t and t' are to be treated identically'.[52] Minkowski thus sided with Einstein on the relative nature of time, rather than with Poincaré.

Minkowski did not take up the question of the form of an electromagnetic pulse for a source in motion, a topic earlier broached by Poincaré (see Section 5.4), but his Cologne lecture carefully illustrated the Liénard–Wiechert potential on a three-dimensional spacetime diagram. In all of these considerations, Minkowski avoided reference to the luminiferous ether; he mentioned it only once, and then in a critical way, saying that the Lorentz contraction should not be thought of 'as something like a consequence of ether drag'.[53]

In Minkowski's view, the ether constituted an obstacle to the general acceptance of his theory of spacetime. He thus concluded his Cologne lecture with an appeal to 'those for whom the abolition of familiar views is unappealing or distressful', suggesting in effect that, although the electromagnetic ether was no longer tenable in modern physics, four-dimensional spacetime was in fact the 'true kernel of an electromagnetic world view'.[54] The idea that the electromagnetic world view could be pursued successfully by means of the spacetime theory was, at best, an instance of wishful thinking on Minkowski's part but, as a rhetorical ploy, it effectively situated spacetime theory at the forefront of theoretical physics.

5.9 MINKOWSKI SPACETIME AS AN ETHER STAND-IN

The pursuit of the electromagnetic world view implied a microphysical reduction and, for Minkowski, this reduction was to be carried out in the arena of four-dimensional spacetime. He was unable to advance this project very far himself, as he perished in January 1909, as previously mentioned. Other theorists stepped in to continue Minkowski's research programme, including his assistant in Göttingen, Max Born, and his boyhood friend, Arnold Sommerfeld, the professor of theoretical physics at the Ludwig-Maximilian University of Munich.

Both Born and Sommerfeld were instrumental to the success of spacetime theory among theoretical physicists. Most importantly in this respect, Sommerfeld reformulated

Minkowski's elegant but unfamiliar matrix calculus as a four-dimensional vector analysis, which appealed to physicists familiar with ordinary three-dimensional vector analysis. When presenting the latter formalism to readers of the *Annalen der Physik*, Sommerfeld formulated an argument that was implicit in Minkowski's Cologne lecture (see Section 5.8), to the effect that spacetime, or the 'absolute world' in the Minkowskian vernacular, was a substitute for the ether: 'The absolute world appears in place of the older theory, that is, the connection of space and time via the velocity of light *c*, the immutability of which now constitutes the absolute substrate of electrodynamics.'[55]

The 'older theory' to which Sommerfeld refers here is that of Maxwell and Hertz, which assumed an ether at rest as the substrate of electromagnetic wave propagation with velocity *c* in matter-free space. The immutability of the latter velocity was postulated by Einstein in 1905, as noted in Section 5.5, and also, in turn, by Minkowski, when he formulated the theory of spacetime.

In the years leading up to Minkowski's discovery of spacetime, Sommerfeld had established his reputation as a leading electron theorist, and a critic of Einstein's theory of relativity. The same was true of the Göttingen *Privatdozent* Max Abraham. A former doctoral student of Max Planck, Abraham was the author of an influential treatise on the electromagnetic theory of radiation. In the 1908 edition, which appeared shortly after Minkowski's spacetime theory, Abraham adopted an approach midway between Sommerfeld's enthusiastic embrace of Minkowskian relativity, and outright rejection of the principle of relativity.

For research in the electrodynamics of moving media, Abraham considered that both Lorentz's theory and that of Minkowski represented attractive alternatives, provided that electrons were considered as point charges, in virtue of the ease of calculation these two theories offered. For electrons in empty space, on the other hand, Abraham felt that Lorentz's theory was worthless. Not only was this theory inconsistent from a formal standpoint, in that it gave an improper value for electron rest mass (unless, like Poincaré, one admitted a non-electromagnetic binding potential), but it was also disconfirmed by Walter Kaufmann's cathode-ray deflection experiments.

Earlier, we saw how Poincaré had no truck with mobile clocks and preferred to keep time with clocks at rest in the ether (see Section 5.4). Abraham shared this view with Poincaré, and consequently dismissed Einstein's notion of path-dependent time, although he acknowledged that this was not ruled out by experiment. Time was absolute, in Abraham's view, such that the difference between the rate of an ideal clock in motion and an equivalent clock at rest in the ether could be detected in principle, if not in practice. Space was also absolute for Abraham and, in light of the null result of the Michelson–Morley experiment (1887), he readily admitted that all material bodies contract in the direction of their motion with respect to the ether.[56]

Both Sommerfeld and Abraham began to favour relativity following Minkowski's contributions, and Alfred Bucherer's experimental confirmation of the relativistic dynamics of charged particles. Sommerfeld argued that spacetime was a replacement for the ether, while Abraham devised an alternative to Minkowski's relativistic theory of moving media, giving rise to over a century of debate over which formulation was to be preferred.[57] The theory of moving media proposed by Abraham was no less relativistic than Minkowski's theory. Yet, Abraham held that Einstein's light postulate was incomprehensible without an ether, arguing that electromagnetic waves and fields could not subsist without a substrate.[58]

Max Abraham was not alone in his rejection of Einstein's relativity, not even in Göttingen. Walter Ritz (1878–1909) had been a former student of Minkowski in Zürich, a former doctoral student of Woldemar Voigt and, from 1909, a *Privatdozent* in Göttingen. Ritz found fault both with Einstein's relativity, for its postulation of universal light-speed invariance, and with field theory in general. He also dismissed the ether in favour of an emission theory, which performed well in comparison to Lorentz's electron theory, although it failed to account for the optics of moving bodies, including Fizeau's demonstration that the propagation speed of light in running water changes with flow rate.[59] Ritz's theory of electrodynamics was even more radical than Einstein's in some respects and, while Einstein paid attention to it, most physicists did not. When Ritz succumbed to tuberculosis in 1909, his theory died with him.[60]

While Ritz dismissed the luminiferous ether, Emil Wiechert sought to defend it. Director of the Göttingen Institute for Geophysics, Emil Wiechert was, like Ritz, unsatisfied with Einstein's relativity. At the University of Königsberg in the 1890s, Wiechert developed a theory of charged particles which shared certain features of Lorentz's theory, including an ether at rest. Unsurprisingly, Wiechert found much to admire in Lorentz's electron theory and, like Lorentz, he felt that Einstein was wrong to do away with the ether. Along with many of his contemporaries, including Max Abraham (as previously mentioned), Wiechert felt that the electromagnetic field was not self-standing but required a substrate. In addition, in his paper entitled 'The Principle of Relativity and the Ether', Wiechert deplored Einstein's neglect of a model of electron shape and charge distribution, which meant for Wiechert that Einstein's theory was not realistic, by default.[61]

It is hard to imagine that Wiechert's criticism would have bothered Einstein much. Unlike other electron theorists, Einstein neither needed nor offered a model of electron structure. Nonetheless, following Einstein's own example, his contemporaries often referred to the 'Lorentz–Einstein' theory, in virtue of a common recourse to the principle of relativity, and they considered the theories of Lorentz and Einstein to be empirically equivalent.

This conflation of theories may be explained by recalling a certain conceptual drift of the idea of ether among electrodynamicists, dating from the 1890s. There were those, like Henri Poincaré, who considered the ether to be real but undetectable in principle. Others, like Paul Drude, imagined that the properties of the electromagnetic field were actually properties of space, and not of the ether. And, while Poincaré and Drude both sought a microphysical reduction of electromagnetic phenomena, one theorist, Emil Cohn of the Kaiser-Wilhelm University of Strasbourg, eschewed both electrons and the ether in favour of a phenomenological theory featuring bare field equations, which compared well with that of Lorentz.[62]

Wiechert's criticism of relativity focused on the ether. The prospect of replacing the luminiferous ether with Minkowski spacetime did not please Wiechert, because spacetime seemed to him to allow bodies to propagate with velocities greater than the speed of light, while no such hyperlight phenomena had been observed. The possibility of hyperlight phenomena destroying the principle of relativity had been examined by Poincaré, who noted that the propagation velocity of gravitational action had been calculated by Laplace to exceed the speed of light a million times over.[63] Once Poincaré had discovered the Lorentz group, however, he showed that, if the speed of gravitation were equal to that of light, this would be no less consistent with astronomical observations than Newtonian gravitation. As one might guess, Wiechert's defence of the ether did not mention Poincaré's argument in favour of relativity.

Among those who admired Wiechert's paper was the geodesist Friedrich Helmert, who cited it approvingly in his letter of nomination of Wiechert as a corresponding member of the Prussian Academy of Sciences; the letter was co-signed by Max Planck and Walther Nernst.[64] One has to wonder how closely these scientists read Wiechert's paper. In response to Wiechert's article, Max Laue remarked the logical fallacy involved in deducing the existence of a class of preferred frames of motion (such as that of the luminiferous ether) from the inexistence of a different class of frames (in Wiechert's case, frames with hyperlight velocity).

Laue went on to suggest that, in the absence of evidence for the existence of hyperlight phenomena, there was 'no shame' in banning from physics further discussion of the existence of the ether and absolute time.[65] Seven years earlier, Einstein had deemed the ether superfluous to physics (as mentioned in Section 5.5), but now Laue was recommending that the topic itself be censored, along with that of absolute time.

Educated in Berlin and Göttingen, Laue was, at the time of his critique of Wiechert's view, a modest *Privatdozent* at the Ludwig-Maximilian University of Munich, attached to Sommerfeld's Institute of Theoretical Physics. He had recently published the first German textbook on the principle of relativity and it soon became the standard work of reference in this domain.[66] Six months after Laue pointed out the logical shortcomings of Wiechert's defence of the ether, Sommerfeld communicated to the Munich Academy a paper written by Walter Friedrich, Paul Knipping and Laue on the interference of X-rays by crystals, a discovery for which Laue was awarded the Nobel Prize in Physics in 1914. That same year, Einstein became a member of the Prussian Academy of Sciences, on the strength of Planck's letter of nomination, which emphasised Einstein's revision of the notion of time and how this revision led to Minkowski's spacetime theory.[67] It is telling that Planck's letter described Minkowski's spacetime theory as a consequence of Einstein's discoveries and that it neglected to mention Einstein's criticism of the ether. Planck must have known that the members of the Prussian Academy were not ready to renounce the ether. The recognition of the two most vocal critics of the ether and absolute time—Laue and Einstein—by the Royal Swedish Academy of Sciences and the Prussian Academy of Sciences, respectively, sent a strong signal to physicists young and old that they, too, could do without the ether.

Yet, even among physicists who promoted the Einstein–Minkowski theory of relativity, Einstein and Laue's anti-ether campaign met with resistance. As observed above, Sommerfeld and Max Abraham did not do away with the ether altogether. Likewise, when lecturing on the theories of Einstein, Minkowski and Sommerfeld at the Collège de France in 1910–11, Paul Langevin deplored Einstein's view of the ether. Lecture notes by Léon Brillouin record Langevin's remark on the subject: 'The very notion of the ether loses its meaning, says Einstein—this is an exaggeration.' Langevin went on to point out that, while it is physically impossible to determine velocity with respect to the ether, 'we can determine accelerations and rotations'.[68]

5.10 CONCLUSION

When Einstein announced that his theory of the electrodynamics of moving bodies had no use for the luminiferous ether and that this ether was consequently 'superfluous' to the theoretical domain of electromagnetism, he must have known it would capture the attention of his peers. And so it did, although few physicists were ready to banish talk of the ether from physics.

The electron-theoretical origins of relativity theory guided its development, even as new conceptual tools were brought into play. Poincaré's discovery of the Lorentz group enabled him to reinterpret Langevin's conception of velocity waves and acceleration waves propagating in the ether, and to recover Hertz's demonstration of the production of electromagnetic waves by an oscillator. Such waves, he showed later, could be used to demonstrate in principle the deformation of temporal intervals for observers in frames of reference in motion with respect to the ether. The mechanism for this deformation remained mysterious, although, for Lorentz, at least, it was no more mysterious than the contraction of bodies in their direction of motion with respect to the ether. Both time dilation and length contraction issued from the same velocity-based modification of dynamical laws, in Lorentz's view.

Directly opposed to the latter 'dynamical' approach to relativity, Einstein's relativistic kinematics had no need of an ether, and did not introduce one. Einstein took care, nonetheless, to provide an argument for the logico-mathematical compatibility of his twin postulates of relativity and universal light-speed invariance. This argument, as we have seen, did not convince electron theorists, including Lorentz, Poincaré, Max Abraham and Wiechert, to forgo the ether.

Soon after the theories of Poincaré and Einstein appeared on the scene, a third approach to relativity proved effective in attracting electron theorists, and many others besides, to the relativistic fold. Minkowski's spacetime theory featured a powerful blend of ideas drawn from a variety of disparate sources, including Hertz's theory of the electrodynamics of moving bodies; Lorentz–Poincaré electron theory; Einsteinian kinematics and the theory of continuous transformation groups. Where Einstein offered in exchange for the ether only a pair of postulates, the logical consistency of which was suspect, Minkowski and his followers proposed spacetime as a conceptual substitute.

From the fact that neither Einstein's theory nor that of Minkowski found immediate success, it may be gathered that, in general, physicists did not find the idea of renouncing the ether to be particularly compelling. Nonetheless, most theoretical physicists adopted spacetime theory, including several theorists encountered in this chapter: Einstein, Cunningham, Abraham, Sommerfeld, Langevin and Lorentz. Many mathematicians, who had no prior attachment to the ether, joined them in this endeavour.[69]

By the end of the first decade of the twentieth century, relativity and spacetime theory were poised to dominate theoretical physics. To some extent, the rise of these theories came at the expense of ether theories, which continued nonetheless to appeal to theorists like Wiechert. The Lorentz–Poincaré electron theory survived the onslaught of relativity, via mathematical reformulation and conceptual adaptation to the principle of relativity. In a lecture delivered in Salzburg in September 1909, Einstein remarked that Lorentz's theory was the only electron theory that was useful and had clear foundations.[70] By the end of 1911, however, electron theory was seen to be incapable of explaining black-body radiation. This incapacity was not due to the theory's fixed-ether foundation but rather on its assumption that bodies emit and absorb energy continuously.[71]

ACKNOWLEDGEMENTS

I am grateful to the Dibner Library of the History of Science and Technology for granting me a residence in November 2016, and to the Niedersächsiche Staats und Universitätsbibliothek Göttingen for authorising publication of the diagram in Fig. 5.3.

NOTES

1. The emergence of electron theory in the 1890s is described by Jed Buchwald, *From Maxwell to Microphysics* (Chicago: University of Chicago Press, 1985), while the development of electron theory has been studied in detail up to the time of Einstein's and Poincaré's theories of relativity (1905–6) by Arthur I. Miller, *Albert Einstein's Special Theory of Relativity: Emergence (1905) and Early Interpretation* (Reading, MA: Addison-Wesley, 1981), and Olivier Darrigol, *Electrodynamics from Ampère to Einstein* (Oxford: Oxford University Press, 2000). For the period up to 1920, works of reference include Russell McCormmach, 'Einstein, Lorentz, and the Electron Theory', *Historical Studies in the Physical Sciences* 2 (1970): ix–xx; Bruce R. Wheaton, *The Tiger and the Shark: Empirical Roots of Wave–Particle Dualism* (Cambridge: Cambridge University Press, 1983); Helge Kragh, 'The Electron, the Protyle, and the Unity of Matter', in Jed Z. Buchwald and Andrew Warwick, eds., *Histories of the Electron: The Birth of Microphysics* (Cambridge, MA: MIT Press, 2001), 195–226; Ole Knudsen, 'O. W. Richardson and the Electron Theory of Matter,

1901–1916', in Buchwald and Warwick, *Histories of the Electron*, 227–54; Walter Kaiser, 'Electron Gas Theory of Metals: Free Electrons in Bulk Matter', in Buchwald and Warwick, eds., *Histories of the Electron*, 255–304; and Jaume Navarro, *A History of the Electron: J. J. and G. P. Thomson* (Cambridge: Cambridge University Press, 2012).

2. Edmund T. Whittaker, *A History of the Theories of Æther and Electricity, Volume 2: The Modern Theories, 1900–1926* (London: T. Nelson, 1953).
3. Stanley Goldberg, 'In Defense of Ether: The British Response to Einstein's Special Theory of Relativity 1905–1911', *Historical Studies in the Physical Sciences* 2 (1970): 89–125, p. 114.
4. For a thoughtful review of the literature, see Oliver Darrigol, 'The Mystery of the Einstein–Poincaré Connection', *Isis* 95 (2004): 614–26.
5. Peter Galison, *Einstein's Clocks and Poincaré's Maps: Empires of Time* (New York: Norton, 2003).
6. On Poincaré's participation in the Lorentz *Festschrift*, see his correspondence with Heike Kamerlingh Onnes: H. Poincaré to Heike Kamerlingh Onnes, in Scott A. Walter et al., eds., *Henri Poincaré Papers*, doc. 2-63-1, http://henripoincarepapers.univ-nantes.fr/chp/text/kamerlinghonnes190009.html, accessed 15 November 2017; H. Poincaré to Heike Kamerlingh Onnes, in Walter et al., *Henri Poincaré Papers*, doc. 2-63-2, http://henripoincarepapers.univ-nantes.fr/chp/text/kamerlinghonnes190010.html, accessed 15 November 2017; and H. Poincaré to Heike Kamerlingh Onnes, in Walter et al., *Henri Poincaré Papers*, doc. 2-63-3, henripoincarepapers.univ-nantes.fr/chp/text/kamerlinghonnes19001110.html, accessed 15 November 2017.
7. A. Liénard, '*La théorie de Lorentz et celle de Larmor*', *Éclairage électrique* 16 (1898): 320–34, 360–5.
8. Henri Poincaré, '*La théorie de Lorentz et le principe de réaction*', *Archives néerlandaises des sciences exactes et naturelles* 5 (1900): 252–78, p. 272.
9. See H. A. Lorentz to H. Poincaré, 20 January, 1901, in Walter et al., *Henri Poincaré Papers*, doc. 2-38-1, http://henripoincarepapers.univ-nantes.fr/chp/text/lorentz-1901-01-20.html, accessed 15 November 2017. According to H. A. Lorentz, *Versuch einen Theorie der elektrischen und optischen Erscheinungen in bewegten Körpern* (Leiden: Brill, 1895), the local time t' was a linear function of the ether time t and the product of the distance x and the frame velocity v, divided by the square of the velocity of light in the ether, c: $t' = t - vx/c^2$. Lorentz noted that his formula was valid up to a first-order approximation in v/c.
10. Henri Poincaré, '*La mesure du temps*', *Revue de métaphysique et de morale* 6 (1898): 1–13. English translation in Henri Poincaré, *The Value of Science: Essential Writings of Henri Poincaré*, ed. by Stephen Jay Gould (New York: Random House, 2001), p. 220ff.
11. Joseph Larmor, *Æther and Matter* (Cambridge: Cambridge University Press, 1900).
12. Joseph J. Thomson, *Notes on Recent Researches in Electricity and Magnetism: Intended as a Sequel to Professor Clerk-Maxwell's Treatise on Electricity and Magnetism* (Oxford: Clarendon, 1893).
13. G. F. C. Searle, 'On the Steady Motion of an Electrified Ellipsoid', *Philosophical Magazine* 44 (1897): 329–41.
14. O. Heaviside, 'On the Electromagnetic Effects Due to the Motion of Electrification through a Dielectric', *Philosophical Magazine* 27 (1889): 324–39. Thomson, *Notes*.
15. Paul Langevin, '*Sur l'origine des radiations et l'inertie électromagnétique*', *Journal de physique théorique et appliquée* 4 (1905): 165–83.
16. A. Liénard, '*Champ électrique et magnétique produit par une charge concentrée en un point et animée d'un mouvement quelconque*', *Éclairage électrique* 16 (1898): 5–14, 53–9, 106–12. E. Wiechert, '*Elektrodynamische Elementargesetze*', *Archives néerlandaises des sciences exactes et naturelles* 5 (1900): 549–73.

17. For details on Langevin's paper, see Arthur I. Miller, 'A Study of Henri Poincaré's *"Sur la dynamique de l'électron"*', *Archive for History of Exact Sciences* 10 (1973): 207–328.
18. See Langevin's notebook, box 123, and the letter to his wife of 26 September 1904, box 3, Fonds Langevin, Library of the École supérieure de physique et de chimie industrielle, Paris.
19. Paul Langevin, 'The Relations of the Physics of Electrons to the Other Branches of Science', in H. J. Rogers, ed., *Congress of Arts and Science, Universal Exposition, St. Louis, 1904, Volume 4: Physics, Chemistry, Astronomy, Sciences of the Earth* (Boston: Houghton, Mifflin & Co., 1906), 121–56.
20. Henri Poincaré, *'Sur la dynamique de l'électron'*, *Rendiconti del Circolo matematico di Palermo* 21 (1906): 129–76, p. 149. On Hertz's solution, see Darrigol, *Electrodynamics*, p. 251.
21. Henri Poincaré, *'Sur la dynamique de l'électron'*, *Comptes rendus hebdomadaires de l'Académie des sciences de Paris* 140 (1905): 1504–8.
22. H. Poincaré to H. A. Lorentz, Walter et al., *Henri Poincaré Papers*, doc. 2-38-3, http://henrip-oincarepapers.univ-nantes.fr/chp/text/lorentz3.html, accessed 15 November 2017.
23. Shlomo Sternberg, 'Imagery in Scientific Thought by Arthur I. Miller', *Mathematical Intelligencer* 8 (1986): 65–74.
24. H. Vergne, notebook 2, p. 52, François Viète Center, Nantes.
25. A. Einstein, *'Zur Elektrodynamik bewegter Körper'*, *Annalen der Physik* 322 (1905): 891–921, p. 892; translation by the author. For line-by-line commentary, see Miller, *Albert Einstein's Special Theory*, p. 392ff, and the editorial notes in Albert Einstein, *The Collected Papers of Albert Einstein, Volume 2, The Swiss Years: Writings, 1900–1909*, ed. John Stachel, David C. Cassidy, Jürgen Renn and Robert Schulmann (Princeton: Princeton University Press, 1989), pp. 275–310.
26. Notable exceptions include the Cambridge electron theorist Joseph Larmor, who remarked on the 'difference of time reckoning' of orbiting electrons (see Joseph Larmor, 'A Dynamical Theory of the Electric and Luminiferous Medium, Part 3', *Philosophical Transactions of the Royal Society A* 190 (1897): 205–300, p. 229), and Poincaré, who gave an operational definition of local time in *'La théorie de Lorentz'*, p. 272.
27. For a clear exposition of Einstein's kinematics, see Alberto A. Martínez, *Kinematics: The Lost Origins of Einstein's Relativity* (Baltimore: Johns Hopkins University Press, 2009).
28. For a more detailed presentation, see Scott A. Walter, 'Figures of Light in the Early History of Relativity (1905–1914)', in David E. Rowe, Tilman Sauer and Scott A. Walter, eds., *Beyond Einstein: Perspectives on Geometry, Gravitation, and Cosmology in the Twentieth Century*, Einstein Studies 14 (Basel: Birkhäuser, 2018), 3–50.
29. Let this factor be ϕ, and coordinates in the frame S' be defined as $\xi = \phi(v)\,\gamma(x - vt)$, $\eta = \phi(v)\,y$, $\zeta = \phi(v)\,z$, and $\tau = \phi(v)\,\gamma(t - vx/c^2)$, where $\gamma = (1 - v^2/c^2)^{-1/2}$, and the coordinate origins of the two frames are assumed to coincide. If we set $\phi(v) = 1$, these transformations are equivalent to the Lorentz transformations.
30. See Section 3 of Einstein, *'Zur Elektrodynamik'*, p. 901.
31. See Goldberg, 'In Defense of Ether', and Bruce Hunt, 'Experimenting on the Ether: Oliver J. Lodge and the Great Whirling Machine', *Historical Studies in the Physical Sciences* 16 (1986): 111–34.
32. C. G. Neumann, *Ueber die Principien der Galilei-Newton'schen Theorie* (Leipzig: Teubner, 1870). E. Cunningham recalled this fact, without mentioning Neumann, in E. Cunningham, 'The Principle of Relativity', *Report–British Association* 81 (1911): 236–45.
33. H. Minkowski, *'Raum und Zeit'*, *Jahresbericht der deutschen Mathematiker-Vereinigung* 18 (1909): 75–88, p. 79, and M. Laue, *Das Relativitätsprinzip* (Braunschweig: Vieweg, 1911), p. 33.

34. E. Cunningham, 'On the Electromagnetic Mass of a Moving Electron', *Philosophical Magazine* 14 (1907): 538–47, and Max Abraham, *Theorie der Elektrizität, 2: Elektromagnetische Theorie der Strahlung* (Leipzig: Teubner, 1905), p. 205. Cunningham's conclusion agrees with that reached later by Enrico Fermi; see Fritz Rohrlich, *Classical Charged Particles* (Singapore: World Scientific, 3rd edn, 2007), p. 17, and Michel Janssen and Matthew Mecklenburg, 'From Classical to Relativistic Mechanics: Electromagnetic Models of the Electron', in Vincent F. Hendricks, Klaus F. Jørgensen, Jesper Lützen and Stig A. Pedersen, eds., *Interactions: Mathematics, Physics and Philosophy, 1860–1930* (Dordrecht: Springer, 2006), 65–134.
35. Cunningham, 'On the Electromagnetic Mass', p. 540.
36. Goldberg, 'In Defense of Ether', p. 114.
37. Cunningham, 'On the Electromagnetic Mass', p. 544.
38. See Section 3 of Albert Einstein, *'Über das Relativitätsprinzip und die aus demselben gezogenen Folgerungen'*, *Jahrbuch der Radioaktivität und Elektronik* 4 (1907): 411–62, reprinted in Einstein, *The Collected Papers of Albert Einstein*, 432–488. Cunningham's paper appeared in the October 1907 issue of *Philosophical Magazine*, and Einstein's review article was submitted for publication in Johannes Stark's *Jahrbuch der Radioaktivität und Elektronik* on 4 December 1907.
39. E. Cunningham, *The Principle of Relativity* (Cambridge: Cambridge University Press, 1914), p. 193.
40. Arnold Sommerfeld, *Atombau und Spektrallinien* (Braunschweig: Vieweg, 1919), p. 319.
41. Albert Einstein, *'Über einen die Erzeugung und Verwandlung des Lichtes betreffenden heuristischen Gesichtspunkt'*, *Annalen der Physik* 322 (1905): 132–48, reprinted in Einstein, *The Collected Papers of Albert Einstein*, pp. 149–69.
42. For details on the lectures, see Jeremy Gray, *Henri Poincaré: A Scientific Biography* (Princeton: Princeton University Press, 2013), David E. Rowe, *A Richer Picture of Mathematics: The Göttingen Tradition and Beyond* (New York: Springer-Verlag, 2018) and Scott A. Walter, 'Poincaré-Week in Göttingen, in light of the Hilbert–Poincaré correspondence of 1908–1909', in Maria Teresa Borgato, Erwin Neuenschwander and Irène Passeron, eds., *Mathematical Correspondences and Critical Editions* (Cham: Springer, 2018), 189–202.
43. Henri Poincaré, *Sechs Vorträge über ausgewählte Gegenstände aus der reinen Mathematik und mathematischen Physik* (Leipzig/Berlin: Teubner, 1910), p. 51.
44. Peter Galison, *Einstein's Clocks and Poincaré's Maps*, p. 276.
45. Poincaré, *Sechs Vorträge*, pp. 54–5.
46. Henri Poincaré, *'La mécanique nouvelle'*, *Revue scientifique* 12 (1909): 170–7, p. 173.
47. Such systems correct for several motional effects, including special-relativistic time dilation, the latter effect being offset by gravitational blue shift; for a clear explanation, see Neil Ashby, 'Relativity and the Global Positioning System', *Physics Today* 55 (2002): 41–7.
48. For a detailed explanation, see Scott A. Walter, 'Poincaré on Clocks in Motion', *Studies in History and Philosophy of Modern Physics* 47 (2014): 131–41.
49. H. A. Lorentz, *The Theory of Electrons and its Application to the Phenomena of Light and Radiant Heat* (New York: Columbia University Press, 1909), p. 226.
50. Other students receiving grades from Minkowski included Louis Kollros, Marcel Grossmann and Jacob Ehrat; undated autograph, Minkowski Papers, Jewish National and University Library.
51. On the electron-theory seminar, see Lewis Pyenson, 'Physics in the Shadow of Mathematics: The Göttingen Electron-Theory Seminar of 1905', *Archive for History of Exact Sciences* 21 (1979): 55–89; on Minkowski's career in physics, see Scott A. Walter, 'Hermann Minkowski's Approach to Physics', *Mathematische Semesterberichte* 55 (2008): 213–35.

52. Cited by Scott A. Walter, 'Minkowski, Mathematicians, and the Mathematical Theory of Relativity', in Hubert Goenner, Jürgen Renn, Jim Ritter and Tilman Sauer, eds., *The Expanding Worlds of General Relativity*, Einstein Studies 7 (Basel: Birkhäuser, 1999), 45–86.
53. Minkowski, '*Raum und Zeit*', p. 80.
54. Minkowski, '*Raum und Zeit*', p. 88.
55. A. Sommerfeld, '*Zur Relativitätstheorie, I: Vierdimensionale Vektoralgebra*', *Annalen der Physik* 337 (1910): 749–76, p. 749.
56. A. A. Michelson and E. W. Morley, 'On the Relative Motion of the Earth and the Luminiferous Ether', *American Journal of Science* 34 (1887): 333–45; Max Abraham, *Theorie der Elektrizität, 2: Elektromagnetische Theorie der Strahlung* (Leipzig: Teubner, 1908), pp. 368–9.
57. For an overview, see Tomás Ramos, Guillermo F. Rubilar and Yuri N. Obukhov, 'Relativistic Analysis of the Dielectric Einstein Box: Abraham, Minkowski and Total Energy-Momentum Tensors', *Physics Letters A* 375 (2011): 1703–9.
58. Max Abraham, '*Die neue Mechanik*', *Scientia (Rivista di Scienza)* 15 (1914): 8–27.
59. H. Fizeau, '*Sur les hypothèses relatives à l'éther lumineux, et sur une expérience qui paraît démontrer que le mouvement des corps change la vitesse avec laquelle la lumière se propage dans leur intérieur*', *Comptes rendus hebdomadaires de l'Académie des sciences de Paris* 33 (1851): 349–55.
60. On Ritz's contributions to electrodynamics, see Alberto A. Martínez, 'Ritz, Einstein, and the Emission Hypothesis', *Physics in Perspective* 6 (2004): 4–28, and Oliver Darrigol, 'Electrodynamics in the Physics of Walther Ritz', in Jean-Claude Pont, ed., *Le destin douloureux de Walther Ritz (1878–1909), physicien théoricien de génie* (Sion: Vallesia, 2012), 207–40.
61. E. Wiechert, '*Relativitätsprinzip und Äther*', *Physikalische Zeitschrift* 12 (1911): 689–707, 737–58, p. 748.
62. Darrigol, *Electrodynamics*, p. 366.
63. H. Poincaré, '*L'état actuel et l'avenir de la physique mathématique*', *Bulletin des sciences mathématiques* 28 (1904): 302–24.
64. C. Kirsten and Hans-Günther Körber, eds., *Physiker über Physiker: Wahlvorschläge zur Aufnahme von Physikern in die Berliner Akademie 1870 bis 1929* (Berlin: Akademie-Verlag, 1975), p. 198.
65. M. Laue, '*Zwei Einwände gegen die Relativitätstheorie und ihre Widerlegung*', *Physikalische Zeitschrift* 13 (1912): 118–20.
66. M. Laue, *Das Relativitätsprinzip* (Braunschweig: Vieweg, 1911).
67. Kirsten and Körber, *Physiker über Physiker*, p. 201.
68. Léon Brillouin, '*Cours de relativité au Collège de France* 1910–1911', Léon Brillouin Papers, box 7, folder 8, American Institute of Physics, Niels Bohr Library, College Park, MD.
69. For quantitative details on the disciplinary reception of relativity theory, see Walter, 'Minkowski, Mathematicians'.
70. McCormmach, 'Einstein, Lorentz', p. 79.
71. Thomas S. Kuhn, *Black-Body Theory and the Quantum Discontinuity, 1894–1912* (Chicago: University of Chicago Press, 1978), p. 134.

6

Making Space for the Soul

Oliver Lodge, Maxwellian Psychics and the Etherial Body

Richard Noakes

6.1 A MAXWELLIAN APPROACH TO PSYCHICS

Historians have shown that Oliver Lodge was one of many late Victorian physicists who took it upon themselves to interpret the complex theories of electricity and magnetism of the Scottish natural philosopher James Clerk Maxwell.[1] As a 'Maxwellian', Lodge accepted not only the widely shared view that some kind of invisible material, space-filling medium was required to explain how waves of light travelled across 'empty' space at finite speed, but also Maxwell's view that the same medium was the seat of electrical and magnetic energies—an identification underpinning Maxwell's major claim that light was an electromagnetic vibration in the ether. In the wake of Maxwell's premature death in 1879, Lodge emerged as one of the most conspicuous of all Maxwellians by virtue of the number of lectures, articles and books in which he interpreted Maxwell's ideas for specialist and general scientific audiences.

From the early 1880s onwards, however, Lodge's interpretations of the ether also embraced his relatively new interest in psychical research. Over the next few decades, this most conspicuous British champion of the ether developed a 'philosophy' in which the increasingly troubled entity became inextricably entangled with his views on telepathy, the afterlife and Christianity. Several historians have studied Lodge's multifaceted etherial philosophy. In 1971 David Wilson argued that it was essentially conservative, insofar as Lodge's ether embodied Victorian scientific principles and values that he felt were being lost, including the possible compatibility of religious and scientific beliefs, the uniformity of nature and the virtue of physical and mechanical models.[2] Eight years later Brian Wynne offered a provocative but problematic case for seeing Lodge as one of many late Victorian physicists for whom evincing a spiritual domain via ether physics and psychical research provided the scientific basis for conservative political and religious ideologies.[3] Most

Noakes, R., 'Making Space for the Soul: Oliver Lodge, Maxwellian Psychics and the Etherial Body' in *Ether and Modernity*, edited by J. Navarro. © Oxford University Press 2018. DOI: 10.1093/oso/9780198797258.003.0006 recently, Peter Bowler and Courtenay Raia have analysed the profound significance of the ether in Lodge's attempts to intervene in debates on the relationships between mind and matter and God's relationship to the physical universe.[4] The following chapter moves beyond this work by examining the debts of Lodge's increasingly daring thinking to new ideas about the ether between the 1880s and 1930s and by revealing the largely positive reception of such thinking in constituencies beyond those of professional physicists whose scepticism towards Lodge's ideas has sometimes been taken as a measure of his scientific marginality *per se.*

One of the earliest hints that Lodge contemplated a psychic function for the ether occurred in a public lecture that he gave in 1882. The lecture consolidated the work of Maxwell, William Thomson, George Gabriel Stokes and other prominent ether architects in representing the medium as a 'perfectly homogenous and continuous' universal 'substance' whose irrotational movements enabled it to transmit radiation and whose rotational motions were the possible basis of the atomic constituents of gross matter.[5] Yet, Lodge went much further than Maxwell in maintaining the role of an etherial continuum in countering the idea of the direct action of one body on another at a distance, an idea which he believed to be 'absolutely unthinkable' to a humanity that boasted a 'mass of experience' of mechanical action mediated by some kind of 'connecting medium'.[6] Lodge was so convinced of the virtues of such a medium in explaining action between two distant bodies that he was prepared to contemplate its existence in a situation that was not ostensibly physical and which Maxwell had certainly not considered in his writings on the ether or action at a distance: the power of acting on a dog by mesmeric impressions.[7]

The confidence with which Lodge referred to mesmerism owed much to the fact that earlier in 1882 the intellectual credibility of investigations into mesmerism had received a significant boost with the foundation of the Society for Psychical Research (SPR). With the renowned moral philosopher Henry Sidgwick as its first president, and the statesman Arthur Balfour, classicist Frederic W. H. Myers and physicist W. F. Barrett among its leading members, the SPR sought to apply the methods of experimental science and critical literary and historical analysis to mesmerism, spiritualism, ghosts, haunted houses and other phenomena that the organisation grouped together as 'psychical'.[8] Although they distanced themselves from those whose religious and emotional prejudices seemed to undermine their capacity to make scientific judgements of psychical phenomena, the SPR's founders were partly motivated by the possibility of answering profound religious and philosophical questions such as the existence of the future life and the efficacy of prayer.

Lodge's reference to mesmerism also reflected an interest in psychical phenomena that had been stimulated by a friendship with Myers and the psychological writer Edmund Gurney, whose evidence for the capacity of one person to transmit a vivid psychological impression to another at a distance Lodge had initially disbelieved.[9] Gurney and Myers seem to have been instrumental in persuading Lodge to undertake his first piece of psychical research which took place in 1884 when he was a young professor at University College Liverpool. His research focused on a local example of the 'thought reading' craze then sweeping Britain and the United States of America. Drawing on methods developed by the SPR, Lodge tested the apparent ability of two young women to correctly identify or draw an object or picture on which one or more persons, unknown to the women but specially invited by Lodge to participate, were focusing their gaze. Convinced that collusion and trickery was impossible, Lodge interpreted his positive results as evidence of the capacity of one mind to communicate impressions to another, independently of the recognised sensory channels—a capacity that Myers had christened 'telepathy' in 1882 and of whose reality he, Lodge and other SPR members remained convinced for the rest of their lives.[10]

Lodge's interpretation of his telepathy tests for the SPR's publications reveals the extent to which he was prepared to apply Maxwellian physics to an ostensibly psychological question. As a Maxwellian, Lodge was strongly preoccupied with the question of how electrical and magnetic energies were transferred and transformed and this was doubtless the reason why he paid critical attention to the assumption that telepathic feats involved a kind of transfer of thought. It was not clear to him that transfer was appropriate in this case because it assumed that mind was located in the brain rather than being distributed in the space around the person. To illustrate this latter possibility—one that he insisted that psychologists could not deny—he mobilised a 'rough and crude analogy' from Maxwellian physics.[11] Just as Maxwellian physics located the energy of an electrical charge in the space around a conductor rather than on it,

> so it may be that the sensory consciousness of a person, though apparently located in his brain, may be conceived of as also existing like a faint echo in space, or in other brains, though these are ordinarily too busy and preoccupied to notice it.[12]

Although the ether was not named as the region outside material objects in which energy could be stored or 'echoed', Lodge's known acceptance that space was filled with the etherial medium suggests that he already saw it as a possible physical realm of mental energy or consciousness.

Lodge's stress on the crudity of his Maxwellian analogy would have pleased Gurney and Myers, who had already drawn analogies between the psychological 'resonance' suggested by telepathy and the resonant behaviour of physical systems (e.g. tuning forks and sensitive flames) but warned against treating them as explanations.[13] Yet, for all its tentativeness, Lodge's analogy reflected his growing confidence in the possibility of extending the 'modern views' of electricity and the ether into other areas of scientific enquiry. In 1889, following Heinrich Hertz's detection of electrical waves in free space, and his own experimental evidence for the propagation of electrical waves along wires, he notoriously declared that Maxwell's theory of light as a form of vibration in an electromagnetic ether had been 'utterly and completely verified' and that the 'whole domain of Optics' was 'now annexed to Electricity, which has thus become an imperial science'.[14] Not long after this, Lodge literally sought to extend the electrical empire into physiology by building a table-top 'electric eye' in which retinal responses to light were simulated by the spark gaps, electric spheres and other components used to generate and detect electromagnetic waves.[15] Lodge did not think this or any other form of physical resonance provided more than a useful analogy for telepathy but his continued use of the language of physical resonance in the psychical context suggests that he was still prepared to accept the possibility that telepathy might be a form of etherial vibration.[16]

By the time Lodge had built the electric eye, the psychical research that he managed to fit in around his commitments to 'physical' research, teaching, administration and writing had persuaded him of the possibility that the ether's functions might be extended further beyond the purely physical realm than he had contemplated in the early 1880s. In 1889 he participated in tests of the American spiritualist medium Leonore Piper. During seances with her, Lodge claimed to be able to speak to disembodied personalities professing to be deceased relatives of his and who revealed information about themselves that Piper could not have obtained even by fraudulent methods, that Lodge had never known (thus reducing the possibility that Piper simply read his mind via telepathy) and that he later verified with those relatives' living descendants. Convinced that he had taken every precaution against fraudulence and self-deception, Lodge declared publicly in 1889 that while some of Piper's revelations could be explained in terms of telepathy, others were achieved by 'none of the ordinary methods known to Physical Science'.[17] He later revealed that the Piper seances convinced him of the survival of the human personality following bodily death but, for over a decade, Lodge kept such convictions private, undoubtedly to avoid professionally damaging charges that he was a deluded convert to spiritualism.[18]

In the aftermath of the Piper seances, Lodge's ether thinking took on a more philosophical and religious aspect. This is evident from a semi-popular article on the ether, published in 1893 and which was mainly preoccupied with familiar physical arguments for the ether's existence and the difficulties physicists had encountered in explaining it (as well as electricity, magnetism and light) in terms of those 'primary objects of experience and consciousness': force and motion.[19] The frustrating ability of the ether to escape scientific attempts to grasp or move it suggested the possibility that it might turn out to be neither material nor something to which 'simple mechanics' applied. For Lodge, accepting the ether as something having a more complex dynamical basis would force the foundations of physics to be 'enlarged' and make it easier for physics to 'annexe' the vital and mental 'processes' that had eluded the science for so long.[20] Moreover, the ether's likely immaterial nature would help illuminate a 'region of the universe which Science has never entered yet, but which has been sought from afar, and perhaps blindly apprehended, by painter or poet, by philosopher or saint'.[21] A better understanding of the ether would make it easier for the sciences to accommodate the insights of psychical research into the nature of artistic inspiration and religious ecstasy, and thereby help reconcile scientific and religious understandings of the world.

By the turn of the twentieth century, Lodge's belief in survival and telepathy was compounding a problem for which the ether seemed to be a promising solution. Evidence of mind existing *independently* of matter merely dramatised the problem of how mind (whether human or Divine) interacted with matter and whether this required a mediating mechanism that was neither wholly material nor spiritual in nature. The use of an ether in these philosophical and theological problems had a long history but, as we shall see later on, Lodge's approach was indebted to two particular interventions in the debate.[22] Lodge's personal belief in survival also seems to have drawn his attention to the need for a better scientific understanding of the Christian idea of the soul, a need that he would explore in his speculations on the etherial body and which underpinned his more general preoccupation with the ways in which the sciences could render Christian theology more credible.

6.2 THE DEBTS TO MYERS AND *THE UNSEEN UNIVERSE*

Lodge's relatively new preoccupation with the possible religious significances of the ether owed much to Myers, a man he later venerated as 'among the chief influences of my life' and who seemed to show many of the 'faculties of instincts of a man of science' despite lacking a formal scientific education.[23] With Myers's guidance, Lodge gradually rejected the philosophical materialism that underpinned the education he had from scientific naturalists such as T. H. Huxley, John Tyndall and W. K. Clifford and he embraced the wider philosophical, religious and spiritual uses to which the results of scientific, and in particular psychical, investigation might be put. Lodge's acceptance of a theistic interpretation of the cosmos inevitably fuelled his preoccupation with the relationships between science and religion, not least the means by which Divine mind could guide the development of the cosmos without violating its physical laws. Lodge would spend the rest of his meditating on the possibility that the ether was the key to explaining how mind interacted with matter.

In an intense private correspondence lasting nearly twenty years, Lodge and Myers often discussed the ether, and this resulted in Myers's etherial thinking becoming better informed by Lodge's vaster knowledge of physics, and Lodge's etherial thinking becoming enriched by Myers's greater understanding of psychology and religion.[24] One idea that probably stimulated Lodge was Myers's notion of a 'metetherial' environment. Building on the relatively new understanding of the dependence of human life on the etherial as well as material environment, and on psychical research's evidence of survival, Myers proposed that life also depended on a third, metetherial environment. Identical to the 'world of spiritual life', this environment was entirely independent of the material world and was both 'within' and 'beyond' the ether.[25] It was the 'spiritual energy' from this environment that was the likely source of genius, inspiration and religious ecstasy and of phenomena studied by psychical research such as telepathy, clairvoyance, spirit communication and apparitions of the dead.[26] Myers never resolved the question of whether the metetherial environment was continuous with or distinct from the ether of physics, the former possibility being fuelled by Lodge's increasing conviction that the ether would turn out to be more accommodating of spiritual functions.

Another idea of Myers's that stimulated Lodge's religious thinking about the ether related to the 'cosmic record of all things'.[27] Drawing on Platonic and neo-Platonic ideas of an *anima mundi*, this represented the capacity of a higher cosmic consciousness to perceive the past, present and future and was something to which clairvoyants and seers appeared to have access. In 1890, only a year before Myers published an early form of the idea, Lodge discussed it with fellow physicist and SPR member Barrett who referred him to a similar idea in the anonymous *Unseen Universe; Or, Physical Speculations on a Future State* (published in 1875), a work that Lodge later called a 'jumping-off' point for his own etherial speculations.[28] One of the bestselling popular scientific works of the late Victorian period, the *Unseen Universe* challenged the materialist and secularist uses to which Tyndall and other scientific naturalists had put the physics of matter, energy and ether. Written by the Scottish physicists Balfour Stewart and Peter Guthrie Tait, the book was the most elaborate and controversial, but by no means the first, attempt by leading British physicists to turn the ether into a reconciliation strategy between scientific and theistic understandings of the cosmos.[29] Only two years before the *Unseen Universe* appeared, for example, Stewart and Tait's close ally Maxwell had insisted that the 'infinite continuity' of the 'wonderful medium' ensured that no part of the universe was devoid of the 'symbols of the manifold order of His Kingdom'.[30]

The *Unseen Universe* had far greater ambitions for the ether. The medium played a key role in the Stewart and Tait's scientific argument for the plausibility for the Pauline doctrine of the spiritual body. They speculated that the ether could be regarded as the luminiferous medium *and* an unseen universe intimately connected to the visible universe which, following evidence of its apparent absorption of starlight, absorbed and stored energy from visible universe. It enabled the disappearance of energy from the visible universe to be reconciled with what the authors proclaimed as a principle of continuity—a major driver of intellectual progress that ensured that humanity would always be able to reconcile states of the universe with immediately antecedent states. On this basis, the discontinuity of energy flow was only apparent because energy merely transited into the unseen part of the total universe. Crucially, among the energies that were dissipated from the visible universe were those associated with thought and the authors believed these arose from the motions of brain molecules. Following the work of William Thomson and others, Stewart and Tait argued that molecules were probably vortex structures in the ether, and so thoughts could be regarded as etherial motions. But the authors also implied that these thoughts could be communicated by an infinite sequence of vortex structures in increasingly rarefied, energetic and robust ethers, which culminated in a truly spiritual body.

Although Lodge shared Myers's interest in the *Unseen Universe*'s suggestion that the ether's storage of thoughts made it a kind of cosmic record, his later writings suggest that he was more interested in the book's connections between the ether and the Christian idea of a spiritual body and between continuity and cosmic intelligibility.

Lodge was certainly aware of one weighty criticism of the book. In a lecture on the ether in 1908, he quoted Maxwell's objection that the 'question' of whether the ether was the 'material organism' of higher order beings was one 'far transcending the limits of physical speculation'.[31] For Lodge, Maxwell's view that Stewart and Tait had taken ether physics too far was less important than the fact that he had considered it worth discussing at all and thereby legitimated further speculation along these lines.

6.3 ETHER PROBLEMS AND NEW PSYCHIC POSSIBILITIES

By the time he gave his 1908 lecture, Lodge had established himself as a leading British authority on electrical science and was equally renowned as the Principal of the fledgling Birmingham University. When in 1900 he resigned his Liverpool chair for the Midlands varsity, he engineered an intellectual profile that differed from that he had developed in Liverpool. A bibliography of Lodge's publications testifies to the shift in his attention away from experimental research in physical and psychical subjects towards writing and lecturing on a much wider range of subjects—including wireless telegraphy, Christian theology, philosophy, education and politics—addressed to specialist and general audiences.[32] Lodge's decision to change how he appeared before the world was partly a pragmatic decision arising from the difficulty of combining research with his burdensome administrative duties but it also reflected a recognition that his 'special function' was in 'writing and calculating', as he remarked in 1925.[33]

Not surprisingly, the ether became one of the subjects to which Lodge devoted considerable literary energy from the early 1900s. Much of this material suggests that he enjoyed writing for publications that permitted more latitude than elite scientific societies normally granted on speculations on the ether's possible constitution and functions. A good example is his review of nineteenth-century physics, published in 1901. Echoing many fin-de-siècle reflections on the limitations of long-cherished physical laws and concepts, the review judged that the most significant tendency in physics was the rise of the electromagnetic and ethereal rather than strictly Newtonian view of the physical world.[34] Despite the successes of Newtonian mechanics, it was now clear that the 'really fundamental dynamics' needed an 'ethereal and not a material basis'.[35] Maxwellian electrodynamics required the luminiferous and electromagnetic medium to be constitutionally very different from the elastic solid model that had been proposed for the ether in the previous century. But, in elaborating the extraordinary dynamics of a physical entity that seemed to be unlike any known material substance, physicists also stood a better chance of addressing questions traditionally excluded from their young discipline:

> Inasmuch as the laws of ordinary dynamics have shown themselves apparently incompetent to deal fully with the phenomena of light and electricity, so much the more it is unlikely that these laws will have any satisfying or comprehensive application in the regions of life and

> mind. Nevertheless, the attempt to include life in the more general scheme of physical science is one that should be strenuously made, and in the long run it may conceivably turn out successful.[36]

Lodge was also hopeful that this 'general scheme', which he clearly believed would have an etherial basis, might also embrace mind but denied that this was Tyndallic materialism by another name. Rather than reducing mind to matter, the encompassing of mind by etherial dynamics was encouraging physicists towards the idealist position of elevating 'matter and all existence to the level of mind and spirit'.[37]

Lodge's remarks represent some of the more optimistic interpretations of what most physicists accepted as the ether 'problem'.[38] The problem was that experimental and theoretical attempts to elucidate the ether's constitution and structure over the past few decades had proven inconclusive, although most physicists maintained that some kind of ether was necessary. The elaborate interferometric experiments of Michelson and Morley in the 1880s had yielded inconclusive evidence for Augustin-Jean Fresnel's theory that the earth moved relative to a stationary ether, while, in the following decade, Lodge's experiments at Liverpool seemed to contradict this, insofar as they suggested that the ether was not dragged by gross matter moving rapidly past it. One reason why it was difficult to relinquish the idea that the ether was physically real was that the achievements of Guglielmo Marconi, Lodge and others in wireless telegraphy suggested that the medium could be manipulated for the purposes of long-distance communication.

Lodge was surrounded by colleagues such as George Francis FitzGerald and Joseph Larmor, who shared his profound commitment to the ether and who devoted much of the 1890s to the task of devising mathematical models of the medium that would explain how it could sustain transverse vibrations (which required it to have some rigidity); how it could appear to present no resistance to the passage of gross matter; how it gave rise to the electrical and magnetic fields demanded by Maxwell's theory and how it related to gross matter. Among the most important results of work on the last of these questions were the electron theories of matter. Devised by Larmor, Hendrik A. Lorentz and others, these proposed that gross matter was built from microscopic, electrically charged particles christened 'electrons', whose movements gave rise to electric currents. They were truly electrical theories of matter because they proposed that the crucial attribute of an electron's mass—inertia—was electrical or etherial in origin. Electron theories proved remarkably effective in explaining a host of experimental discoveries made in the 1890s (notably J. J. Thomson's evidence of the corpuscular nature of cathode rays) but they only compounded the problem of the ether's constitution. Lorentz's version effectively sidestepped the problem by drawing a sharp distinction between the matter and a stationary ether; Larmor's version blurred the distinction by proposing that electrons were localised modifications in the ether, but represented the ether as an abstract dynamical continuum.

Larmor's version of electron theory was more useful to Lodge since it made the immaterial ether antecedent to gross matter and, diverging from Larmor's own inference, showed why the ether needed to be more substantial than previously assumed.

According to estimates made by Lodge and J. J. Thomson in the early 1900s, electron theory required that the density of the ether (which stored most, if not all, of the mass and energy of gross material bodies) needed to be enormously larger than previously thought (10^{12} g cm^{-3}).[39] Although the ether's possibly enormous density baffled some physicists, Lodge turned its differences with gross matter into yet another reason why the medium was a potential space for non-physical attributes of the cosmos.[40] 'We know that matter has a psychical significance', he explained in 1908,

> since it can constitute brain, which links together the physical and the psychical worlds. If anyone thinks that the ether, with all its massiveness and energy, has probably no psychical significance, I find myself unable to agree with him.[41]

Uttered in the same lecture where he alluded to the *Unseen Universe*, this was one of many examples of how Lodge's psychic interpretations of the ether had become far bolder since his move to Birmingham.

6.4 THREATS TO ETHER AND SPIRIT

In 1902 Lodge delivered the first of several presidential addresses to the SPR. Having recently suffered the deaths of Sidgwick and Myers, the organisation needed somebody of Lodge's scientific distinction at the helm to symbolise its claims to represent an intellectually respectable approach to the controversial phenomena of psychical research. His 1902 address is significant because it contains his first public reference to the concept of an 'etherial body', on which he would write at length after 1918.[42] Building on ideas in the neo-Platonic writings of classical antiquity and in later Christian and esoteric philosophies, the idea of an 'etherial', 'ethereal' or 'astral' vehicle of the spirit, independent of the mortal body, gained wide circulation in spiritualistic and modern Theosophical discourses of the nineteenth century. Lodge's interest in the idea needs to be seen in the context of his increasingly conspicuous alliances with early twentieth-century modernising and liberalising trends in Christian theology.[43] The possibility of updating the Christian idea of the soul or etherial body in the light of ether physics was no less important to him than rendering the efficacy of prayer more credible via scientific evidence of telepathy.

It was not until the end of World War I, however, that Lodge felt a substantial intervention on the conception of the etherial body was necessary. I suggest that there were at least two reasons for this. First, Lodge had tragic personal reasons for interpreting the ether as something giving physical substance to the psychological and spiritual essence of an individual. In September 1915 his son Raymond was killed on the Western Front and, to cope with his grief, Lodge attended numerous seances with the British spiritualist medium Gladys Leonard, and these eventually persuaded him of Raymond's post-mortem existence. Although Lodge had openly declared his personal belief in survival as early as 1902, his experiences of Raymond's personality prompted him to write a book that he hoped would console the myriad families who had lost relatives in the global conflict. Entitled *Raymond or Life and Death*, the book was an immediate bestseller and caused Lodge to suffer criticism and enjoy praise for what he believed to be a substantial case for survival.[44] One of the reasons why *Raymond* brought comfort to so many was because it gave empirical and theoretical reasons for thinking that death did not mean the annihilation of the individual. Death was 'not a word to fear, any more than birth is' because it was likely to be merely a change of state: at birth we entered a 'world of air and sense and myriad existence', but at death we probably passed into a region called ether and a 'still more myriad existence'.[45] Those with loved ones whose bodies had been blown to pieces in the military conflict may well have been comforted by the distinguished physicist and pioneer of etherial telegraphy suggesting that the personalities of the dead might still have bodies, albeit etherial bodies that, while eluding our sight and touch, were still substantial.

The second reason why Lodge was prompted to publish on the etherial body related more directly to the threats posed to the idea of the ether, and the intellectual and moral values that he and others attached to it. Even before Raymond's death, World War I would have given Lodge cause to defend the wider symbolic significances of the ether. In his angry *The War and After* (1915), Lodge attacked Germany for its 'war spirit and war caste' and for descending into a civilisation without morality, 'loftiness of soul' or a capacity to serve anything higher than the state.[46] For Lodge, Germany's failure to prevent the material overpowering the spiritual was not surprising given the 'specifically scientific philosophy of German physicists' that neglected the ether.[47] German physicists' disbelief in and denial of the medium was an 'allegory' of the morally dangerous 'larger scheme' in which the immaterial had been abandoned for the material.[48]

The German 'scientific philosophy' to which Lodge was so hostile was, of course, that associated with the Special Theory of Relativity. By the outbreak of World War I, he had established himself as the most outspoken British defenders of the ether in the face of Einstein and other proponents of relativity who regarded the medium as a superfluous concept in understanding the electrodynamics of moving bodies.[49] Lodge had no doubt that the acceptance of principle of relativity would be the 'death-blow to the Ether' because it abandoned Newtonian conceptions of absolute space and time, which he associated with a stationary ether.[50] Although motion relative to this stationary reference frame remained experimentally unproven, Lodge insisted that it was 'not meaningless' or as 'undiscoverable' as relativity's architects supposed and that the alternative—abandoning absolute space and time—was intellectually more perilous.[51]

Lodge was certainly not alone in thinking that the search for ether drift was worthwhile or that abandoning the ether involved more than giving up a problematic physical hypothesis.[52] In the early 1900s, Larmor insisted that abandoning the ether, qua dynamical continuum, removed a proven source of progress in physics and of cosmic intelligibility, while J. J. Thomson invested the medium with providential significance in declaring that, since it enabled humans to benefit from the sun's 'gifts', physicists had a 'duty' to study it.[53] Larmor and Thomson certainly shared the views expressed in Lodge's notorious plea made in his 1913 address to the British Association for the Advancement of Science.

He urged a 'conservative attitude' towards new theories that were 'uprooting' and removing the successful 'landmarks' of the sciences.[54] Implicitly recalling the *Unseen Universe*, Lodge's chief landmark was the idea of the '*ultimate* Continuity' of the cosmos, which both relativity and quantum theories of radiation seemed to be undermining.[55] Unable to imagine the 'exertion of mechanical force across empty space', he saw the ether as 'at least' the

> great engine of continuity. It may be much more, for without it there could hardly be a material universe at all. Certainly, however, it is essential to continuity; it is the one all-permeating substance that binds the whole of the particles of matter together. It is the uniting and binding medium without which, if matter could exist at all, it could exist only as chaotic and isolated fragments.[56]

The ether was also the 'engine of continuity' in non-material domains. In one of the most discussed portions of his address, Lodge outlined his position on psychical research and affirmed his personal belief in the continuation of the human personality in the post-mortem 'etherial existence'.[57] Given the scientific calibre of his audience, Lodge was wise to eschew further speculation on the ways in which the ether might furnish a habitat for the soul, but this did not stop him from suggesting that ether physics and psychical research were connected in yet another sense. Both stood and fell by the scientific attitude to invisible, intangible and elusive phenomena which, if it turned from the 'cold light of reason' to 'dogmatic negations' and denials, spelled doom for the ether and psychic manifestations.[58]

6.5 LODGE'S ETHERIAL BODY

Lodge's British Association address embodied a common rhetorical strategy in his writings on the ether and psychical research. Here physics was turned into a parable for the study of psychics. This was particularly striking in the article of 1918 where he presented his fullest account of the etherial body.[59] The place of publication—the *Hibbert Journal*—was sagacious because it was a leading forum of philosophical and religious debate that had long engaged with the arguments of Lodge and others for a liberalised form of Christianity whose doctrines were more consistent with, and safeguarded by, new developments in the sciences, not least physics and psychical research. For Lodge, the emergence of electron theory furnished a 'sort of parable' to be applied to Christianity and psychical research.[60] The theory had turned electricity from 'something vague, mysterious and immaterial' into something more definite and material.[61] The Christian conception of the soul stood in the same problematic position that electricity once had and, given the public indifference or hostility to the conception, it was in dire need of being rendered more 'concrete and tractable'.[62]

Just as electricity probably comprised 'specks of modified ether', so Lodge hypothesised that the soul might turn out to be a structural feature of the same universal medium.[63] Lodge's reasoning was as follows. The ether appeared to play such a critical part in material bodies—in producing atoms and mediating the cohesive forces between them—that it was possible to imagine that all bodies had material and etherial components. In the case of animate bodies, it was possible that the vital, mental and spiritual qualities that were commonly associated with the material component were also attached to the etherial component. But it was precisely because the ether's constitution appeared to be so radically different from gross matter that it struck Lodge as a possible way of giving much-needed substance to the conception of the soul. Since the etherial component did not suffer from the fatigue, imperfect elasticity, viscosity, decay and other processes that blighted its material component then it was the 'more permanent' vehicle of the spirit that survived bodily death and enabled spirit to lead a 'less abstracted and livelier existence'.[64] On this basis, the ether was the more robust home of our personality. To distinguish himself from the dogmatism that he deplored in recent scientific pronouncements on the ether and psychic phenomena, Lodge insisted that his hypothesis that the etherial component could carry mental and spiritual qualities after death was purely a matter of evidence. Just as current scientific research had brought the ether 'partially brought into the scheme of physics', so future research would help bring it 'into the scheme of psychics' and this would render 'intelligible' the 'obscure communications and strange movements' of spiritualistic seances as well as the Christian idea of the soul.[65]

Lodge was not content, however, to relinquish the task of bringing the ether into psychics entirely to future 'generations of men'. From 1919, when he retired from Birmingham and freed himself from onerous administrative work, until his death in 1940, he wrote, lectured and broadcasted voluminously on the possible role of the ether in solving the fundamental question of how mind interacted with gross matter. In much of this writing, Lodge often drew upon new developments in physics to strengthen lines of reasoning he had been developing for decades. He was predictably critical of the General Theory of Relativity for being a 'blindfold mathematical method' that sidestepped questions of physical cause and dynamical reality and he upheld the virtue of an objective ether in giving physical meaning to potential energy, elasticity, cohesion and the finite speed of light.[66] Yet he also saw General Relativity as an extension of a fruitful scientific tradition privileging space over matter. Einstein's 'real achievement' had been in extending the work of Thomas Young, Fresnel, Michael Faraday and Maxwell in bringing gravity within the ether's embrace and so strengthened hopes of including within its functions psychic 'forms of existence'.[67] Lodge was generally more excited by wave mechanics because it was an aspect of quantum theory that appeared to reinstate the importance of the ether as a 'physical reality'.[68] He also seized on it as a promising new, etherial way of speculating about the mind–matter interaction problem with which he had been struggling for decades. Wave mechanics represented a material particle as a 'group wave' and proposed that the group wave was guided by a 'form wave' with which it was interlocked. Unlike group waves, form waves carried no energy, were undetectable and could travel significantly faster than the speed of light. For Lodge it was not impossible that form waves constituted a 'physical basis' of life and mind, insofar as they showed how those ineffable qualities might interact with and guide matter without violating energy conservation.[69]

When in 1933 Lodge published his final book, *My Philosophy*, his etherial 'philosophy' had not changed substantially in over a decade.[70] Like so many of Lodge's books, it rehashed many articles and other texts he had published elsewhere. Despite upholding the ether's objective reality, Lodge was as modest as ever in accepting that its constitution was still largely unknown, that his basis for believing in it was partly instinctive and that, in describing it in terms of elasticity, inertia and other physical properties ascribed to gross matter, he was 'speaking apologetically'.[71] He held out the hope that the ether would turn out to be some kind of dynamical medium whose drift relative to moving matter would be eventually be detected and which still represented the most attractive 'path of the progress of discovery' regarding the interaction of the immaterial and material domains.[72] As a synthesis of all his writings on the ether, *My Philosophy* highlighted the significant extent to which the medium had become relevant to the 'progress of discovery' in all his major intellectual endeavours. The ether had become his driver of intelligibility in physics, electrical engineering, psychical research and the modernisation of Christian theology. He was certainly not alone among late nineteenth- and early twentieth-century physicists in regarding arguments against the ether's reality as threats to cherished scientific, philosophical and religious values attached to the idea of a dynamical continuum, but he was rare in championing the ether as passionately as he did and in how many aspects of his life's work stood to lose by its abolition.

6.6 THE RECEPTION OF LODGE'S ETHERIAL IDEAS

In apologising for using mechanical terms to describe an entity whose constitution and objective reality was uncertain, Lodge engaged with a difficulty that he and other proponents of ether theory had struggled with for decades. In 1889 the mathematician Karl Pearson spoke for many readers of Lodge's *Modern Views of Electricity* when he warned that explaining the ether in terms of mechanism was putting 'the cart before the horse'.[73] Twenty-six years later, in a review of Lodge's *Atoms and Rays* (1924), the physicist Edward Andrade echoed more widely shared difficulties with Lodge's popularisation of outdated etherial conceptions. He warned the 'lay reader' to whom the book was aimed that Lodge's medium was really the 'good old mechanical ether' that no 'orthodox physicist of today' believed in because they accepted relativity's argument that the position and speed of the space-filling medium were undetectable.[74] However, for the dwindling number of physicists who, like Lodge, grew up with the 'good old mechanical ether', Andrade's criticisms would not have constituted decisive reasons to abandon an objective ether *per se*.[75]

What bothered many of Lodge's critics was his insistence on combining increasingly outdated conceptions of the ether with the controversial results of psychical research. In response to Lodge's British Association address of 1913, the devoutly Christian American physicist Edwin Hall upheld relativity's case against the ether and the inconclusiveness of psychical research before expressing fear of the day when people abandoned communion with God for intercourse with 'spirits clad in "ethereal bodies"'.[76] Even some advocates of psychical research found the ether to be a liability in interpreting evidence of a world beyond gross matter. In 1918, one of the SPR's leading figures, Eleanor Sidgwick, warned Lodge that a 'good many physicists' would question his ideas about the ether and judged his hypothesis of the etherial body 'too materialistic to be probably true'.[77] Lodge may well have recognised that, for all its constitutional difference from Stewart and Tait's ether, his conception of the ether was still too material to be immune to the charges of materialism once levelled at the *Unseen Universe*'s idea of the abode of spirit.[78]

Much to his critics' chagrin, the books, articles, lectures and radio broadcasts where Lodge presented his controversial etherial ideas enjoyed a large following.[79] Not surprisingly, spiritualists were among the most voracious consumers of what he wrote on survival and psychical questions generally and, by the 1920s, they widely regarded him as a champion of their cause, despite his objections to the spiritualist label.[80] Neither was it surprising that spiritualists should have welcomed Lodge's writings on the ether because they lent a form of scientific credibility to ideas about the etherial domicile of mind and spirit that they already accepted. Although some spiritualists were acutely aware of the debate about the ether's constitution and objective reality, most found that Lodge's ether writings provided enough of the scientific support they required for the general idea that human existence was not limited to the material domain.[81]

By far the most instructive responses to Lodge's psychic interpretations of the ether came from a scientific constituency with which he closely identified himself in the interwar period: professional and amateur wireless telegraphers. A survey of British and American wireless textbooks, periodicals and other texts published in the 1920s and 1930s testifies to the survival of the concept of the ether long after it was supposed to have been killed off by relativity and quantum theory. For many in this scientific constituency, the quotidian practices of wireless telegraphy were far more intelligible with an ether than without it. In his 1924 textbook, for example, the British wireless telegrapher P. J. Risdon warned readers that the only 'reasonable course' to follow in a situation where the etherial implications of relativity and quantum theories had caused 'dissension among scientists' was to assume that space was filled with an ether.[82] Given Lodge's pioneering contributions to wireless telegraphy, he was inevitably summoned by wireless telegraphers as a weighty scientific dissenter from the view that the ether did not exist.[83]

Wireless telegraphers were rather less unsympathetic to Lodge's psychic and religious ideas about the ether than were his physicist colleagues.[84] Risdon was one of many wireless telegraphers who upheld the old arguments for the ether's existence and helped promulgate a psychic interpretation of the ether from which even Lodge dissented. In the same textbook where he defended the ether, he speculated that some kind of ether might carry 'thought waves' arising from the electrical activity in the brain.[85] He was hardly the first to suggest that telepathy might have such an electrical and etherial basis:

brain wave theories of telepathy had been circulating since the 1880s and enjoyed a renewed popularity in the wake of the successes of wireless telegraphy. Yet, by the early 1900s, Lodge regularly expressed severe doubts about the theories. He questioned whether telepathy behaved like an etherial or any other physical form of vibration (e.g. because the intensity of telepathic impressions did not decline with distance) and argued that it was more likely to be a purely psychical form of communication.[86] For some of the wireless telegraphers who read Lodge's books and heard his radio broadcasts, however, the physicist's arguments that the ether was the 'true vehicle of life and mind' seemed to contradict this argument and inspired numerous speculations on and experimental investigations into the possibility that minds of the living and the dead might utilise the ether to commune with the living.[87]

The foregoing survey of the reception of Lodge's etherial ideas suggests that the post-relativistic lives of the ether and its most vociferous British champion were a good deal rosier than historians have claimed. If we follow many historians and confine our attention to physicists, then Lodge and his objective ether do indeed look increasingly outdated and marginal; but, if we broaden our sample to include practitioners of other scientific enterprises, then they are far from marginal. Wireless telegraphers played an important part in sustaining a place for the ether and for Lodge's already widely circulated ideas about it, in the popular literature of the sciences. Historians have only begun to study this literature, a literature to which Lodge and many of his critics and champions made notable contributions. Paradoxically, a recent study of the British market for this literature implies that the ether had been killed off by the 1920s, even though its greatest champion, Lodge, was one of the most respected scientific expositors of the period.[88] One way to address this apparent paradox is to better understand the extent to which popular scientific literature engaged with the ether and gave currency to the wider philosophical and religious arguments for its place in the cosmos. Given the significance of these kinds of texts in the early education of scientists, this investigation may help explain why, long after Lodge's death, the ether never entirely disappeared from the discourses of professional physicists.[89]

ACKNOWLEDGEMENTS

The author would like to thank Linda Henderson, Jaume Navarro, Richard Staley and Aaron Wright for their comments on earlier versions of this chapter. Permission to quote from unpublished material held in their collections has been granted by Senate House Library, University of London and the Syndics of Cambridge University Library.

NOTES

1. Jed Buchwald, *From Maxwell to Microphysics: Aspects of Electromagnetic Theory in the Last Quarter of the Nineteenth Century* (Chicago: Chicago University Press, 1985); Bruce Hunt, *The Maxwellians*

(Ithaca: Cornell University Press, 1991); Peter Rowlands, *Oliver Lodge and the Liverpool Physical Society* (Liverpool: Liverpool University Press, 1990).
2. David B. Wilson, 'The Thought of Late-Victorian Physicists: Oliver Lodge's Ethereal Body', *Victorian Studies* 15 (1971): 29–48.
3. Brian Wynne, 'Physics and Psychics: Science, Symbolic Action and Social Control in Late-Victorian England', in *Natural Order: Historical Studies of Scientific Culture* (Beverly Hills, CA: Sage, 1979): 167–87. For a critique of Wynne, see Richard Noakes, 'Ethers, Religion and Politics in Late-Victorian Physics: Beyond the Wynne Thesis', *History of Science* 43 (2005): 415–55.
4. Peter J. Bowler, *Reconciling Science and Religion: The Debate in Early-Twentieth Century Britain* (Chicago: Chicago University Press, 2001); Courtenay Raia, 'From Ether Theory to Ether Theology: Oliver Lodge and the Physics of Immortality', *Journal of the History of the Behavioural Sciences* 43 (2007): 19–43, 95–101. Other studies include G. N. Cantor, 'The Theological Significance of Ethers', in *Conceptions of Ether: Studies in the History of Ether Theories*, ed. G. N. Cantor and M. J. S. Hodge (Cambridge: Cambridge University Press, 1981), 135–55, and Janet Oppenheim, *The Other World: Spiritualism and Psychical Research in England, 1850–1914* (Cambridge: Cambridge University Press, 1985), pp. 378–85.
5. Oliver Lodge, *Modern Views of Electricity* (London: Macmillan and Co., 1889), p. 358.
6. Lodge, *Modern Views*, p. 331.
7. Lodge, *Modern Views*, p. 335.
8. For the SPR's foundation, see Alan Gauld, *The Founders of Psychical Research* (London: Routledge, 1968).
9. Oliver Lodge, *Past Years* (London: Hodder and Stoughton, 1931), p. 271.
10. Oliver Lodge, 'An Experiment in Thought Transference', *Nature* 30 (1884): 145.
11. Oliver Lodge, 'An Account of Some Experiments in Thought-Transference', *Proceedings of the Society for Psychical Research* 2 (1884): 189–200, p. 191.
12. Lodge, 'Account of Some Experiments', p. 191.
13. Edmund Gurney, F. W. H. Myers and Frank Podmore, *Phantasms of the Living* (London: Society for Psychical Research, 1886), p. 112.
14. Lodge, *Modern Views*, p. 307.
15. Oliver Lodge, 'Electrical Radiation from Conducting Spheres, an Electric Eye, and a Suggestion Regarding Vision', *Nature* 41 (1890): 462–3, p. 462.
16. Oliver Lodge, 'Thought Transference: An Application of Modern Thought to Ancient Superstition', *Proceedings of the Literary and Philosophical Society of Liverpool* 46 (1892): 127–45.
17. Oliver Lodge, 'Account of Sittings with Mrs Piper', *Proceedings of the Society for Psychical Research* 6 (1889–90): 436–659, p. 436.
18. Lodge, *Past Years*, p. 279.
19. Oliver Lodge, 'The Interstellar Ether', *Fortnightly Review* 53 (1893): 856–62, p. 856.
20. Lodge, 'Interstellar Ether', p. 862.
21. Lodge, 'Interstellar Ether', p. 862.
22. Cantor, 'Theological Significance'.
23. Oliver Lodge, 'The Life Work of My Friend F. W. H. Myers', *Nature* 144 (1939): 1027–8, p. 1027; Oliver Lodge, 'A Scheme of a Vital Faculty', *Nature* 68 (1903): 145–7, p. 145.
24. See letters between Myers and Lodge from 1889–1900: SPR.MS.35/1298–1572, Oliver Lodge Papers, Society for Psychical Research Archive, Cambridge University Library (hereafter OJL-SPR).
25. F. W. H. Myers, *Human Personality and the Survival of Bodily Death* (London: Longmans, Green and Co., 1903), vol. 1, p. 215.

26. Myers, *Human Personality*, p. 219.
27. F. W. H. Myers, 'The Principles of Psychology', *Proceedings of the Society for Psychical Research* 7 (1891–2): 111–33, p. 120. Lodge's interest in this is evident in Oliver J. Lodge, 'Address', *Report of the Sixty-First Meeting of the British Association for the Advancement of Science* (London: John Murray, 1892): 547–57, p. 552.
28. W. F. Barrett to Oliver Lodge, 15 October 1890, SPR.MS.35/60, OJL-SPR. [Balfour Stewart and Peter Guthrie Tait], *The Unseen Universe; Or, Physical Speculations on a Future State* (London: Macmillan and Co., 1875); J. Arthur Hill, ed., *Letters from Sir Oliver Lodge* (London: Cassell and Co., 1932), p. 33.
29. On *Unseen Universe*, see Crosbie Smith, *The Science of Energy: A Cultural History of Energy Physics in Victorian Britain* (London: Athlone Press, 1998), pp. 253–5.
30. James Clerk Maxwell, 'On Action at a Distance', *Nature* 7 (1873): 323–5, 341–3, p. 343.
31. Oliver Lodge, 'The Ether of Space', *Contemporary Review* 93 (1908): 536–46, p. 540. Lodge was quoting James Clerk Maxwell, 'Ether', in *Scientific Papers of James Clerk Maxwell*, edited by W. D. Niven, vol. 2 (Cambridge: Cambridge University Press, 1890), 763–75, p. 775.
32. Theodore Besterman, *A Bibliography of Sir Oliver Lodge* (Oxford: Oxford University Press, 1935).
33. Oliver Lodge to Harry Price, 9 May 1925, HPC/4B/150, Harry Price Collection, University of London.
34. John L. Heilbron, 'Fin-de-Siècle Physics', in Carl-Gustav Bernhard, Elisabeth Crawford and Per Sèrböm, eds., *Science, Technology, and Society in the Time of Alfred Nobel*, (Oxford: Pergamon Press, 1982): 51–73.
35. Oliver Lodge, 'Scope and Tendencies of Physics', in *The 19th Century: A Review of Progress* (London: G. P. Putnam's Sons, 1901), 348–57, p. 352.
36. Lodge, 'Scope and Tendencies', p. 354.
37. Lodge, 'Scope and Tendencies', p. 355.
38. Olivier Darrigol, *Electrodynamics from Ampère to Einstein* (Oxford: Oxford University Press, 2000), pp. 314–50; Loyd Swenson, *The Ethereal Aether: A History of the Michelson–Morley–Miller Aether Drift Experiments, 1880–1930* (Austin: University of Texas Press, 1972), pp. 98–120.
39. Oliver Lodge, *The Ether of Space* (London: Harper and Brothers, 1909), p. 150.
40. H[orace] L[amb], 'The Ether of Space', *Nature* 82 (1910): 271.
41. Lodge, 'Ether of Space', p. 543.
42. Oliver Lodge, 'Address by the President', *Proceedings of the Society for Psychical Research* (1901–3), 37–57, p. 47.
43. Bowler, *Reconciling Science and Religion*, p. 100.
44. Oliver Lodge, *Raymond, or, Life and Death* (London: Methuen and Co., 1916). On *Raymond*'s impact, see Jay Winter, *Sites of Memory, Sites of Mourning: The Great War in European Cultural History* (Cambridge: Cambridge University Press, 1995), pp. 61–3.
45. Lodge, *Raymond*, p. 298.
46. Oliver Lodge, *The War and After* (London: Methuen and Co., 1915), pp. 65, 91.
47. Lodge, *War and After*, pp. 15, 18.
48. Lodge, *War and After*, p. 18.
49. Stanley Goldberg, 'In Defence of the Ether: The British Responses to Einstein's Theory of Relativity', *Historical Studies in the Physical Sciences* 2 (1970): 89–125.
50. Oliver Lodge, 'The Ether of Space and the Principle of Relativity', *Science Progress* 6 (1911–12): 337–44, p. 339.
51. Lodge, 'Ether of Space and the Principle of Relativity', p. 343.

52. Swenson, *Ethereal Aether*, pp. 171–89; Milena Wazeck, *Einstein's Opponents: The Public Controversy about the Theory of Relativity in the 1920s* (Cambridge: Cambridge University Press, 2014), pp. 136–47.
53. Andrew Warwick, '"That Universal Aethereal Plenum": Joseph Larmor's Natural History of Physics', in Kevin Knox and Richard Noakes, eds., *From Newton to Hawking: A History of Cambridge University's Lucasian Professors of Mathematics* (Cambridge: Cambridge University Press, 2003), 343–86; J. J. Thomson, 'Address', in *Report of the Seventy-Ninth Meeting of the British Association for the Advancement of Science* (London: John Murray, 1910), 3–29, p. 15.
54. Oliver Lodge, 'Continuity', in *Report of the Eighty-Third Meeting of the British Association for the Advancement of Science* (London: John Murray, 1914), 3–42, p. 19.
55. Lodge, 'Continuity', p. 15.
56. Lodge, 'Continuity', p. 27.
57. Lodge, 'Continuity', p. 40.
58. Lodge, 'Continuity', pp. 5–6.
59. For a fuller discussion see Wilson, 'Thought of Late-Victorian Physicists'.
60. Lodge, 'Ether, Matter and the Soul', p. 257.
61. Lodge, 'Ether, Matter and the Soul', p. 252
62. Lodge, 'Ether, Matter and the Soul', p. 259.
63. Lodge, 'Ether, Matter and the Soul', p. 255.
64. Lodge, 'Ether, Matter and the Soul', p. 258.
65. Lodge, 'Ether, Matter and the Soul', pp. 259–60.
66. Oliver Lodge, 'Einstein's Real Achievement', *Fortnightly Review* 110 (1921): 353–72, p. 366.
67. Lodge, 'Einstein's Real Achievement', p. 372.
68. Oliver Lodge, 'The Nineteenth Kelvin Lecture, "The Revolution in Physics"', *Journal of the Institution of Electrical Engineers* 66 (1928): 1005–20, p. 1014.
69. Oliver Lodge, *Beyond Physics, or, The Idealisation of Mechanism* (London: George Allen and Unwin, 1931), p. 144.
70. Oliver Lodge, *My Philosophy. Representing My Views on the Many Functions of the Ether of Space* (London: Ernest Benn, 1933).
71. Lodge, *My Philosophy*, p. 16.
72. Lodge, *My Philosophy*, p. 235.
73. [Karl Pearson], 'Recent Works on Physical Science', *Athenaeum* (February 1893): 132–3, p. 133.
74. E[dward] N. Da C[osta] A[ndrade], 'A Veteran's View of Modern Physics', *Nature* 114 (1924): 599–601, p. 600; cp. [J. W. N. Sullivan], 'Ether and Reality', *Times Literary Supplement*, 14 May 1925.
75. J. A. Fleming, *Fifty Years of Electricity: The Memories of an Electrical Engineer* (London: Wireless Press, 1921), p. 303; J. J. Thomson, *Recollections and Reflections* (London: G. Bell, 1936), p. 432.
76. Edwin H. Hall, 'Sir Oliver Lodge's British Association Address', *Harvard Theological Review* 8 (1915): 238–51, p. 251.
77. Eleanor Mildred Sidgwick to Oliver Lodge, 4 October 1918, SPR.MS.35 / 2255, OJL-SPR.
78. C. G. Knott, *The Life and Scientific Work of Peter Guthrie Tait* (Cambridge: Cambridge University Press, 1911), p. 239.
79. 'Professor Lodge's Theology', *Church Times*, 4 December 1908.
80. Lodge to J. A. Hill, 8 October 1917, published in Hill, *Letters from Sir Oliver Lodge*, p. 141.
81. H. A. Dallas, 'The Etherial Body: Its Nature and Scope', *Light* 44 (1924): 116; Arthur Findlay, *On the Edge of the Etheric* (London: Psychic Press, 1931), pp. 33–47.

82. P. J. Risdon, *Wireless* (London: Ward, Lock and Co., [1924]), p. 34.
83. Ralph Stranger, *The Outline of Wireless* (London: George Newnes, 1932), pp. 203–4.
84. See e.g. 'The Ether and the Soul', *Popular Radio* 8 (1925): 280; 'Ether and Reality', *Wireless World* 17 (1925): 16.
85. Risdon, *Wireless*, pp. 347–51.
86. Oliver Lodge, *The Survival of Man: A Study in Unrecognised Human Faculty* (London: Methuen and Co., 1909), p. 35.
87. Jack Frost, 'Radio and Psychic Science', *Light* 48 (1926): 221; E. J. G. Lewis, 'Spirit Communication and the Ether', *English Mechanic*, 12 March 1926. Citation from Oliver Lodge, *Ether and Reality* (London: Hodder and Stoughton, 1925), p. 166.
88. Peter Bowler, *Science for All: The Popularisation of Science in Early Twentieth Century Britain* (Chicago: University of Chicago Press, 2009), p. 36.
89. P. A. M. Dirac, 'Is There an Aether?', *Nature* 168 (1951): 906–7; Geoffrey Builder, 'Ether and Relativity', *Australian Journal of Physics* 11 (1958): 279–97.

7

Lenard's Ether and Its Vortex of Emotions

Between Accommodating and Fighting Modern Physics with *Äther* and *Uräther* in the German Political Context

Arne Schirrmacher

7.1 INTRODUCTION: PHILIPP LENARD IN (ANGLO-AMERICAN) PERSPECTIVE

On Wednesday, 18 May 1932, Philipp Lenard was awarded the Franklin Medal in Philadelphia, which was at that time the highest recognition for science in the United States. Like the Nobel Prizes, but with a much longer tradition and wider scope, the medallists honoured by the Franklin Institute comprised the technological, industrial and scientific elite worldwide. Lenard was an exemplar of a great scientist whose 'life-work [was] devoted to fruitful research in physics, in the course of which he added greatly to scientific knowledge and especially showed that it was possible for cathode rays to exist outside the generating tube'; as well as discovering the nature and the laws of the photoelectric effect.[1] Thus, Lenard stood in a line of the great physicists of the twentieth century, together with Hendrik Antoon Lorentz, Joseph John Thomson, Albert Abraham Michelson, Niels Bohr, George Ellery Hale, Max Planck, Charles Thomson Rees Wilson, William Henry Bragg and James Jeans, while Albert Einstein, who had just received a frantic welcome to the United States, would be admitted to the club of Franklin medallists only in 1935.[2]

At the age of almost 70, Lenard did not travel to Philadelphia and his paper was read by an American friend from the early stages of his career. His 'Some Scientific Views to the Franklin Institute', which was published in the institute's journal, appear to be Lenard's only presentation of his ideas on the ether and the distinction between *Äther*

Schirrmacher, A., 'Lenard's Ether and Its Vortex of Emotions: Between Accommodating and Fighting Modern Physics with *Äther and Uräther* in the German Political Context' in *Ether and Modernity*, edited by J. Navarro.
 DOI: 10.1093/oso/9780198797258.003.0007

and *Uräther* in English. It is also one of few instances where reference is made to his extensive writings on the ether from the previous two decades; these writings had otherwise not found much international attention, besides being read as obstinate attacks on Einstein's theory of relativity.

For many, it may seem awkward to see Philipp Lenard honoured in this way by an American institution during the early 1930s. Had Lenard in his recollections not just given ample display of his disdain for Americans and American scientists? He had, for instance, stated that Millikan claimed priority for researches Lenard himself had earlier already published, which 'shows how shabby in general the mindset of representatives of natural research had become, even in the prestigious academic circles'.[3] Furthermore, was Lenard not also the obtuse foe of modern physics—one who, since World War I, had become an ideologue, discarding foreign research and condemning first British, French and American and later all scientists deemed Jewish? How could he be a worthy representative of good science, when he would soon become one of 'Hitler's scientists', whom he had endorsed and promoted as early as 1924?[4] Had the seeds not yet developed visibly enough, which few years later became so drastically explicit, when he proclaimed in the preface to a four-volume physics textbook in 1936 that the idea of internationalism in science 'is always rooted in an error', since only the 'Nordic race' would be capable of true natural research, which should be called 'German Physics' or 'Aryan Physics'.[5]

This chapter deals with German conceptions of the ether in the twentieth century and the wider uses of this notion. It is from this perspective that Philipp Lenard becomes a central figure, perhaps comparable to Oliver Lodge in Britain. It is besides the scope of the present paper to unfold a fuller account and explanation of the intricacies and inconsistencies of Lenard's life, career, understanding of natural research, national ambivalences and religious, political and ideological views; it often seems easier to think of Lenard taking up various different personae.[6] Whenever I have to touch upon these complexities, I shall only do it insofar as it is relevant to his views on the ether. The disentanglement of the complicated overlap of epistemic, disciplinary, political and religious/ideological factors in Lenard's ways of natural research (he tried to forgo the term 'science', or *Wissenschaft*) has still to be attempted.[7]

The ceremony of the Franklin Institute may help us see the ambiguous and multifaceted landscape of recognised fields of scholarship and the extent to which scientific, ideological and political factors interfere. The variety of prizes and medals given by the institute bear witness to its role as a national institution that helped bring about the 'making of an industrial America'.[8] Inventors, engineers and scientists were treated on equal footing, as success, both in industry and in science, was guaranteed only when they worked in collaboration. Moreover, the committee of the Franklin Institute wanted to achieve a balance of national and foreign medallists, and it appears that Lenard complemented rather ideally the other recipient of the medal that year, the American self-made engineer Ambrose Swasey, an archetype of a successful machinist, entrepreneur, philanthropist and amateur astronomer who constructed the mountings of the best American telescopes. The German professor, *Geheimrat* and Nobel Prize winner who had just retired from active duty appeared to be a deserving candidate. With his fame as an experimentalist, he resonated well with the practical and applied focus of the institution. Indeed, objects as well as physical entities were carrying his name: Lenard tubes, windows and rays.

Lenard's paper for the ceremony was read by Charles B. Thwing, who had made a career in selling and producing scientific measuring instruments, after he had given up the post of head of department of physics at Syracuse University. In a sense, he personified the relation of American physics, in general, and the Franklin Institute, in particular, with Lenard. As Thwing explained before reading Lenard's paper, his contact and friendship with Lenard went far back to his time as the last doctoral student of Heinrich Hertz at Bonn, where Lenard was Hertz's assistant and made his first steps in university teaching. With an untiring attitude, Lenard had introduced innovative teaching, when he complemented his lectures with informal discussions and demonstrations of instruments, always keen to get feedback from his students. This was the beginning of a 'regular friendship' between two aspiring physicists of roughly the same age, and resulted in a transatlantic correspondence for years. Their closeness included Thwing asking Lenard for the privilege of naming his second son after him, a request which was cordially granted.

This persona of Lenard—the international, open-minded one, so to speak—was also visible in 1889 when Lenard applied for a job at the University of Lincoln, Nebraska, as career chances in Germany seemed dim and he felt that his teachers were not role models.[9] Later, he would turn to Britain, as I will discuss in Section 7.2.2. The American Philip Lenard Thwing, however, personified in another way the end of the transatlantic friendship as, in World War I, he was fighting the Germans: 'I could not write to Dr. Lenard and tell him that his namesake was fighting against his country. Our correspondence ended.'[10]

Apparently, in the same way that Thwing had lost contact with Lenard due to World War I, the Franklin Institute and with it most of the American scientific world had disconnected from him. Many remembered his work from around the turn of the century on cathode rays and the photoelectric effect, results that had had tremendous impact on technological innovation. Instrument makers and technical physicists, who had a prominent standing in the Franklin Institute, may have preserved the memory of Lenard's earlier work. Some could probably recall the 1920 confrontation between Lenard and Einstein; but few, if any, would have followed Lenard's research in the 1910s and 1920s, due to the war, the subsequent boycott of German science or the transformation of relevant fields of physics.

Lenard's paper, which Thwing had translated and read in March 1932, attempted to 'state briefly what today appears as the most essential result of the views established at that time, and in the course of subsequent considerations', that is Lenard's ether conceptions.[11] When, around 1943, Lenard prepared the fourth volume of his collected works, he included this paper as it was 'the first wholly summarizing publication of the new conceptions... on ether and energy', which had been published in three books during the 1920s[12] and which he later had carried out further in his *Deutsche Physik*, that very textbook, which in its preface proclaimed a German or 'Aryan' physics as opposed to a 'Jewish' one.[13]

In order to exhibit the all-pervading role of the ether in Lenard's career and thinking, I will discuss his experiences of the ether, the experiments he envisaged to prove the ether, and the emotions that came from the ether and which impacted his scientific, philosophical and political approach.

7.2 EXPERIENCES OF THE ETHER

7.2.1 Crookes, or, The fourth state of matter

William Crookes's 1879 British Association lecture, which was translated into German, had a lasting impact on Lenard when he was a Pressburg soon-to-be high-school graduate and it influenced his whole academic life.[14] The plethora of new colourful experiences from electricity in diluted gases revealed phenomena of the ether. Not only did this define the material world of Lenard's future laboratories, it also instilled a whole attitude into the incipient *Naturforscher*. This attitude is nicely summarised in Crookes's final lines of *On Radiant Matter*, which in its German edition carried the subtitle '*or the Fourth State of Matter*':

> We have actually touched the border land where Matter and Force seem to merge into one another, the shadowy realm between Known and Unknown which for me has always had peculiar temptations. I venture to think that the greatest scientific problems of the future will find their solution in this Border Land, and even beyond; here, it seems to me, lie Ultimate Realities, subtle, far-reaching, wonderful.[15]

It turned out that also Lenard's physics teacher, whom he assisted in private experimentation in the school laboratories on weekends, had been electrified by the idea of a fourth state of matter, and he had already ordered a number of tubes like the ones Crookes had described. Soon, they started making their own tubes and, although they failed to make it to the fourth state, they still enjoyed huge glowing lights.[16] Lenard would henceforth strive for 'ultimate realities' in a number of ways reminiscent of Crookes, extending his physics towards more general meanings of ether and a world of spirits, without, however, subscribing to Crookes's spiritualism.[17]

It took Lenard thirteen years and much struggle—with family, with academia, particularly professors from Vienna, Budapest and Heidelberg, as well as with himself, that is, with figuring out a (national) identity—until he found satisfaction in having made progress 'after Crookes had advanced in this direction for the last time'.[18] As an assistant of Heinrich Hertz in Bonn, Lenard strived to free the cathode rays from the messy conditions within the tube and let them escape through a 'Lenard window', as it was soon called. Thus, in 1893, it became possible to investigate what Crookes radiant matter was: was it radiant energy, a stream of fast-moving 'hurled' molecules, effects in the ether, or something completely novel?

Generally speaking, British scientists favoured the existence of an all-pervasive ether. Lord Kelvin, in particular, based some of his ideas on the hydrodynamic theories of Hermann von Helmholtz, in which vortex rings, twists and knots could appear, and so did J. J. Thomson. French physicists would prefer corpuscular models, while Dutch and German voices like Lorentz and Wiechert were favouring non-mechanical variants or some world ether. Meanwhile, Lenard stayed closer to Crookes's stream of radiant matter, or hurled molecules. In Bonn, Lenard had learned a lot about mechanical and hydrodynamic ether theories, both from Hertz, whose *Prinzipien der Mechanik* he completed and edited after Hertz's sudden death in 1894, and from the Norwegian physicist Vilhelm Bjerknes. The latter, one of Lenard's peers and a later pen pal, would be extending the hydrodynamic theories for electromagnetism and gravitation of his father, Carl Anton Bjerknes, to a hydrodynamic understanding of the ether.[19]

Lenard's main occupation, however, was not yet theories of the ether but observations that could be recognised as phenomena related to the ether. He wanted to prove with 'pure' experiments that cathode rays were hurled electricity in the ether, rather than matter. While he was doing that, others like Thomson and Jean Perrin made progress that was often perceived as more groundbreaking, even though, from Lenard's point of view, they did not care enough about obvious interfering influences, which rendered their findings inconclusive or 'impure'. Still, they could claim priority for demonstrating the negative charge of the rays and the electron as its constituent.

7.2.2 Lodge, or, International exchange, ether and spirits

The British experts of the ether quickly became interested in Lenard's work on cathode rays and his window tube. Lord Kelvin invited Lenard to the 1895 Ipswich British Association meeting and, when Lenard declined, Kelvin and Oliver Lodge then invited him to the 1896 Liverpool assembly, proposing that he stay at Lodge's home and asking him to bring his tube to demonstrate it in Lodge's laboratory; this time, Lenard accepted. Actually, before this, in 1890, Lenard had tried to emigrate to England, frustrated with his Heidelberg professor, and had worked for six months at the City and Guilds of London Institute. With help of some of his British friends, he had been introduced to Lodge and saw his laboratory (Lenard took detailed notes of its equipment).[20] But it was only the collaboration with Hertz and the window tube of 1892 that would promote the German researcher to the rank of a respectable scientist in Britain.

In 1896, Lenard enjoyed Lodge's splendid hospitality in Liverpool, appreciated the attention and decided spontaneously to give his talk at the meeting in English; the talk was greeted by an audience of some 500, including most of the leading British physicists, with only Crookes missing, to Lenard's regret.[21] It was here that Lenard developed cordial relations with Lodge and, to some degree, also with Kelvin, although less so with J. J. Thomson, who was clearly the strongest competitor in cathode-ray research at this time.[22]

As much as Crookes had triggered Lenard's interest in cathode rays as phenomena in the ether, Lodge eventually became responsible for Lenard's turn to problems on the existence and the qualities of the ether, when he sent him copies of his books, including his 1908 *Substance and Faith*. Lenard's diary reads: 'Comes across as a prophet for me; of much impact.'[23] Lenard started experiments on ether drag at his Heidelberg institute, as I shall discuss in Section 7.3, and he began to make the ether a topic of his publications from 1910 on. His *Über Äther und Materie* [*On Ether and Matter*] grew out of a plenary lecture to the Heidelberg Academy of Sciences and saw a second amended printing in 1911. This meant the first departure from a 'rough-mechanical analogy' for Lenard. The structure of the ether should be understood as 'a spatially discontinuous, moving, and all-pervading ether', which he had derived both from his experiments with cathode rays and from Bjerknes's hydrodynamic works.[24]

Lodge's *Substance and Faith*, with its broad visions that went beyond physics proper, made a discernible impact on Lenard, particularly the following far-reaching analogy in the chapter entitled 'Immanence', in the section entitled 'Mind and Matter':

> An omnipresent, uniform, immanent Purpose, running through the whole of existence . . . is beyond our experience; and, like every other uniformity, is difficult to detect or realise. As an instance of this difficulty, I need only cite the long-delayed discovery of an all-embracing medium like the terrestrial atmosphere. An intelligent deep-sea creature would find it most difficult to become aware of the existence of water. Similarly humanity has existed all along in a pervading and interpenetrating ether, of which to this day men have for the most part no cognisance; although it is probably the fundamental substratum of the whole material world, underlying every kind of activity, and constituting the very atoms of which their own bodies are composed.[25]

The first footnote of Lenard's published 1910 lecture refers to Lodge's *Life and Matter* of 1905,[26] from which he took the speculation that sufficiently complex atomic structures like protoplasm may 'hold what we call spirit' and they may 'become the carriers of the wonderful phenomena of life' otherwise out of reach of the scientist.[27] In this way analogical reasoning along the lines of Lodge allowed for a first glimpse into the *Geisterwelt* ('world of the spirits'), which Lenard would invoke much more forcefully in later times. However, unlike Crookes and Lodge, who were members of the more secretive Ghost Club as well as the more rational Society for Psychical Research, Lenard would not follow them as far as to link attempts at understanding the fundamental structures of nature with secretive sessions or public spectacle of spiritualism and telepathy.[28]

Despite these excursions into the world of spirits, Lenard mostly stuck with the more physical ether concepts of Maxwell, Hertz, Kelvin and, in particular, Bjerknes and his hydrodynamic theory. Lenard argued that 'cathode rays, which are pure negative electrons, are processes in the ether'. Therefore 'electrons [can] be understood as parts of the ether inasmuch they are the end points of ether vortex lines'. Moreover, the concept of light quanta, which had been introduced by Planck and Einstein, could be understood as a 'transverse flow' related to vortex threads.[29]

Throughout his scientific life, Lenard had tried to grasp the new and unknown—the fourth state of matter, the empty atom, the photoelectric effect, and so on—using an experimental approach which proceeded step by step (reflected by a continuous numbering in his notebooks and publications) and which led to an understanding in basic, mostly mechanical pictures. The human mind, according to Lenard, was 'set up for comprehension, not just for mathematical description'.[30] And this is what his *Über Äther und Materie* was about—to show in a Hertzian way how nature could be comprehended:

> I do not think that difficulties should prevent us from extending and maintaining the existing image [*Bild*], for then we would forgo any such image and the mechanical intelligibility of nature altogether. I do not think that this will happen, even if—in order to explicate the mechanics of the ether—we have to introduce beside or behind the ether and its parts another ether.[31]

Ten years after this first mention of a possible second kind of ether, Lenard introduced the *Uräther* and conceded that mechanical intelligibility needed to be transcended. By then, the world of physics had undergone dramatic changes.

World War I had a severe impact on Lodge and Lenard, due to their painful experiences during its course. After Lodge's son Raymond was killed in action in 1915, the mourning father took console in spiritualist sessions in which he allegedly connected with his dead son through a medium. Based on his extraordinary experiences, he developed his views on the afterlife in his 400-page-long *Raymond, or Life and Death*, published in 1916. Reference to the ether was made in multiple instances, some of which were highly criticised, such as the claim of heavenly laboratories which 'were able to manufacture what looked like a cigar... out of essences, and ethers, and gases'.[32] Still, the more general idea that 'we change our state at death and enter a region of—what? Of Ether, I think' found more popular acceptance, as did its underlying assumption that the ether of the physicists was a particularly 'recalcitrant' one which could not be explained in terms of pure dynamics alone and, hence, foreshadowed a wider theory which might include spirits and the realm of the dead.[33]

While Lodge consolidated his spiritualism, Lenard turned political. This change he shared with many German scholars, albeit in a more passionate way. Already in August 1914, he had written a pamphlet entitled *England and Germany at the Time of the Great War*, an essay which was republished during World War II under the fitting title *Ideelle Kontinentalsperre*, meaning 'Ideational, or Spiritual, Continental Blockade'. Like Lodge, Lenard had lost his only son, although only after the war. He, however, saw his son as a casualty of the battles, too, as the war's privations had wrecked his son's already weak health. His son's kidney problems had prevented him from reaching the battlefield and the honour attached to it; instead his miserable state of agony had tested Lenard's soul, with little opportunity of heroic transfiguration.[34]

7.2.3 Simpson and Thomson, or, Competition, betrayal and spying

For Lenard, the war meant a dramatic break, especially with the British scientists, two of whom particularly attracted his ire: J. J. Thomson and George Simpson. While Lenard's rage at the British when they entered the World War I followed the general sentiment, from 1916 on his rhetoric changed and became the most radical among his German physics colleagues. The Rumford Medal, which he had received in 1896 from the Royal Society, was now only worth to Lenard the value of the gold it contained and so was donated to help support dependants of the war dead.[35] More important for his scientific views, however, were the years of his son Werner's suffering, from 1920 to 1922, which saw growing fights between a more traditional and experimentally grounded fraction of German physics, which Lenard led, and a more avant-garde and theoretical one around Albert Einstein in particular.[36] The complete reversal of the mutual esteem which Lenard and Einstein had expressed during the preceding two decades became apparent in the heated discussion on the limitations of relativity theory at the *Naturforscherversammlung* in 1920, and intensified thereafter, with ideological and racial undertones.[37]

Gradual changes in Lenard's way of dealing with competition can be found in his main three papers on waterfall and thunderstorm electricity, a field which was also pursued by the meteorologist George Simpson (1878–1965).[38] In 1892, in a groundbreaking paper containing forty-five numbered paragraphs, Lenard described the basic phenomena of waterfall electricity as well as the experimental techniques of its investigation, thus opening up a research field which he himself would constantly pursue for decades. In hindsight, Lenard thought his early results were the obvious seed for a 'thunderstorm theory', which he himself rather carefully tried to establish, first between 1898 and 1904, but successfully only in 1914 and 1915. In the meantime, however, 'the Englishman Simpson flouted the difficulties and published a "thunderstorm theory"... on the basis of observations, which turned out not to be reliably reproducible'.[39] In his 1915 paper, Lenard dealt explicitly with the young competitor. One of his students had replicated Simpson's experiments but could not reproduce the results; instead, it turned out that a much more complex interrelation of water-drop tearing, electricity and air flow had to be considered. In a footnote, Lenard remarked on Simpson's 1909 thunderstorm theory, claiming that Simpson took an easy way out by eliminating an apparent contradiction 'while he was expecting of his readers in an inappropriate way to deem some parts of my preceding observations to be incorrect. In this way, however, a satisfactory understanding cannot be obtained.'[40] He also opposed J. J. Thomson's claim that hydrogen was electrified differently than air:[41] in one paper, Lenard criticised Thomson's generally accepted but 'incorrect theory' on vapour condensation and he subsequently wrote a detailed 'historical account' of the 'fallacies' in Thomson's work.[42] Thus, Lenard did not shy away from clarifications where his results corrected or improved the findings of others and, when necessary, from defending the validity of his own results.

A third paper, which appeared in 1921, brings us to the post-war Lenard—best known for his confrontation with Einstein at the September 1920 Bad Nauheim meeting. It contains the talk he gave at this event, on the waterfall theory of thunderstorms; the contrast between Lenard's nineteenth-century topic of electrification through fluids and Einstein's new theory of general relativity could hardly have been more striking. In this paper, Lenard addressed points in question with Simpson, with the discussion no longer buried in the footnotes but instead carried out prominently in the main text (and probably also during his talk), pointing out Simpson's flawed observations and conclusions.[43] Similarly, in the 1920 edition of his Nobel lecture, he 'exposed' Thomson, detailing how Thomson had cited not the first version of a paper but a reprint and thus had suggested a misleading chronology which, as Lenard later noted, had 'severely struck' him.[44]

Shortly before the war, Lenard had become director of one of the largest German physics institutes in Heidelberg, and this status, combined with the war experiences, turned him into a more combative persona. The research programme of his institute with quite a number of students, assistants and heads of sub-departments consisted mainly of expanding on Lenard's long-lasting topics: electrons; the electricity of thunderstorms, waterfalls and flames; electric arcs; phosphors and the ether. There were also some experimental activities in connection with the modern physics of ether and relativity, as well as with quantum phenomena, especially those carried out by Jakob Laub, Rudolf Tomaschek and Carl Ramsauer.[45]

In the course of the 1920s, Lenard's criticism of physical theories and their proponents grew fiercer, paralleling his political radicalisation, as did his writings and the comments that flowed into his biographical recollections and the editing of his collected works in the 1930s and early 1940s. J. J. Thomson had been a rival from the beginning, but Lenard now spent considerable effort in substantiating various allegations of the Briton's playing foul. Was it not Thomson who was rejecting, manipulating and deceiving from the outset, for example, when he cut short a discussion of Lenard's talk in Liverpool (as Lenard would claim only in 1943) or when he carefully studied Lenard's experiments at that occasion and one year later published his own experimental results, which were done without sufficient care and hence not validated, just to pre-empt Lenard?[46] Somewhere between interpretations and fabrications, these reproaches exhibited Lenard's later emotions, rather than being reliable accounts of the course of events. They reflect his growing frustration with the more efficient publication strategy of the British, when even his own students found the presentation in Thomson's rather popular books on electricity, cathode rays and matter more appealing than studying Lenard's rather formal and complicated works. In a work report from his Heidelberg institute in 1916, Lenard complained that much 'confusion' about basic knowledge and essential sources of experimental error was due to relying excessively on 'compilations (mostly translated from the English), which prejudiced many inasmuch as they dispensed with the use of the original literature'.[47]

The peculiar combination of the ideal of experimental rigour and the ideology of conspiracy theories becomes particularly visible in the composition of Lenard's autobiographical writings. In a first version of his *Erinnerungen* in 1931, for example, a footnote refers to Thomson's determination of the speed of electrons in cathode rays: 'His experiments must have been "impure". At this time, there occurred strange, disturbing visits of a young Englishman, who even intruded at night into the institute, which I was not authorised to lock; once, he destroyed a discharge tube. His intentions remained unfathomable.' Further on, he added: 'Later he appeared to me—please excuse this suspicion after a number of experiences—like a spy sent by J. J. Thomson in order to find out what I was up to.' For the final 1943 version, this sentence was reduced to 'he looked like a spy', but the preceding characterisation of Thomson's experiments was boosted to 'highly impure... (which often was the case with him).'[48] Ironically, one may add, Lenard did never realise that Thomson's experimental approach, as well as his distance from modern theory and adherence to the concept of ether, was much akin to his own understanding.

7.3 EXPERIMENTS ON THE ETHER

7.3.1 Philip Lenard himself, or, From ether to atomic theory

When the Nobel laureate of 1905 delivered his lecture entitled 'On Cathode Rays' in Stockholm, he tried to portray the development the images of electricity and matter had undergone from the perspective of his own experiences and, rather than talking about the 'fruits', he intended to focus on the 'trees' that bore fruit; in modern language, his 'trees' would equate to experimental systems and research technologies.

The nature of cathode rays, which exhibited a wealth of mysterious phenomena within the tubes with diluted air or gases, was long disputed: were they processes of matter, the ether or even a kind of light ray? With the 'Lenard tube' of 1894, the place of the creation of cathode rays could be separated from the place of its investigation, as the 'Lenard window' allowed the rays to escape into another tube of vacuum or substance of the experimenters' choice. In this way, Lenard created an experimental system that helped demonstrate both that cathode rays were processes in the ether (amounting to 'pure electricity' without carrier) and that they could be used to probe the structure of matter. And, rather than speculating about the wave or corpuscular nature of the rays—be it 'ether waves' or 'elementary quanta of electricity'—Lenard delved into the microcosm, which, when investigated with cathode rays, appeared as vacuous as the cosmos of the stars. From scattering experiments conducted between 1895 and 1903 Lenard could determine the degree to which the atom was empty, so that its material building blocks, which he called 'dynamids', had to be tiny.[49]

Cathode-ray tubes became a research technology, which led to the experimental proof of Bohr's atomic model, via the experiment of Franck and Hertz, or to oscilloscopes and television (during the Nazi times Lenard's students, which probably excelled more in fascist thinking than physics skills, tried to attribute the invention of television to their master). With the corpuscular electron, cathode rays, which would penetrate many layers of atoms easily, and conceptions of matter made up from tiny building blocks leaving the space mostly empty, the space-filling ether remained both omnipresent in Lenard's physics and, at the same time, rather irrelevant. Besides the already mentioned waterfall electricity, it was the photoelectric effect and phosphorescence—two fields that concerned the interactions of electrons and light—that kept Lenard busy in his newly won position and institute at the University of Heidelberg, while experimental work on the ether was left to his staff.[50]

7.3.2 Jakob Laub, or, Detecting ether jolts

One of Lenard's assistants who was supposed to do experimental work on the ether was Jakob Laub. He had studied and worked in Würzburg, where he had completed his experimental dissertation on secondary cathode rays with Wilhelm Wien in 1906. Three years later, he went to Berne to see Einstein, with whom he extended his 1905 seminal paper introducing special relativity. Only a month later, however, he became Lenard's assistant in Heidelberg—the Swiss patent office clerk and the Heidelberg professor were at this time on cordial terms, and Laub appeared to be the ideal link between their fields of expertise, as he had a strong background in experimentation and was into Einstein's electrodynamics of moving bodies. His first Heidelberg publication, 'On a Theory of Longitudinal Magneto-Optic Effects in Luminous Gases and Vapours', was based on relativistic computations, and he later completed a survey article 'On the Experimental Foundations of the Relativity Principle'. Both papers had to be approved by Lenard, as he did with all publications of his institute.[51]

Laub's arrival coincided with Lenard's new interest in the re-conceptualisation of the ether, triggered by Bjerknes's 1909 book on the hydrodynamic ether concept,[52] which, as we saw earlier, led to his *Über Äther und Materie* lecture in June 1910. He found that electromagnetic and hydrodynamic theories of the ether might predict slightly different properties with respect to ether density and distribution, as Bjerknes's theory would entail an additional contribution to Maxwell's equations. After his talk, he asked Laub to study the differences of the theories and to start experiments for the determination of ether densities, which he hoped could be detected in arrangements of combined strong electric and magnetic fields. Switching fields on and off would produce jolts in the ether, and these might be detectable by galvanometers.[53] In July, Lenard wrote to Bjerknes about Laub giving a lecture on the issue and about making him report about the book to ensure he had understood it fully. He is 'very smart and has the most anti-mechanical attitude at the institute', Lenard wrote; 'therefore I selected him to engage him in this, up to the possible refutation or proof by experiments'.[54] The experiments were, however, unsuccessful. In the amended second printing of his academy talk, dated February 1911, Lenard mentioned in a footnote that all the results to detect ether jolts in Laub's experiments were negative; to Bjerknes he confided at the same time that he realised that Laub was 'to such extent captured by the relativity principle, that I fear all the time, it won't add up altogether'.[55] When Laub left for a professorship in Argentine in 1911, his relation with Lenard had already soured; regarding empirical evidence of the ether—and possibly the existence of 'another ether' mentioned at the end of his 1910 lecture or of a 'meta-ether' as speculated about already in a letter with Bjerknes in October 1909—Lenard was left empty handed.[56]

During the war, Lenard returned to the problems of the ether after Einstein's theory of gravitation had appeared. In *On the Relativity Principle, Ether and Gravitation*, published in 1918, he developed a rather substantive and unpolemical critique of the principles of general relativity and tried to find alternatives from classical physics.[57] This meant to resist Einstein's call for the elimination of the ether while, at the same time, to demonstrate that ether physics could still provide a more plausible approach. Eventually, Lenard argued, one would have to accept the conjecture of a bifold ether, where the regular ether, which was understood as dynamic, was complemented with a 'meta-ether' or *Uräther*. While the regular ether appeared discontinuous and moving with matter and hence evaded detection through ether-drift experiments, the *Uräther* reinstated an absolute background for all physics distinct from matter.[58] This paper laid the basis for the 1920 Bad Nauheim dispute, and its main ideas were advanced considerably into a new treatise *Über Äther und Uräther*, which appeared between 1920 and 1922 in three editions.

7.3.3 Rudolf Tomaschek, or, Finding the *Uräther* in the light of stars

Despite his more ideological approach against general relativity after 1920, Lenard still hoped to substantiate the ether by objective experiment. After the 1919 solar eclipse results were interpreted to corroborate Einstein's general relativity—probably more a media success than a scientific one[59]—Lenard hoped for a similar proof of the existence of the ether, with help from the stars. As mentioned in *Über Äther und Uräther*, starlight should keep its velocity with respect to the *Uräther* as long as it did not interact with matter. Therefore, it could show a way out of the so far futile Michelson–Morley experiments, and Lenard announced that such experiments were now on their way at his Heidelberg institute.[60] This effort got additional inducement when in 1922 Dayton Miller reported measurements at the Mount Wilson Observatory indicating ether drag effects decreasing with altitude (see Chapter 9).[61]

The research was done by Rudolf Tomaschek, another assistant, who was observing light from sun, moon, planets and stars at the Königstuhl observatory, residing high above the city, from April to July 1923. In a publication of 1924, he meticulously presented the recorded deviations of interference patterns produced by apparatus including a heliostat from the 1914 German eclipse expedition in Norway and other equipment funded by the *Notgemeinschaft*. Tomaschek concluded that the light quanta from extraterrestrial sources would not anymore propagate in the *Uräther* of outer space, but underwent an aberration in the ether at rest relative to the earth. He claimed that one could recognise a property of light quanta: that they change their velocity component which is pointing in the direction of propagation when entering the ether, while the transversal component remains, thus causing aberration.[62] It was then left to Lenard in a separate paper to offer a slightly refined interpretation—that of a rotation of the wave front of starlight when entering the ether of the earth—and to combine these results with the gravitational aberration of starlight. This was, however, not pursued in consonance with Einstein's theory of general relativity, but rather with reference to old mechanical theories of gravitation that he had dug out and mobilised against general relativity, which now appeared 'only artificially and feignedly interwoven with the result'.[63]

As falsification could always be evaded by adjustments to the preferred theory, Lenard went on and attacked whatever would undermine the ether and with it the human comprehensibility of nature, or at least large parts of it, as provided by mechani(sti)cal conceptions in the line of Hertz.

7.4 EMOTIONS FROM THE ETHER

7.4.1 The Lenard conundrum

Lenard's personality did not allow him to separate research experiences and experimental undertakings from his emotional life. The *Naturforscher* was used to smelling chemical substances, tasting weak electricity or hearing whether the pump was producing a proper vacuum; just as his senses and his body were part of his scientific persona, so was his mind with its emotions. Lenard took physics personally, and he grew more and more convinced that it might explain the world even beyond the boundaries that had been drawn in the modern organisation of disciplines. He had struggled to be accepted in the academic world, found it difficult to find true friends and renewed his interest in religion exactly at the time when he felt most challenged by the new theoretical physics as epitomised in the heated 1920 Einstein exchange.

Lenard's recollections, which in parts go back to drafts from the 1920s but mainly were written in 1931, reworked in 1943 and finalised only after the war in 1947, are full of tales of academic disrespect: Austrian and Hungarian universities did not allow him to study in line with his abilities; when he was performing groundbreaking research in Bonn, the humanities faction of the philosophical faculty made him waste precious time to bone up on Latin, which was then prerequisite for becoming eligible as professor in any field; when both he and Röntgen were candidates for the Nobel Prize in 1901, the academy opted to award the prize to the latter, relegating Lenard to receiving the prize in 1905; British writers, and especially J. J. Thomson, omitted Lenard's contributions in their semi-popular books, which sold widely in Germany, so that even Lenard's students did not get things right; his hope of attracting the brightest students to his brand-new Heidelberg institute did not come to fruition; when Lenard was trying to defend himself against Einstein's assault in a newspaper article, it was Einstein, not Lenard, who played the racial card first; and so on.[64]

More than this, Lenard felt betrayed, for example, when Laub defected before accomplishing his task; when Lenard found his aspiring assistant Emil Rupp forging results for experiments which Einstein had suggested; when Heisenberg popularised the forged results in his 1930 book and never corrected the error in later editions; and when excellent and generally esteemed collaborators like Carl Ramsauer, Carl Krauch and others, left academia for industry positions.[65]

Of course, and historical research has shown this very clearly, Lenard cannot be painted as a product of circumstances or as a victim, for he was much the culprit and active aggressor himself, for example, by defying state orders with his open approval of the assassination of Walther Rathenau and Matthias Erzberger and by denigrating his foes from the lectern and in infamous writings like his *England und Deutschland zur Zeit des großen Krieges*, an agitation against British science during World War I, or his *Mahnwort an deutsche Naturforscher* of 1921.[66]

In order to understand the extent to which Lenard's adherence and advancement of the concept of ether was not only a traditional, conservative and internalised pillar of physics but also the result of an emotional interpretation of the reach of natural science into the realms of living matter, agency and spirit, it may help to examine his recollections and some of his later writings. These often transcend both rationality and the limits of physical analysis, and display various emotional registers deeply linked to the ether. Lenard's turn to a number of irrational ideas can be interpreted, indeed, as rather a deliberate and rational choice, particularly when viewed in social and political contexts, as well as in his presumptuous zeal to teach about living matter, free will, mind and spirit, a move that he deemed feasible on the ground of an even broader interpretation of the ether.

7.4.2 A 'rational' irrationalist?

It is hard to date exactly when Lenard compromised his physics by taking up questionable interpretations that were, however, easier to sell and fitted better his position as a *Naturforscher* of fame. Having realised his limited resources to compete in research, he tried to find a simplified account of nature that would also extend to the questions of living matter, muscular force, the human mind and the relation of science with religion. All this would let him appear like a more complete *Naturforscher* in public. Along these lines, he proclaimed that basic laws of nature must be simple, provided the appropriate concepts are chosen. In addition, Lenard took up a new role as teacher of physics—and beyond—for large audiences of students of all fields: his lectures in Heidelberg became famous for turning the demonstration of basic phenomena of physics into big shows and, after World War I, his lectures were attended by over 500 students, making it necessary to add makeshift benches to the lecture room.[67] His success came from the fact that he tried to combine accessible mechanical concepts à la Hertz with results, mostly experimental, from modern physics—although clearly keeping up traditions and views deeply rooted in nineteenth-century physics.

Lenard's intensified turn to the ether from 1910 became increasingly difficult to defend. Bjerknes's hydrodynamic theory, for which Laub did not provide the desired experimental results, lost earlier supporters, and it became questionable whether there could be a measurable result serving as an *experimentum crucis* between standard and hydrodynamic electrodynamics at all.[68] In a review of Lenard's *Über Äther und Uräther*, published in the widely read journal *Die Naturwissenschaften* in 1923, Hans Thirring concluded that the text had 'sprung from the author's desire to find an explanation for various, albeit, mutually contradicting findings of ether physics, while avoiding those consequences of the theory of relativity that do not appear plausible to him' and that the solution Lenard presented 'easily goes astray into a thicket of hypotheses on further elaboration'.[69]

Since it would be too easy an explanation to say that Lenard just became stubborn or crazy as he continued to advance his conceptions of ether and meta-ether, even at a time when there was more and more evidence of insurmountable problems, could it be that he was a rather 'rational' irrationalist, that is, that he made a deliberate decision? Clearly, by doing so he lost credibility with respect to modern physics, but could this have been the lesser drawback when compared to the potential loss of his position as the well-known Heidelberg *Naturforscher*? This way, he could remain, in society's view, a respected 'great man' of science, well known for his famous lectures. With the nationalist or *völkisch* movements in German universities in the 1920s, the opportunity arose to become a nationalist leader of science while forgoing the hope of attracting brilliant research students.

Historically, real or feared loss of resources, respect or recognition has often been a force which contributes to exclusion, nationalisation and the creation of ideologies. The bourgeois anti-Semitism of the nineteenth century may be largely seen as having been produced by fears of loss among wide strata of the middle class in the face of industrialisation and modernisation.[70] Lenard's way of coping with disappointment and a looming loss of significance was to seek out new intellectual and social circles which could bring renewed recognition and which, with their offers of community (*Gemeinschaft*) and youth cult(ure), may have provided both an intellectual refuge for Lenard and a substitute for scientific disciples.

This perspective renders Lenard's deliberate change from a 'nationalist scientist' to an 'irrational nationalist' plausible. Already in May 1925, as Karl Wien wrote to his father Wilhelm, who used to be in Lenard's camp with respect to Einstein, that it was hard to tell 'whether one becomes *völkisch* first and then a doctoral candidate, or vice versa', and that Lenard's institute was 'quite homogeneously' right wing.[71] This seems consistent with Lenard's early inclination towards Hitler as, at that time, he, together with Johannes Stark, wrote a newspaper article entitled '*Hitlergeist und Wissenschaft*' ('Hitler Spirit and Science') in which they 'praised Hitler's simple and natural way of thought'.[72]

It remains difficult to tell the extent to which ideology compromised science in Lenard's Heidelberg institute, and also to assess, for example, the impact of his membership—as the only science representative—in Alfred Rosenberg's Combat League for German Culture, which is infamous for its campaigns against 'degenerate art', although it did not develop a comparable crusade on science. In 1933, however, his statements on Nazism and science were unequivocal. In a letter to Hitler, dated 21 March, Lenard offered to help bring about the 'renewal' of university staff after the Nazis seized power:

> There are not as many really scientific high-fliers living at the same time as university professors are needed; but there are enough thoroughly German people with the necessary scientific education and teaching qualifications who could be role models for the German youth in the lecture halls and unselfish administrators of the university institutions, who, however, have by no means been sought after for a long time.[73]

However, nothing came out of this offer.

7.4.3 Haunted Lenard and spirited ether

It was a bizarre event when, on 7 June 1939, a bust of Philipp Lenard was inaugurated in the garden of the Heidelberg physics institute, and a *Geisterrede* (literally, a 'spirit's speech') was read after high-ranking representatives of the university, the state ministry and the brown-shirted students had opened the ceremony from a swastika-decorated platform. As monuments are only erected for the dead, 'spirits may well come [here] occasionally', the speech explained, and speak about who the person sculpted in stone was. The speech had been written by Lenard himself, who, however, stayed away from the ceremony, and it self-indulgently described his life path as that of one who aspired to study 'remote, mysterious things' to obtain scientific knowledge about them and apply it, although he often did not receive adequate recognition for this. Although the speech contained passages of self-pity and crude anti-Semitism (coming in scientific guise), here particularly aimed at Einstein and at Hertz's Jewish ancestry, Lenard's 'spirit' also acknowledged the support Lenard had received from a number of 'pure-blood' Jewish teachers against the lukewarm response of the 'major Aryans'. Without Leo Königsberger, for example, he would have attained neither his Heidelberg position nor adequate working conditions. In this way, Lenard, or rather, the spirit speaking about him, demonstrated a certain independence and staunchness which distanced Lenard somewhat from the many Nazi representatives at the event. Afterwards, however, most of them met Lenard at home to celebrate his seventy-seventh birthday.[74]

Although it is not easy to make sense of this event, a few points may be highlighted. Lenard had finally received proper recognition via the monument, but the author of *Deutsche Physik* and an early promoter of 'Aryan physics' did not present himself as the leader of a Nazi-endorsed rally against 'Jewish science'. Rather, he suggested that Aryan scientists had not performed well enough, and neither had the active proponents of the movement, by pushing rather mediocre candidates into professorships meant to promote the alternative racial interpretation of science. And, whether one interprets this stance as a critique of the leadership, which had obviously failed to foster good science, or simply as Lenard's insight that he had, once more, failed to find a proper place where he could feel at home scientifically and emotionally, the spirit both reproached the audience and haunted Lenard. In different ways, he would invoke this spirit also in his funeral eulogy, which he wrote himself in 1938, and in the preface, which he dictated to his daughter, to his collected works, which were to be published posthumously.[75]

Besides these rather mundane uses of spirits, there was, however, an increasingly more scientific invocation of ghosts, and this developed much in parallel with Lenard's theories about the ether. Simply put, Lenard tried to bridge the 'world of matter' with the 'world of spirits' via a 'world of ether', a scientific programme long in the making but only laid out fully in his *Deutsche Physik*. As we have seen, his 1932 Franklin Institute paper presented some such aspects of this programme, and more appeared a year later in the English translation of another publication about which Edward Andrade[76] said in his preface to it that 'all English-speaking men of science will rejoice to greet... this work'.[77] In *Great Men of Science*, Lenard also dealt with Darwin and his investigation into the evolution of life, giving his own views on the subject. From his point of view, one could actually identify 'the seat of life' or 'centres of life' in the largest molecules found in living substances and which 'appear to form that part of the cell most important to life'.[78] In a footnote, he further explained that, already around 1910, it had occurred to him that the physics of matter and ether might lead to the understanding that 'the new formation of life' was necessarily linked to 'a suitable spirit'. And he clarified that 'spirit is here a name for that which is obviously necessary, apart from material, to life'. This third quality of nature—perhaps, like the fourth state of matter in his youth—was a thrilling concept, or rather

> a name that enables the concept thus grasped, the thing discovered—unknown but still existing—to be firmly retained in order that it may be investigated as far as possible by further experiment. It follows from this that we must assume as existing in space in an available form spirits of many different kinds (derived from dead living beings), and that these have the property of uniting with molecules which suit them, as soon as they find such, whereby these become living beings or the germs of them, which then develop in a suitable environment in accordance with the spirit and guided by it.[79]

His idea was to extend the concept of the ether by the addition of a second one, the *Uräther* (meta-ether), and to connect the latter, however loosely, to spirits.

In the troubled times around 1920, Lenard had turned also to religion, or rather, to developing his own ideas about it after he had disappointedly left the church. In his 'Three Pieces on Religion', he tried, possibly for the first time, to establish a connection between religion, natural sciences and political leadership by means of a 'world of spirits' (*Geisterwelt*).[80] Nobody has realised more clearly than the savants of nature, he wrote, that there exists a 'universal spirit' (*Allgeist*) 'infinitely more comprehensive and sublime than all the natural laws we have ever have fathomed, and the whole of all spirits—souls—life, which have ever inhabited the earth'. This 'universal spirit' would also explain the evolution of animal and man, determined by the 'linkage of the world of spirits with the universal spirit that gives life in the first place'.[81]

Lenard's *Deutsche Physik* was a comprehensive four-volume textbook which appeared in 1936–7. It has fallen into disrepute, mostly because of its infamous polemical, racist and anti-Semitic preface; the body of the work, however, tried to develop a closed, comprehensible and tradition-conscious approach to the description of nature and summarised Lenard's world view as it developed in the 1910s and 1920s and became ideologised in the 1930s. It was here that Lenard made a more coherent attempt at explaining the interrelations between the world of matter (or rather, energy, as he stressed later on), the world of ether, which is the companion of energy and determines gravitation, and the world of spirits, which govern living matter.

Already on the first page of the introduction to the first volume on mechanics, the material world is juxtaposed with the immaterial one, which we experience inside of us and which is studied by the humanities (*Geisteswissenschaften*). The living part of nature is guided by the world of spirits, which cannot be determined spatially, although they connect to bodies like parts of the ether that relate to certain parts of matter.[82] The ether lacks the tangibility and impenetrability of matter, and so do spirits. Even in basic mechanics, where one can forgo the ether, Lenard observes the influence from the world of spirits when he discusses muscular force, which is related to mental causation and free will: 'It is my spirit that guides my will and thus also my muscles.' After identifying nerve tracts and protoplasm as important connections, he discusses in a footnote the old Germanic wisdom that all things have life, that is, that each atom carries spirit: 'The "spirit" of dead matter would be ether in our mode of expression, in distinction to the unfathomable spirits of the creatures.' Volume 2 on acoustics and thermodynamics only mentions that, while life processes do not change atoms individually, it is the rather stable large molecules that are able to capture spirit. Volume 3 focuses on the physics of the ether, that is, optics and electromagnetism, but sticks mostly to the traditional concept of ether as the necessary carrier of electromagnetic waves. It introduces the *Uräther* and the discussion of the experiments and the evasion of the negative ether-drift experiments, but mentions also that there are kinds of 'further efficaciousness' to be discussed later.

It is in the last section of Volume 4 and after the exposition of his general theory of energy, gravitation and ether that Lenard turns to 'ether and the world of spirits'. There are limits to the extent to which the human mind can comprehend nature, and this applies to the ether, which is 'largely withdrawn from the senses; it is almost transcendental'. However, it is an outstanding achievement of the human mind that it can recognise the existence of the ether from observations of inanimate nature alone and, given this difficult quest, it is no wonder that it took some 250 years from Huygens's first work until 'again experience led to new ideas on the nature of the ether'. From the hypothesis of the even distribution of ether in space—and thus being equivalent to it —over to the disproof of absolute movement and the recognition of the ether as the 'companion atmosphere' of matter, or rather all energy, one found a second kind of ether that describes the fact that any amount of energy extends via the ether in all space. This non-local coexistence of all ethers, more or less evenly distributed, now defines the *Uräther*. And, from this, Lenard tried to move on to spirits and life:

> The ether has accordingly in its fine composition some similarity with the world of spirits, which is also not directly accessible through the senses and moreover completely out of reach by quantitative analysis. There is no doubt that the same space that contains all ethers also somehow contains the spirits, which we find in creatures tied to matter; space-objects these spirits are in as far as they remain bound to the animate body, which during its life constantly produces new matter. In this way they exhibit a remarkable similarity with the ether of the body. We have not been able to attribute spirit as a companion to all matter, but only to the very large molecules. The spirit of a living organism is part of the universal spirit, onto which its large molecules can hold on.[83]

While the spirit remains after death, it is only when the spirit is bound to matter that 'there appears to be the possibility of the transformation and rearrangement of spirits . . . under the involvement of matter'. And this interplay of spirits and matter Lenard extended to heredity, envisaging that, in the mechanisms of heredity, new, spontaneously arising traits could spread when larger molecules evolved, which then were able to incorporate new parts of the world of spirits and thus enable the development of a new kind of living organism.

Writing this in 1937, at a time when he was receiving a great deal of recognition from the Nazi regime and when he producing numerous speeches and articles against 'Jewish science'—with the *Geisterrede* just a year later —Lenard did not draw a connection between his ultimate insights as a *Naturforscher* and his otherwise pervasive talk of leadership principle, hereditary selection or ideas of the new man. He rather tried to create a physics as intriguing and mysterious as Crookes's 1879 book on a 'fourth state of matter' had appeared to him. By constantly suggesting that behind the world of ether there was a world of spirits in which he dared to enter—quite in contrast to the self-restraint of other physicists—he promised his readers much more than a typical physics textbook could deliver. Labelling this work then as 'Aryan physics' in the preface and claiming that 'natural laws are always simple' and that simplicity would come automatically with the appropriate concepts,[84] Lenard's textbook quickly went through four editions by 1944. As adopting the Nazi ideology and hammering it out in the preface to *Deutsche Physik* was probably a rational decision, so was his reluctance to interpret his physics in ideological terms within the four volumes. Or, as the motto opening his opus magnum says: 'The preface comes from the fight of the day; the work seeks values of eternity.'[85]

NOTES

1. 'Medal Day Meeting', *Journal of the Franklin Institution* 214 (1932): 346–56, p. 353. Other prizes that year went to various engineers, entrepreneurs and research scientists, ranging from G. P. Thomson from Imperial College for his 'brilliant experimental demonstration of the wave properties of electrons', to Percy Bridgman from Harvard for his high-pressure experimentation, to Charles Fortescue from Westinghouse Electric for network management, to John B. Whitehead from Johns Hopkins for his work on dielectrics.
2. 'Medal Day Meeting', p. 353f.
3. Arne Schirrmacher, ed., Philipp Lenard: Erinnerungen eines Naturforschers. Kritische annotierte Ausgabe des Originaltyposkriptes von 1931 / 1943 (Heidelberg: Springer, 2010), p. 108.
4. Alan D. Beyerchen, Scientists under Hitler. Politics and the Physics Community in the Third Reich (New Haven: Yale University Press, 1977).
5. Philipp Lenard, Deutsche Physik (Munich: J. F. Lehmanns Verlag, 4th edn, 1944), vol. 1, p. ix.
6. For some preliminary attempts, see the introduction in Schirrmacher, Philipp Lenard: Erinnerungen; Andreas Kleinert, 'Von Preßburg nach Heidelberg: Philipp Lenard (1862–1947) und die Schwierigkeiten einer Biographie' in Peter Zigman, ed., Die biographische Spur in der Kultur—und Wissenschaftsgeschichte (Jena: IKS Verlag, 2006), 195–203; and Arne Schirrmacher, '*Ein Leben in Experimenten. Philipp Lenard zwischen Naturforschertum und moderner Physik*' in Wilhelm Füßl and Johannes-Geert Hagmann, eds., *Konstruierte Wirklichkeit. Philipp Lenard 1862–1947* (Munich: Deutsches Museum, 2012), pp. 18–27.

7. This would entail combining writings, diaries, laboratory notebooks, and so on, which have only in recent years been deposited in archives.
8. See Stephanie A. Morris, ed., The Franklin Institute and the Making of Industrial America, (Bethesda: CIS Academic Editions, 1987), and Michael McMahon, '"Bright Science" and The Mechanic Arts: The Franklin Institute and Science in Industrial America, 1824-1976', Pensilvania History 47 (1980): 351–68.
9. Philipp Lenard to DeWitt B. Brace, 22 July 1889, reprinted in Schirrmacher, Philipp Lenard: Erinnerungen, pp. 103–4.
10. 'Medal Day Meeting', p. 356. On Thwing (1860–1946), see John W. Leonard, Who Is Who in Engineering 1922–1923 (New York: John W. Leonard, 1922), and James McKeen Catell, American Men of Science (New York: Science Press, 5th edn, 1933).
11. Philipp Lenard, 'Some Scientific Views to the Franklin Institute on the Eighteenth of May 1932', Journal of the Franklin Institute 214 (1932): 257–63, p. 257; reprinted in Philipp Lenard, Wissenschaftliche Abhandlungen, vol. 4 (abbreviated as WA4 hereafter), edited and critically commented by Charlotte Schönbeck (Berlin: Verlag für Geschichte der Naturwissenschaften und der Technik, 2003), 575–79.
12. The three books were Philipp Lenard, 'Über Kathodenstrahlen' (Berlin: De Gruyter, 1920); the second enlarged edition of his Nobel lecture of 1906, 'Über Äther und Uräther' (Leipzig: S. Hirzel, 1921), which appeared in the second edition in 1922 with a 'word of admonition' (Mahnwort) towards the German Naturforscher; and 'Über Energie und Gravitation' (Berlin: Springer, 1929); the latter two were reprinted in WA4 on pp. 471–532 and 549–73, respectively.
13. Philipp Lenard, Deutsche Physik, Vol. 1, p. ix.
14. Schirrmacher, Philipp Lenard: Erinnerungen, p. 77. In the following, I will rely to some extent on ego-documents, such as Lenard's recollections (written in 1931–43), diaries and letters, which may seem especially problematic in this case since Lenard was an infamous nationalist, racist and anti-Semite. As I have argued in the introduction to Schirrmacher, Philipp Lenard: Erinnerungen, his ideals of experimental accuracy, however, let him render the facts rather consistently, while going astray or turning ideological with a number of its interpretations.
15. William Crookes, On Radiant Matter: A Lecture Delivered to the British Association for the Advancement of Science, at Sheffield, Friday, August 22, 1879 (London: E. J. Davey, 1879), p. 30.
16. Schirrmacher, Philipp Lenard: Erinnerungen, pp. 78f.
17. See Courtenay Raia-Grean, 'Picturing the Supernatural: Spirit Photography, Radiant Matter, and the Spectacular Science of Sir William Crookes', in Minsoo Kang and Amy Woodson-Boulton, eds., Visions of the Industrial Age, 1830–1914 (Aldershot: Ashgate 2008): 55–80.
18. Schirrmacher, Philipp Lenard: Erinnerungen, p. 118.
19. Heinrich Hertz, Die Prinzipien der Mechanik in neuem Zusammenhange dargestellt (Leipzig: Johann Ambrosius Barth, 1894), with a preliminary note by Philipp Lenard; Vilhelm Bjerknes, Vorlesungen über Hydrodynamische Fernkräfte nach C. A. Bjerknes Theorie (Leipzig: Johann Ambrosius Barth, 1900, 1902) and Vilhelm Bjerknes, Die Kraftfelder (Braunschweig: F. Vieweg, 1909).
20. Diary, reprinted in Schirrmacher, Philipp Lenard: Erinnerungen, pp. 105–6, and unpublished diary book T11, Deutsches Museum Archives, Munich.
21. See the detailed account in his diary, reprinted in Schirrmacher, Philipp Lenard: Erinnerungen, pp. 171–3.

22. Schirrmacher, Philipp Lenard: Erinnerungen, pp. 173–9.
23. Diary 1904–8, transcript 480, 'Wirkt als Prophet für mich; von viel Einfluss', Archives Deutsches Museum, Munich.
24. Philip Lenard, Über Äther und Materie (Heidelberg: Winter, 2nd amended edn, 1911), p. 3.
25. Oliver Lodge, The Substance of Faith Allied With Science. A Catechism for Parents and Teachers (London: Methuen & Co, 1907), p. 73.
26. Oliver Lodge, Life and Matter (London: Putnam, 1905). A further influence on Lenard's speculation about ether and cosmos in his 1910 lecture may have been Lodge's theory of spiritual evolution as proposed in Oliver Lodge, Man and the Universe (London: Methuen & Co., 1908). On Lodge's Christian spiritualism, see Janet Oppenheim, The Other World: Spiritualism and Psychical Research in England, 1850–1914 (Cambridge: Cambridge University Press, 1988).
27. Lenard, Äther und Materie, p. 8.
28. The Society for Psychical Research, founded in 1882 to scientifically investigate phenomena called mesmerism and spiritualism, counted among its members not only key figures such as Crookes and Lodge but also J. J. Thomson and Lord Rayleigh. Heinrich Hertz became a corresponding member in 1891. Max Planck showed sympathy with Lodge's research in telepathy and mediumship when he received his copy of Lodge's Phantom Walls in 1929, as he himself was trying to bridge science and religion. See Andreas Sommer, 'Oliver Lodge, Psychical Research and German Physicists: Heinrich Hertz and Max Planck', History of Science Society Newsletter 43 (2014): 13.
29. WA4, pp. 394, 395.
30. WA4, p. 383.
31. Lenard, Äther und Materie, p. 51.
32. Oliver Lodge, Raymond, or, Life and Death (London: Methuen & Co., 1916), p. 197.
33. Lodge, Raymond, pp. 286, 298.
34. Life-saving dialysis did not arrive before 1924. On Werner Lenard, see Schirrmacher, Philipp Lenard: Erinnerungen, p. 221.
35. Schirrmacher, Philipp Lenard: Erinnerungen, pp. 220f.
36. See Stefan Wolff, 'The Establishment of a Network of Reactionary Physicists in the Weimar Republic', in Cathryn Carson, Alexei Kojevnikov and Helmuth Trischler, eds., Quantum Mechanics and Weimar Culture: Revisiting the Forman Thesis (London: Imperial College Press, 2011), 293–318.
37. From the wide literature on the meeting see e.g. Jeroen van Dongen, 'Reactionaries and Einstein's Fame: "German Scientists for the Preservation of Pure Science", Relativity, and the Bad Nauheim Meeting', Physics in Perspective 9 (2007): 212–30, and Charlotte Schönbeck, Albert Einstein und Philipp Lenard: Antipoden im Spannungsfeld von Physik und Zeitgeschichte (Berlin: Springer, 2000).
38. Lenard's papers are 'Über die Elektrizität der Wasserfälle', Annalen der Physik und Chemie 46, (1892): 584–6; 'Über Wasserfallelektrizität und über die Oberflächenbeschaffenheit der Flüssigkeiten', Annalen der Physik 47 (1915): 463–524; and 'Zur Wasserfalltheorie der Gewitter', Annalen der Physik 65 (1921): 629–39. Reprinted and annotated in WA4; references are made to this edition.
39. WA4, p. 117.
40. WA4, p. 135.

41. WA4, p. 142.
42. Philipp Lenard, 'Probleme komplexer Moleküle, Teil III', Sitzungsberichte der Heidelberger Akademie der Wissenschaften. Math.- naturwissen Kl., Abt. A, vol. 1914, No. 29 (Heidelberg: Heidelberger Akademie der Wissenschaften, 1914). All quotes are from Philipp Lenard, 'IV. Tätigkeitsbericht des Radiologischen Instituts der Universität Heidelberg', Elektrotechnische Zeitschrift 37 (1916): 553–4, p. 554.
43. WA4, p. 174.
44. Lenard, Über Kathodenstrahlen, footnote on p. 41; quotes from Schirrmacher, Philipp Lenard: Erinnerungen, p. 117.
45. Schirrmacher, Philipp Lenard: Erinnerungen, p. 224.
46. WA4, pp. 59f. Lenard's Erinnerungen are full of meticulous reconstructions of priority claims, and so are his annotated collected works.
47. Philipp Lenard, 'IV. Tätigkeitsbericht des Radiologischen Instituts der Universität Heidelberg', Elektrotechnische Zeitschrift 37 (1916): 553–4, p. 553.
48. The different text layers have been marked in Schirrmacher, Philipp Lenard: Erinnerungen, p. 116.
49. See Arne Schirrmacher, 'Das leere Atom. Instrumente, Experimente und Vorstellungen zur Atomstruktur um 1903', in Ulf Hashagen, Oskar Blumtritt and Helmuth Trischler, eds., Circa 1903: Artefakte in der Gründungszeit des Deutschen Museums (Munich: Deutsches Museum, 2003), 127–52.
50. See the list of laboratory notebooks in Schirrmacher, Philipp Lenard: Erinnerungen, pp. 328ff.
51. Jakob Laub, 'Zur Theorie der longitudinalen magnetooptischen Effekte in leuchtenden Gasen und Dämpfen', Sitzungsberichte der Heidelberger Akademie der Wissenschaften, Math-naturwissen. Klasse, Abt. A, vol. 1909, No. 9 (Heidelberg: Heidelberger Akademie der Wissenschaften, 1909 / 10); Jakob Laub, ,'Über die experimentellen Grundlagen des Relativitätsprinzips,' Jahrbuch der Radioaktivität und Elektrizität 7, (1910): 405–63.
52. Bjerknes, Kraftfelder.
53. WA4, p. 337.
54. Lenard to Bjerknes, 9 July 1910, cited from Charlotte Schönbeck, Einstein und Lenard, p. 15.
55. WA4, p. 399; Schönbeck, Einstein und Lenard, p. 16.
56. Lenard to Bjerknes, 27 October 1909, cited from Schönbeck, Einstein und Lenard, p. 15.
57. David E. Rowe, 'Einstein and Relativity. What Price Fame'?, Science in Context 25 (2012): 197–246, p. 218.
58. On the development of Lenard's understanding of the ether, cp. Rudolf Tomaschek, 'Lenards Äthervorstellungen', Zeitschrift für die gesamten Naturwissenschaften 8 (1942): 117–36.
59. Matthew Stanley '"An Expedition To Heal the Wounds of War". The 1919 Eclipse and Eddington as Quaker Adventurer', Isis 94 (2003): 57–89.
60. WA4, pp. 496, 499.
61. Dayton C. Miller, 'Ether-Drift Experiments at Mount Wilson Solar Observatory', Physical Review 19 (1922): 407–8. See also Loyd S. Swenson Jr, The Ethereal Ether: A History of the Michelson–Morley Aether-Drift Experiments, 1880–1930 (Austin: University of Texas Press, 1972), and Roberto Lalli, 'The Reception of Miller's Ether-Drift Experiments in the USA: The History of a Controversy in Relativity Revolution', Annals of Science 69 (2012): 153–214.
62. Rudolf Tomaschek, 'Über das Verhalten des Lichtes außerirdischer Lichtquellen', Annalen der Physik 73, (1924): 105–26, p. 120.

63. Lenard republished Johann Georg Soldner's 1801 paper on gravitational aberration; this paper first appeared in the 'Berlin Astronomisches Jahrbuch', in Annalen der Physik vol. 65 (1921): 593–9, with an introductory note challenging Einstein's much more complicated explanation of the same result with his theory of general relativity; see WA4, p. 465.
64. See Schirrmacher, Philipp Lenard: Erinnerungen.
65. Schirrmacher, Philipp Lenard: Erinnerungen, pp. 109, 300.
66. Schirrmacher, Philipp Lenard: Erinnerungen, p. 255.
67. Diary, June 1919; reprinted in Schirrmacher, Philipp Lenard: Erinnerungen, p. 227.
68. Arthur Korn, 'Weiterführung eines mechanischen Bildes der elektromagnetischen Erscheinungen', Berichte der Deutschen Physikalischen Gesellschaft 13, (1911): 249–56.
69. Hans Thirring, 'Lenard, P., Über Äther und Uräther', Naturwissenschaften 11, (1923): 228–30, pp. 228, 230.
70. See e.g. Detlev Claussen, Grenzen der Aufklärung. Die gesellschaftliche Genese des modernen Antisemitismus (Frankfurt: Fischer Taschenbuch, 1994).
71. Beyerchen, Scientists unter Hitler, p. 138.
72. Philipp Lenard and Johannes Stark, 'Hitlergeist und Wissenschaft', Großdeutsche Zeitung, 8 May 1924, p. 23; Schirrmacher, Philipp Lenard: Erinnerungen, pp. 262, 263.
73. Memorandum, reprinted in Schirrmacher, Philipp Lenard: Erinnerungen, p. 275f.
74. Schirrmacher, Philipp Lenard: Erinnerungen, pp. 289–93.
75. 'Totenrede', Schirrmacher, Philipp Lenard: Erinnerungen, pp. 311–13; 'Vorwort zu den "Erinnerungen"', Schirrmacher, Philipp Lenard: Erinnerungen, pp. 313–14.
76. Andrade, who was Jewish, had completed his doctorate summa cum laude with Lenard in 1911.
77. Philipp Lenard, Great Men of Science. A History of Scientific Progress (London: Bell & Sons, 1933), p. vii. The book was written between 1927 and 1929, when it appeared as Große Naturforscher; it went through four editions until 1941, during which time references to Jewish scientists were gradually reduced and, by a kind of scientific Aryanisation, their contributions were attributed to precursors and non-Jewish contemporaries.
78. Lenard, Great Men, p. 312.
79. Lenard, Great Men, p. 313.
80. Lenard, 'Drei Stücke über Religion', ms. c.1920, Deutsches Museum Archives, Munich.
81. Lenard, Great Men, pp. 311, 312.
82. Lenard, Deutsche Physik, vol. 1, p. 17.
83. Lenard, Deutsche Physik, vol. 4, pp. 267–8.
84. Lenard, Deutsche Physik, vol. 1, p. 8f.
85. Lenard, Deutsche Physik, vol. 1, p. ix.

8

Ether and Wireless

An Old Medium into New Media

Jaume Navarro

8.1 INTRODUCTION

'War in the Ether!' With this alarming headline, some British newspapers prophesised an apocalyptic chaos in radio broadcasting in 1934. The initial impetus of wireless had taken place without much national or international regulation, making it increasingly difficult to tune in the right wavelength without interference from other senders. At the Lucerne conference in the previous summer, most European countries had reached a formal agreement to limit the number of stations and their range, so that broadcasting remained controlled within the limits of national boundaries. A few countries, however, had officially rejected the deal, and some feared that the overcrowded ether would become ever more confused. The sponsored article suggested a solution: buy the new Philco radio set, whose accurate selectivity 'will let you sort out each program from its ether neighbours, to bring every program loud, clear, distinct, free from interference and background overlapping'. This was 'the surest defence against the war in the ether'.[1]

The war *in* the ether about which some newspapers were dramatically reporting was contemporary with another war, a war *about* the ether, also taking place in the public sphere. After the much-advertised 1919 astronomical observations, Einstein's relativity theory became a major cultural subject on which every educated person was expected to have an opinion. Physicists and philosophers engaged with the general public in trying to understand the depth and implications of the new theory; not least among the debates at that time was that on the existence or not of the ether. Following Einstein's famous statement that his theory was neutral regarding this issue,[2] the floor was open for arguments on whether it was convenient to regard the ether as a real entity, keep it as a useful heuristic tool or dismiss the very word 'ether' as part of an old, outdated form of physics.

Navarro, J., 'Ether and Wireless: An Old Medium into New Media' in *Ether and Modernity*, edited by J. Navarro.
 DOI: 10.1093/oso/9780198797258.003.0008

To be sure, reference to the ether in matters related to wireless and radio broadcasting in the 1920s and early 1930s was common in the British press. The word 'ether'[3] became shorthand for many aspects of radio: the undefined medium through which wireless phenomena took place, the dial of the radio set, the space between sender and receiver, the awe surrounding the new technology, and so on. The plasticity that nineteenth-century British physics had granted to first an optical and then an electromagnetic ether was now extended through the spread and popularisation of wireless communications. With the irruption of radio sets in domestic and social spheres, an agonising ether found a new environment in which to stay alive. From amateurs to inventors, from regulators to consumers, a large number of actors helped the ether to endure in the British collective imagery.

In this chapter, I propose to explore the role that wireless and radio broadcasting had in keeping the ether alive at a time—the 1920s and early 1930s—when traditional history of science assumes the imponderable fluid to be mostly dead and buried. The story will begin at the end of World War I, when wireless communications turned into the modern source of entertainment for the middle and working classes. So far, wireless had made a huge impact in maritime communications, as well as in military and imperial activities. It had also become a new toy for the amateur scientist–engineer, without whom the early history of wireless communications cannot be understood. But in 1922, with the creation of the British Broadcasting Company and the generalised accessibility to a radio set, interest in the basic mechanisms of wireless escalated, creating a demand for clear and simple explanations based on the common sense of the time. And the ether was, for such purposes, part of common sense.

After a short introduction on the status of wireless in Britain before and after the creation of the BBC, I shall delve into the ways different publics responded to the new technologies and used them in relation to the controversy on the existence of the ether. As we shall see, amateurs were crucial in the development and spread of wireless and radio broadcasting, and their explanatory tools and social networks shaped their stance in the debates on the ether. Professional inventors and engineers were also instrumental in the revival of the ether. Their distance from the academic debates of highly esoteric physics enabled them to keep the ether alive as a most useful tool in the popularisation of wireless technologies. In the third section, I shall enlarge the picture and situate the ether of wireless amateurs and engineers in the broader context of the war about the ether. Some physicists of the old school saw in the popularity of radio broadcasting a timely tool to promote their views: the success of radio was, for some of them, the best physical proof of the existence of the ether. I shall finally discuss how this episode can be understood as an instance of epistemic pluralism in science and technology—broadly understood—rather than a battle between old and new physics.

The geographical span of this paper is mainly Britain, the traditional home of the ether, and the temporal range is the 1920s and early 1930s, the initial years of radio broadcasting. I have chosen not to follow a strict chronological pattern but rather to focus on a thematic analysis, concentrating on the uses of the ether by different groups of people in diverse settings. My interest is to study and highlight the role of radio broadcasting in the public debates on the existence of an ether during the early years of the BBC, when the novelty of this means of communication and entertainment coincided with the popularity of the theory of relativity. The process by which this renewed interest in the ether declined is beyond the scope of this paper. Following my argument, however, one can infer that the consolidation of radio as a matter-of-fact technology meant that the need for simple explanations of wireless waves in terms of a mysterious medium—the ether—largely decreased, thus contributing to what can be characterised as the *demise* of the ether.[4]

8.2 WIRELESS IN BRITAIN AFTER WORLD WAR I: THE ROLE OF AMATEURS AND THE BBC

In the early days of wireless communication, amateurs were a fundamental driving force in the use, spread and improvement of this technique.[5] *The Wireless World*, an *Illustrated Monthly Magazine for All Interested in Wireless Telegraphy and Telephony*, as the subheading went, was the tip of the iceberg of a large network of amateur and semi-professional practitioners of wireless in the interwar period. First issued in 1911 as a bulletin for Marconi employees under the name of *The Marconigraph*, the publication changed its name in 1913, and in 1922 turned into a weekly magazine with huge commercial success until the outbreak of World War II. This periodical, with its articles and advertisements, is a privileged showcase of the development of wireless technologies, its publics and its practitioners. It reveals a large and transversal army of experts and amateurs working together in the creation of a network of resources, information, parochial concerns and commercial interests. Just to mention a few examples, most issues of *The Wireless World* contain a section with reports on technical problems found by local operators and the way to solve them, a section with accounts of experiences in the practice of wireless by amateurs, and advertisements to encourage young people to train in wireless and secure a job in an area increasingly needed of qualified workforce.

Together with the increase in the use of wireless as an alternative to cable-based telegraphy communications, the awareness and involvement of amateurs and the public at large in matters related to wireless dramatically escalated with the creation of the British Broadcasting Corporation in 1922.[6] Neither the Marconi Company nor the Post Office had initially been very keen on the use of wireless for entertainment and amusement for fear that excessive interference might compromise the more serious side of commercial communications. But the success of the first experimental wireless emissions of music and the business plans of the more entrepreneurial early American broadcasters set the BBC in motion. Throughout the 1920s, the radio receiver would become a common household good in the British middle-class home, hence triggering a new demand for accessible, popular explanations of the mechanisms of wireless. And, as we shall see, the ether came in handy for this task.

Electrical communications without cables had been a theoretical possibility ever since Heinrich Hertz's 1888 experiments had proven Maxwell's theories of electromagnetism. But it was inventors and entrepreneurs like Guglielmo Marconi coming from the fringes of the academic world who actually developed the technology and created the niche markets for the new inventions. Thus, the expertise among enthusiasts of wireless, from engineers and professional scientists to domestic amateurs and small stations' technicians, was diverse enough so as to trigger a wide range of explanatory strategies. Explaining and understanding wireless was a need that had to be fulfilled in different ways due to this diversity in the actors involved. In other words, at a time when the most esoteric circle of academic physics was challenging the very concept of ether as a remnant of pre-relativity science, many in the circles composed of technicians, inventors, engineers and users of wireless embraced the ether as a necessary tool to understand and explain the transmission of electromagnetic waves.

8.3 THE PUBLICS OF WIRELESS AND ETHER

In 1903, Sir Norman Lockyer made an arrangement with Guglielmo Marconi in order to send wireless messages across the Atlantic for the improvement of meteorology and weather forecast in Britain. In analogy to the word 'telegram', Lockyer coined the term 'ethergram' for the new type of messages.[7] The short-lived expression made some sense, since it emphasised the medium through which information was supposedly transmitted. This point was also stressed by the London lecturer in physics and science populariser Walter Hibbert who, in his 1909 book *Popular Electricity*, rejected the widespread use of the word 'wireless' because it was simply a negative term: 'without wires' was giving no hint of what medium was actually used in sending messages. It might mean 'signalling by semaphore, or by flag, or by the sun's rays'. That is why, from his point of view, 'it is easy to see that the correct name is "Etheric telegraphy"'.[8] The problem was not new: action at a distance had been a matter of deep metaphysical concern among scientists and natural philosophers alike, and wireless technology was only putting this problem back on the floor. The ether had played that role: to serve as the medium to explain action at a distance for over a century in the British milieu, and many agreed that it should not be abandoned lightly now that one of its main explanatory roles was most needed.

Like many new technologies, wireless was placed in a limbo between mystery and human ingenuity: mystery, because it proved the dominion of mankind over enigmatic forces of nature, and human ingenuity, because it was providing with new commodities that could only be understood by the connoisseur. The ether served as a good metaphor for both roles at the same time. In a statement in 1922, in the context of advertising the new possibilities opened by the Radio Corporation, Marconi himself used the ether in these two senses: 'In twenty years the mysterious all-pervading ether will be surging with human speech conveyed by ether waves. Whispered conversation with friends in lands as remote as Australia will probably be commonplace, and science having revealed to humanity another wonder of Nature, will have forged thereby a fresh link in the much desired chain of international friendship.'[9] In a similar tone, the famous inventor and populariser of science and technology Archibald Montgomery Low wrote the following as late as 1939 in a chapter called 'Science Conquers the Ether' in one of his more than forty books: 'In addition to the wave-lengths which produce these sensations of light or colour and heat, there are in the ether—that mysterious all-pervading "nothingness" which we postulate—many other waves... At present science only uses those within certain well-defined bands, the best known of which are the waves we use for broadcasting.'[10] It comes as no surprise that such people as Marconi and Low were at times referred to as 'wizards of the ether':[11] they could, almost like magicians, produce new gadgets, and they knew how to control that immaterial medium—the ether—through which all sorts of messages could travel.

Actors involved in this mastery of the ether understood it and explained it in different ways. I shall loosely classify them in three groups of people: the amateurs, the technicians and the commercial and political agents. Certainly, this division may be problematic but it is only a working tool to ease the narrative and analysis of this paper.

8.3.1 Wireless amateurs and their place in the ether

Shortly after the end of World War I, the assistant general manager of the Marconi Company, George E. Turnbull, wrote the following in the editorial of *The Aerial*, a gazette for all the staff in the corporation: 'The spirit of comradeship in the wireless world arising out of, and fostered by, the material as well as *the ethereal nature of our work*, is one of our priceless assets and it must be retained.'[12] The expression shows the *esprit de corps* that the company was trying to promote among its employees, but it also reflects the uses that companies like Marconi made of the passionate and free work of the amateur. *The Wireless World*, with its sections of letters to the editors and news exchange, became a source of information useful for the development of the new techniques. In 1925, an internal memorandum of the Marconi Company praised the work of individuals who regarded their activity as an amusing hobby for their long and dark evenings in winter: 'The advent of long evenings is recalling the amateur wireless experimenters to the fascinating hobby of exploring the ether to add new names to the list of towns and countries with which wireless messages have been exchanged.'[13] This activity was ethereal not only because it dealt with the ether but also because it was uncontrolled, unorganised and—also important—unpaid. Moreover, the amateur was a customer of the Marconi and other wireless providers, so it was only natural that they promoted this activity.

The general public was less sympathetic with amateurs because, when radio sets began to be common in many households, reception was, far too often, hindered by hobbyists interfering with BBC signals. The *Evening News* of 24 April 1926, for example, complained about the 'too keen amateur nuisance' that was 'crowding-out the ether' with their so-called 'research': 'To many listeners the expression "research" seemed to be unnecessarily dignified for a process which appeared to consist merely of the broadcasting of a number of inferior gramophone records, interspersed with the announcement "Hallo! Hallo! Hallo! This is station 5ABCD . . . "'.[14] With the spread of commercial entertainment broadcasting, amateurs increasingly turned into a nuisance and appeals arose to improve the organisation of the ether, like a territory demanding fences.

The 'Amateur's Guide to the Ether' did precisely that. With this headline, *The Wireless World* of 1920 showed all the stations amateurs could tune in in their search for wireless signals.[15] To be sure, the headline of this section changed in the following year to a more accurate and prosaic 'Time Signalling Stations'. But the ether came back in the same magazine and for the same purposes later in the decade: trying to help amateurs and customers of radio to search their favourite stations, the magazine reintroduced a section entitled 'Who Is Who in the Ether', a guide to distant reception.[16]

What was the amateur meant to know about the ether? Certainly, not much. But neither did the more-learned inventor or engineer, although some of them had a patronising attitude towards the amateur. In one of his many books on wireless, Philip J. Risdon advised that 'the amateur . . . is at least, for the time being, relieved of the necessity of trying to investigate or understand the properties of the ether. Our advice is: Do not attempt to do so; be satisfied to assume that space is occupied by a medium of some kind which serves for the propagation of light and heat waves, and of the electromagnetic waves of wireless.'[17] And that was one attitude that many amateurs and practitioners of wireless had taken on board, not dissimilar to the mindset of many nineteenth-century electricians: to assume the existence of an ether about which nothing was known except that it needed to exist in order to serve as the medium for electromagnetic waves.

Risdon's 1924 book on wireless is also typical of this type of publication in the 1920s. It starts with 'The Ether Theory' (the second chapter) and 'Ether Waves and Vibrations' (the third chapter), where the ether is presented as a necessary convention about which 'a great deal of nonsense is talked and written' and which might even not exist but which, 'in theory, it has served a useful purpose, and will continue to do so unless and until its existence is disproved and some better theory is established'.[18] Not surprisingly, however, ether does not appear again in the book, since the terms used to describe the technical aspects of wireless are waves, frequencies, amplitudes, interference, and so on, that is, aspects of the periodicity of the waves but not of the supposed medium in which they travel. Incidentally, the book finishes with the chapters 'Wireless and Telepathy' (Chapter 46) and 'Interplanetary Wireless' (Chapter 47).

The ether was a foundational notion which would seldom appear in the daily work of the amateur, the technician or even the inventor but which, at the same time, should not be lightly abandoned, else the most basic questions would remain unexplained. Laurence B. Turner, an engineer from King's College London who had also worked for the Post Office, made it clear in his very technical book on wireless, which was published in 1931: 'Whether in the new physics it is necessary or nonsensical to think of energy as possessing the attribute of position, and although the aether of Faraday and Maxwell threatens to introduce intellectual difficulties of a kind it was expressly invented to dispel, it seems certain that for many years to come engineers . . . must continue to employ the language and conceptions of the classical mechanics, of Newton and of Maxwell.'[19] In other words, the deep transformations in fundamental physics that were occurring at the time were, in his opinion, irrelevant for the technician and the engineer: the ether could remain as part of the terminology of the wireless practitioner.

Reference to Newton and Maxwell in the last quotation merits some consideration. Wireless technologies were certainly part of modernity in the 1920s, just like the electron, X-rays and radioactivity. As Richard Staley has shown, the divide between classical and modern physics was the result of a complex process of co-creation stretching well into the 1930s.[20] There was no agreement on what was modern and what classical since such divide was still in the making. Thus, when people like Risdon and Turner related the foundations of wireless to the physics of Maxwell, they were certainly not challenging the modernity of wireless but probably using its modernity as a way to safeguard the ether in the tradition of British physics while, at the same time, exonerating the amateur from delving into areas that were beyond their reach. The wireless amateur did not need to worry about the ether in his practical work but could safely keep talking about it. As we shall see in Section 8.4, the defence of the ether by Oliver Lodge, one of the father figures of wireless technologies, only helped in securing a place for the ether in the minds and rhetoric of non-physicists.

8.3.2 Technicians and inventors

One way to get a sense of the use of the ether by technicians, inventors and trainees in the area of wireless is to explore the many available dictionaries, compilations of important terms, yearbooks, and so on. Most of them include the term 'ether' or/and 'aether'. A first disclaimer is common: 'using the word not in its chemical sense of ether the anaesthetic, but with the meaning commonly assigned to it by the physicist'.[21] That is why some dictionaries preferred the spelling 'aether', so as to differentiate it from the chemical compound, although ether became more common in the 1920s. An example of the fluidity between the two types of spelling can be found in *The Year Book of Wireless, Telegraphy and Telephony*, a publication from the Marconi Company: In the first volumes, published during 1913–15, 'aether' is used; in the volumes published during 1916–23, 'ether' is used; and, in the last two editions, which were published in 1924 and 1925, 'aether' is used again.

The definition in this publication also changed slightly in its form but not in its content. Thus, in 1913 we find: 'Aether. The imponderable, elastic, all-pervading medium which cannot be detected by any of our senses, but which is supposed to exist because the Undulatory Theory of Light and of Electro-magnetic Waves (q.v.) based on that supposition, gives a good working hypothesis by which to explain a large number of important phenomena, not only fitting in well with known facts, but even leading to the discovery of new ones.'[22] In 1915, the definition became much shorter: 'The medium assumed by electromagnetic theory in order to explain the translation of energy at finite speed by electromagnetic waves'[23], and it was reformulated in 1924, but without changing the meaning: 'The imponderable, elastic, all-pervading medium which is assumed to exist in order to explain the transmission of energy in the form of electromagnetic waves.'[24]

Two general conclusions can be inferred from the dictionaries and glossaries of terms analysed.[25] First, the ether was a necessary supposition; no dictionary did away with it as irrelevant or falsified. Second, the term 'ether' did not appear in the definition of other more technical terms. To be sure, some of these glossaries also show an inertial tendency to republish the same entries unless some important new fact forced a substantial change, and deleting the ether did not seem to be a priority. For example, as late as the 1938 third edition of Roget's *Dictionary of Electrical Terms. Including Telegraphy, Telephony and Wireless*, the entry 'ether' was still preserved: 'The hypothetical all-pervading medium by which light and all other electric waves are regarded as being propagated, and through which the forces between electrons are exerted.'[26] Incidentally, being a more general dictionary on electricity, the book also explained the Michelson–Morley experiment, which 'appeared to indicate that the earth carries the ether round with it, as no interference in the velocity of light in the two cases could be detected'.[27]

Some inventors and engineers became public faces of the wireless industry and authoritative voices for the young engineer and the public at large. Such was, for instance, John Ambrose Fleming, inventor of the thermionic valve and a key agent in the development of the Marconi Company. His 1902 lectures at the Royal Institution, *Waves and Ripples in Water, Air, and Aether* (reissued in a new edition in 1912) were quoted in many articles, lectures and books throughout the 1920s. The book contained a step-by-step understanding of wave phenomena, first in water, then in air (sound), finishing with electromagnetic waves. His first step was to introduce the medium in which these waves took place since, as he explained, 'the production of a wave of any kind implies . . . two things—first, a medium or material in which the wave exists; and, secondly, some kind of periodic change or movement which is experienced by . . . this medium'.[28] And such a medium—the ether—was, for Fleming, a most real thing and not simply a hypothesis:

> The fact that there is such a space-filling aether is, therefore, only to be deduced by reasoning from experiments and observations, but it is not directly the object of our sense-perceptions in the same way that water or air can be. Nevertheless, there is abundant proof that it is not merely a convenient scientific fiction, but is as much an actuality as ordinary gross, tangible, and ponderable substances. It is, so to speak, matter of a higher order, and occupies a rank in the hierarchy of created things which places it above the materials we can see and touch.[29]

This conceptual structure was repeated in most books for wireless amateurs, trainees and the general public at large, with these books often quoting directly from Fleming's book as one of the most authoritative sources for matters related to wireless communications.

Another such authoritative character was Captain Peter Pendleton Eckersley, the chief engineer of the early BBC. His technical writings do not contain much about the ether, and we find more often than not reference to signals than to waves. This terminology is important because it conceals the problem of action at a distance and, thus, the need for a medium. It is only in more general writings, such as his 1925 book *All about Your Wireless Set*, that the issue arises, and he takes for granted that 'there must be an elastic medium between us and the sun because we reject the idea of action at a distance'.[30] The presence of the ubiquitous medium that permeates everything, both in outer space and within sealed rooms, is important because, without the ether, 'no light (as a wave motion), no heat (again, a wave) and, worse, no Wireless, would come to let us see, to live, and to be amused!' And, sadly for him, as chief engineer of the BBC, the ether was so ubiquitous that it also 'titivate[d] the carefully concealed indoor aerial of a license-avoiding listener',[31] thus making fraud possible.

Wireless communications and radio broadcasting were helping the cause for the ether. The success of popular and semi-technical explanations in terms of waves in the ether went as far as to enable some people to turn the analogy into the analogised. Thus, together with those who needed ripples in water in order to explain that wireless was a wave, there were those who saw in wireless a window open to explaining other mysteries in nature. Such was the case of the physicist and Nobel laureate William H. Bragg, who explained the nature of light in terms of radio sets and not the other way around:

> Transmission by radio, broadcasting as we generally call it in this country, has made us familiar with the idea of a disturbance or condition which travels in waves from a central station and is interpreted by 'receiving sets' near and far. The sun sends out waves just as the central station does: our eyes receive and interpret the wave motions that strike them, just as our wireless sets receive and interpret the waves from the 'studio': there is a difference only in their dimensions. They travel at exactly the same rate; whatever medium carries the one carries the other also.[32]

Bragg was a strong supporter of the ether, and there is no doubt that by 'medium' here he meant the ether, and that he assumed that his audience understood the mechanism of radio better than the nature of light. Similarly, James Rice, a lecturer in physics at the University of Liverpool, used the familiarity of radio sets and wireless technology as the means by which to explain some of the intricacies of modern atomic physics and radiation. In his 1928 *Introduction to Physical Science*, the analogy between radio sets and atoms was very explicit: 'Atoms and molecules with their vibrating electric charges are minute "broadcasting stations", light is the electromagnetic wave sent out, and our eyes are "tuned receivers".'[33] Thus, wireless technologies created a space for the ether to remain rooted in the collective imagery so as to even make it, for people like Bragg and, indirectly, Rice, the entity used to explain some of the novelties in physics.

8.3.3 The ether as a sales strategy

Charles R. Gibson was a very prolific and successful populariser of science in Britain in the early days of wireless and radio broadcasting. His books were often reviewed as exciting, amusing and comparable to books of adventures like those of Jules Verne.[34] In his 1914 survey book *Wireless Telegraphy and Telephony without Wires*, he stressed that 'the real marvel about wireless telegraphy is not the apparatus;...The marvel is that one piece of mechanism can operate another at a great distance without any apparent connection. The idea of "action at a distance" is dead; there must be a real connecting medium, and it is the all-pervading aether of space.'[35] The wonder of wireless resided in the command over an entity, the ether, which had an aura of mystery. A decade later, in 1925 he wrote a book on electromagnetic radiation, from ordinary light to X-rays. The title clearly aimed at larger, popular audiences: *The Mysterious Ocean of Aether*. In it, Gibson explained that there were three oceans in the universe (water, air and ether), each the medium for a specific means of communication (sailing, sound and light). Even at the end of the book, in a series of questions to check whether the reader had grasped the main concepts, we read 'Mention three different kinds of oceans', 'What is the ether made of?' and 'To where does the ether penetrate?'[36] These questions refer only to the first chapter, where answers to these questions can be found. To be sure, however, the book did not say much else about the ether, even though its existence permeated the argument of the whole volume. The point to stress here, however, is that a science popular book with such a title had a secured market.

'Professor' Archibald Montgomery Low, who was another self-appointed populariser, defendant and prophet of science (many of his books delved on the future promises of science and technology), used the ether as a way to legitimise wireless as a serious scientific means and not only synonymous for amusement. Low was not happy with the fact that, after the early success of the BBC and radio broadcasting, 'Many people would

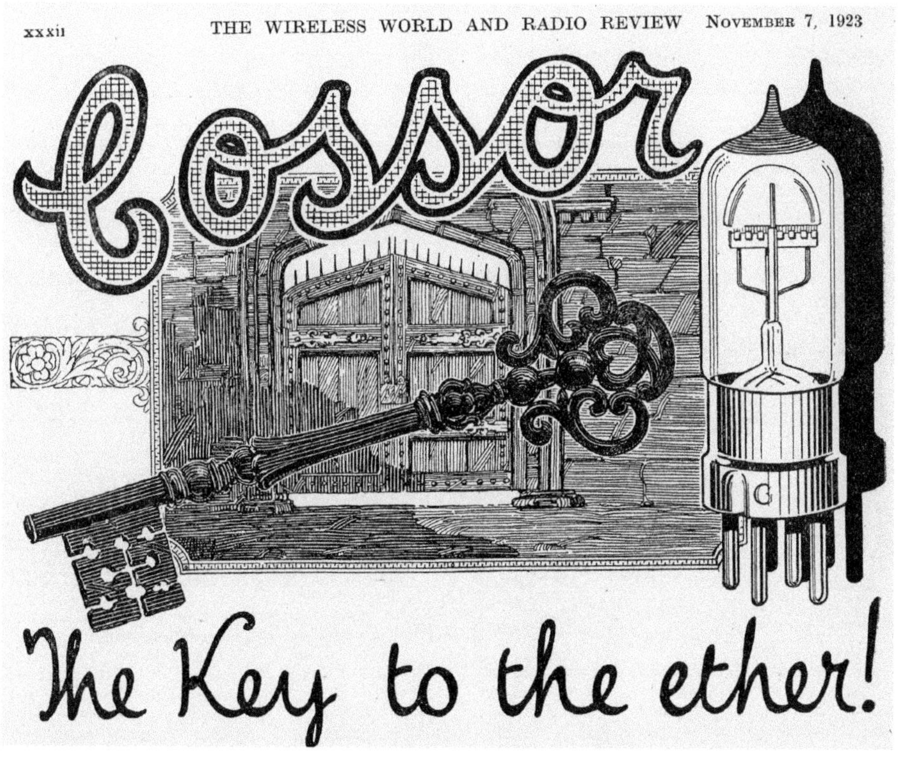

Fig. 8.1 The 'Key to the ether' referred to in this advertisement was a new valve built and commercialised by Cossor. Advertisement in *The Wireless World*, 7 November 1923, p. xxxii. (Reproduced by kind permission of the Syndics of Cambridge University Library)

undoubtedly expect the subject of "wireless" to be dealt with under the heading of "Amusement". It is therefore important to emphasise that broadcasting is perhaps the least important branch of radio.' And he went on: 'It is not true that aether used to be employed as an anaesthetic, and has now "been put to another use" . . . Neither do the suspiciously high-pitched voices of male singers struggling with a negro ditty represent the ultimate value of aetheric radiation.'[37] For him, ether and ether radiations were serious, scientific elements that gave solid ground to the industry of commercial wireless communications and entertainment radio broadcasting. In other words, while 'ether' was a catchword that could grant Gibson sales for his book on radiation, Low used it as the technical element that legitimised wireless as a serious, scientific activity.

The ether could be used, manipulated or domesticated. But could it be owned? Should it be policed? Should it have boundaries? These are three topics we find often in *The Wireless World*. In 1931, for instance, we read the following complaint: 'Nobody seems to care about policing the ether, when we are paying for it! The Postmaster General has a "monopoly of the ether" and the BBC is licensed to broadcast . . . for which there is a fee to the listeners. These should be granted the right to listen without illegal interferences.'[38] Since the public had to pay for the ether, the BBC was keen to ensure people were paying for it rather than letting commercial interests control it. With the headline 'Keep money out of the Ether', the *Daily Telegraph* reported in April 1930 on a statement by Sir John Reith, director general of the corporation: 'The broadcasting service should

6 ADVERTISEMENTS. THE WIRELESS WORLD MARCH 14TH, 1928.

The Ether Master
is ready for use
with your gramophone

Just plug into the jack provided and the Ether Master plays your gramophone records with a fidelity that will amaze you. Intricacies of orchestration and of the singing voice are brought out with full volume and greatly improved tone.

The Ether Master is simplicity itself in use. A new dual drum control of great sensitiveness makes child's play of tuning-in. Without changing coils it brings in all the principal European Stations (high and low wavelengths) at substantial loud-speaker strength.

It uses the latest Screened Grid circuit with none of the disadvantages of neutralising, and all batteries are housed inside the cabinet.

In appearance the Ether Master is a handsome piece of furniture. The pleasing lines and beautiful proportions of the cabinet ensure its looking well in the most luxuriously furnished room.

Ask your dealer to arrange a demonstration.

Mr. E. S. Thomas, of Anfield Road, Liverpool, who got 26 stations in 45 minutes with his Ether Master, says: "The tone of the instrument has positively astounded me and is the envy of all my friends. I can honestly say that no previous purchase of mine has ever given me more satisfaction. If there is anything of value in this letter, by all means use it to your own good. I am delighted."

IN TWO MODELS.

Table Model, less valves, £25 (Royalties, 50/- extra). Console Model (as illustrated), less valves, £30 (Royalties 50/- extra).

EASY PAYMENTS WITH NO DEPOSIT.

The Table Model and Console Model of the Ether Master can be supplied on Easy Terms with No Deposit. Full details on application.

UNIQUE OFFER

The "Ether Master" in Cabinet with electric Gramophone Eliminator, all working from Electric Light Mains, supplied for £5 5 0 down.

Write to-day for full details of this unique offer, marking your envelope "W.W."

ART BROCHURE W.W. ON REQUEST.

PETO-SCOTT Co. Ltd.

77, City Road, London, E.C.1. (Phone: Clerkenwell 9406)

Branches at

62, High Holborn, W.C.1, & 4, Manchester Street, Liverpool.

2056

Fig. 8.2 The 'Ether Master' was a new radio set to which a gramophone could also be attached. Advertisement in *The Wireless World*, 14 March 1928, p. 6. (Reproduced by kind permission of the Syndics of Cambridge University Library)

be established under the auspices of the State, but certainly not conducted by the State.' And he demanded: 'The ether should not be at the mercy of money.'

Perhaps money 'should be kept out of the ether' but, in a press release by the Marconi Company, the ether could also be regarded as a source of income and economic progress in the peak of the Great Depression: 'Wireless, carrying its signals through the ether with the speed of light, over ranges of few miles of the greatest terrestrial distances at

(a)

Fig. 8.3 'The Lure of the Ether' was a slogan used on several occasions to sell a new portable radio set fabricated by L. McMichael, Ltd. (a) Advertisement in *The Wireless World*, 28 November 1928, p. 17. (b) Advertisement in *The Wireless World*, 17 September 1930, p. 5. (Reproduced by kind permission of the Syndics of Cambridge University Library)

(b)

Fig. 8.3 continued

will, is able to serve this industrial age as it could have served none previous.'[39] Certainly, thanks to their dominion over the ether wireless companies, the Marconi Company had made an extraordinary amount of money from wireless, even though the rhetoric of freedom and openness had always been part of their sales strategy.

Finally, since 'ether' became one of the catchwords for wireless and broadcasting in Britain, it was also used in the advertisements for domestic radio sets, although perhaps less often than one might expect: the product name 'Ether Master' and the slogans 'The key to the ether!' and 'The lure of the ether' are examples of the use of the term 'ether' in advertising (see Figs. 8.1–8.3).

8.4 ETHER UNDER THREAT

In an interview with Guglielmo Marconi in 1929, a journalist asked him, 'Did Einstein's discovery of the nature of light and of curved time space affect your research in the realm of power transmission and radio?' Marconi 'smiled' and responded: 'I deal with more practical things. Einstein's discovery does not affect my immediate work. But I realise that it constitutes the most stimulating contribution to the human thought of our age.'[40] Ten years had passed since the famous observations of the deviation of light during a solar eclipse had put Einstein and his relativity theory in the limelight. And one of the topics under serious scrutiny in the public sphere was the fate of the ether with the new theories. In her book *Einstein's Opponents. The Public Controversy about the Theory of Relativity in the 1920s*, Milena Wazeck describes part of the public controversy about relativity in the 1920s, especially in the German and American milieus, where battles for and against the ether became but a part of the larger warfare around the person and the theories of Einstein.[41] Wireless and radio broadcasting easily became a weapon to be used by some of what she calls the 'world-riddle-solvers' who opposed Einstein. One such self-appointed scientist–philosopher, Erich Ruckhaber, claimed as late as 1941 that 'the "abolition" of ether, which caused so much argument some while ago, has been adequately refuted by radio'.[42] For him and for many others, radio broadcasting turned into the most solid proof of the ether.

That very same reasoning had been used in Britain by no other than Oliver Lodge, one of the main champions of the ether, and whose prestige as an icon in telegraphy and wireless and in physics at large gave legitimacy to this argument. In a lecture at the Wireless Society of London in June 1922 he used the opportunity to provide reassurance that, for 'those who are working with the ether and to whom we look for the discoveries of the future', the ether did, in fact, exist. 'It is fashionable sometimes to disbelieve in the ether,' he cautioned. Contrary to what 'some of the newspapers' say, Lodge reminded his audience that the only thing Einstein had done was to stop using the ether. But, he went on, 'a thing does not cease to exist because you refrain from attending to it. Those who daily use the ether for signalling purposes can hardly disbelieve in its existence.'[43] This last sentence epitomises the core of the argument in favour of the ether in the context of wireless communications: the ether was obviously real since, so the argument went, without it, wireless would not be even possible. From his point of view, the ether was a fundamental (albeit only implicit) element of the daily work of engineers, amateurs and users of wireless.

Lodge's case was simple and reassuring for those without knowledge of the intricacies of the theory of relativity. The success of wireless technologies was the best proof that the ether was real. But the argument needed to be repeated often, since popular debates about Einstein's theory seemed to challenge, among other things, the very existence of the ether. In the 1928 edition of *Pitman's Radio Year Book*, this uneasiness found a reply in an article by one J. F. Corrigan, MSc, from the University of Manchester; the article was called 'Radio and Relativity. How the Einstein Theory Affects Our Conception of the

Ether'. The author used the confusion around the theory of relativity and its interpretations to challenge those who denied the existence of the ether: 'Einstein, it is said, with his now celebrated theory of relativity has stripped us of our conception of the ether. At least, that seems to be the popular opinion.' But, he asked, 'is such an opinion true? It is to be doubted, and, although one seldom comes across two people holding the same interpretation of the relativity theory, it is certain that we cannot convict Professor Einstein of having robbed us of our ether.'[44] His answer to the conundrum was: 'It is more correct, perhaps, to say that Einstein has modified the ether into a four-dimensional "continuum", that is to say, a something which is ever-present, and which is continuous, permeating and transcending the material.'[45]

This explanation was close to the interpretations of the ether by Arthur Eddington, who was an early champion of relativity in Britain but was reluctant to abandon the use of the word 'ether' while, at the same time, stressing that relativity had done away with what he called the 'materialistic ether'. Ever since 1918, Eddington had identified the new notion of spacetime with the ether[46] and, by doing so, deprived the ether of any material properties. As a matter of fact, his argument explicitly relied on an old tradition, going back to the mid-nineteenth century, in which matter was regarded as an epiphenomenon of the ether, as a kind of vortex ring or some other structure in the ether.[47] And, thus, the latter should not have any material properties, 'for if matter is vortex-motion or beknottedness in aether, the aether cannot be matter—some state in itself. If any property of matter comes to be regarded as a thing to be explained by a theory of its structure, clearly that property need not be attributed to the aether.'[48] In other words, the old quest for an explanation of the structure of the ether was, according to Eddington, a mistake from the past. Ether should not be thought of in terms of matter if the latter was to be a manifestation or an epiphenomenon of the former, in analogy to the new way matter was understood in terms of the curvature of the spacetime in general relativity.

As a consequence, Eddington went on reasoning, the kind of properties one normally attributed to matter, such as density, rigidity or velocity, made no sense when applied to the ether. In the following extract, we sense his uneasiness with the abandonment of the word 'ether', as he advocates only for a refinement of the notion. Ether and space might be understood as being the same thing, but both words were equally misleading: the former because it might be thought of in terms of some material entity, which it clearly was not, and the latter because it might transmit the idea of totally passive nothingness, which space was not in the new theory of relativity:

> How dense is the aether? Is it fluid like water or rigid like steel? How fast is our earth moving through it? Which way do the particles of aether oscillate when an electromagnetic wave travels across it?...We are, however, convinced that the unanswerableness of these questions is to be reckoned not as ignorance but as knowledge. What we have found out is that aether is not the sort of thing to which such questions should apply. Aether is not a kind of matter. Questions like these could be asked about matter but they could not be asked

> about *time*, for example; and we must reckon aether as one of the entities to which they are inappropriate.[49]

Following his idealist philosophy, Eddington's argument in defence of the ether pivoted on stressing its non-materiality, just as time or other physical concepts were non-material. As a matter of fact, even matter was, in a way, dematerialised by his interpretation of relativity, in which space and the curvature of space played the central role that mass and force had played in Newtonian physics. From this point of view, ether and the new space could be easily identified with each other:

> There is no space without aether, and no aether which does not occupy space. Some distinguished physicists maintain that modern theories no longer require an aether—that the aether has been abolished. I think all they mean is that, since we never have to do with space and aether separately, we can make one word serve for both; and the word they prefer is 'space'. I suppose they consider that the word aether is still liable to convey the idea of something material. But equally the word space is liable to convey the idea of complete negation. At all events they agree with us in employing an army of mathematical symbols to describe what is going on at any point where the aether is—or, according to them, isn't.[50]

Here, Eddington was stressing that debates on the existence of the ether were deeper than most people actually grasped, especially among non-specialists. His point was that one could not do away with the ether while keeping the whole edifice of traditional physics, especially the classical notions of space, matter and force. With his reference to mathematical symbols, Eddington was changing the notion of physics and, with that, the permanence of words like 'ether', 'matter', 'force' and 'space' was simply a question of choice, since their meanings were essentially different. Furthermore, since 'those to whom the word space conveys the idea of characterless void are probably more numerous than those to whom the word aether conveys the idea of a material jelly... aether would seem to be the less objectionable term'. But, in order not to get stuck in nominalisms, 'it is possible to compromise by using the term "field"'.[51]

In the British public sphere, the popularity of Eddington as the expert on relativity was instrumental in reassuring science dilettantes that they could keep the ether, even if they knew little about its properties. In his 1928 bestseller *The Nature of the Physical World*, Eddington reiterated that the new physics did 'not mean that the aether is abolished. We need an aether. The physical world is not to be analysed into isolated particles of matter or electricity with featureless interspace. We have to attribute as much character to the interspace as to the particles.' This insistence on the properties of interspace and his rejection of an absolute void was consistent with the explanatory needs of the wireless populariser. To be sure, wireless technologies were not of great concern to Eddington, but ether enthusiasts could easily invoke his authority in relativity as a strategy to justify keeping the ether.

In 1922, the young physicist Herbert Dingle wrote a popular book, *Relativity for All*, which contained very sharp insights on the new general relativity, partly influenced by the philosophy of Alfred N. Whitehead. Later in life, Dingle would become an opponent of certain principles of relativity and cosmological theories, on the grounds that they were highly speculative and had moved away from the tradition of British empiricism. In his 1922 book, however, he addressed the early controversy on the new status of the ether head on, following the pattern, common at the time, of arguing that the consequences of Michelson and Morley's experiments did not necessarily mean that there was no ether. 'There is still the possibility that the ether possesses some physical property, other than the power to transmit waves, for which no compensation is made by the relativity transformations.'[52] Dingle thus argued that the theory of relativity was indifferent towards the existence of an ether, while, at the same time, stressing the need for an ether in order to explain electromagnetic waves. As a matter of fact, relativity could not account for all aspects of electromagnetic waves and, therefore, the need for some kind of ether should not be lightly abandoned, even if it was only what he called a 'subordinate ether': 'The essence of relativity is the universal character of the event and the subordination of space, time, and matter. A subordinate ether in addition would appear not to be inconsistent with this.'[53]

Thus, for people like Dingle or Eddington, for whom Einstein's theory changed the ontological status of matter, force, space and time, the challenge to the ether was only part of a much bigger reformulation of physics. That is why an ether-focused discussion was alien to their interests since, from their point of view, one could not simply do away with the ether while preserving the whole edifice of traditional physics, since phenomena like electromagnetic waves would lack a solid explanation. What could, therefore, be done was a reformulation of the ether in the new physics, either identifying it with spacetime (Eddington) or as a second-order entity like matter or force (Dingle).

8.5 ETHER, WIRELESS AND PLURALISM

In a short paper called 'The Ether' in the 1924 edition of *Pitman's Radio Year Book*, the author, one Francis M. Denton from the Institute of Electric Engineers, dealt 'with the paradoxes which arise as a result of the acceptance of Einstein's Theory of Relativity'. One of those paradoxes was precisely the fact that while relativity did not seem to need the ether, 'in the theory of radio-transmission the assumption that the ether is a real medium . . . appears to lead to no false predictions of practical importance'.[54] The article suggested that the theories of relativity and wireless transmission were independent and were valid for their respective areas and that, therefore, one could dismiss the ether in the former while keeping it in the latter. As we shall see later on in this section, this duality of attitudes towards the ether may be interpreted as a kind of 'pluralism in science': relativists could choose to dismiss the elusive entity, while radio engineers could keep it as a matter of principle.

Another member of the Institute of Electric Engineers, George G. Blake, argued that there was no way to decide, at least not at that time (1923), whether the ether existed or not. From the engineer's point of view, 'as thoughts of the existence of the ether assist us in our study of radiotelephony, and account for all the facts we so far know concerning this branch of science, we will be of those who take its existence for granted'. But the engineer should also be aware of the fact that 'owing to the work of Einstein, the ether theory, which we radio men feel is almost impossible to do without, is in the melting pot'.[55] A provisional solution to the dispute was to keep both points of view valid in their respective fields, admitting that each theory was valid in one particular area, for one particular set of problems.

As a matter of fact, this plurality of points of view was not unique to the ether in the 1920s. The dichotomy between ether and non-ether was also a reflection of the deeper tension in the physics of the time concerning the nature of radiation. Corpuscular and undulatory explanations of light and radiation were competing in what J. J. Thomson famously called a 'battle between a tiger and a shark, each is supreme in his own element, but helpless in that of the other'.[56] While the corpuscular framework in which light was accounted for in terms of particles or quanta had no need of an ether, the undulatory conceptualisation implicitly demanded a medium in which to support and transmit waves. Since the matter could not be settled among theoretical physicists in the mid-1920s, it comes as no surprise that engineers, wireless amateurs and science dilettantes found multiple and often contradictory ways to cope with the discrepancy.

Indeed, even those who, like James Jeans, were happy to dismiss the ether, they did so with a language full of vagueness and hesitation: 'We need no longer discuss whether light consists of particles or waves. On our days of thinking of it as waves, we may if we please imagine an ether to transmit waves, but this ether will vary from day to day.' Light 'exists in a mathematical formula, and we can picture it as representing waves in three, six or more dimensions whenever we please'.[57] From his point of view, the question whether the ether existed or not was largely irrelevant since the only significant foundations were the mathematical formulae on which physical theories were grounded. Keeping the ether or not was only secondary to the more foundational role of mathematics in physical theories: 'the ether has dropped out of science, not because scientists as a whole have formed a reasoned judgement that no such thing exists, but because they find they can describe all the phenomena of nature quite perfectly without it. It merely cumbers the picture, so they leave it out. If at some future time they find they need it, they will put it back again.'[58] Incidentally, as the historian Gillian Beer described in her seminal work, the language used by Jeans and others had many points in common with that of modernist writers such as James Joyce or Virginia Woolf.[59] The possibility of both keeping and rejecting the ether at the same time was no stranger to the literary devices of the time.

This pluralism was, on occasion, the result of inconsistency or carelessness in one and the same work. In his 1930 bestseller *The Mechanism of Nature*, the physicist and writer Edward N. da C. Andrade explained the rejection of the ether in a chapter on relativity and the Michelson and Morley experiment, while taking it for granted, without much hesitation, in the chapter on Hertz's waves. Thus, following the famous statement by

Lord Salisbury, Andrade reminded that 'the ether was little more than a subject of the verb "to undulate"'[60] and that the experiments with the interferometer had proven that 'there cannot be an ether with the properties of any material body'.[61] But, only thirty pages later, we learn that electromagnetic waves are the result of moving charged particles: 'The moving charge has, as it were, a grip on what, for want of a better term, we call the ether of space, and sends out its waves through this ether.'[62] Certainly, both statements were not black and white, but they reveal the uneasiness with keeping the ether in the context of relativity and the difficulty of doing without it in giving basic explanations of electromagnetic radiation.

Certainly, the kind of pluralism I am hinting at here is related to but different from the one advocated by Hasok Chang in his studies on phlogiston.[63] The main difference is that, while Chang argues that phlogiston was killed prematurely, leaving some phenomena unexplained until the advent of the electron, and that phlogiston and non-phlogiston were valid theories in the same area of chemistry, the ether of wireless was playing an explanatory role with little impact in the daily work of practitioners but with much use in the understanding of that work. As we have seen, many wireless engineers, science dilettantes and the public at large needed an ether only as a matter of principle, and this explanatory role did not directly affect their everyday, more practical work. For some amateurs and the consumers of radio, it was something that helped them make sense of the new, wonderful technology. The long British tradition of explaining electromagnetic waves in terms of an ether could not easily be wiped away unless the question of what medium was undulating in electromagnetic waves was abandoned. And this was a question not to be dismissed easily by the publics I refer to here.[64] For the same reason, however, many fundamental theoretical physicists in the 1920s were increasingly abandoning the ether: they did not need it in their equations and, thus, following some of Einstein's more pragmatic remarks on the matter, they could easily do away with it.

8.6 CONCLUSION

In their chapter on the turn from classical to modern physics in the recent *The Oxford Handbook of the History of Physics*, Graeme Gooday and Daniel Mitchell claim that the historiography about the ether has focused in excess on the category of *belief* rather than actual *use*. Building on the work of Andrew Warwick,[65] they argue that the permanence of the ether in early twentieth-century physics was not so much a question of belief in a metaphysical entity but rather an indirect consequence of the use of certain mathematical tools that took for granted the continuum. From their point of view, 'a fruitful way to investigate alternative metaphysical frameworks to the ether is to consider how they were developed and grounded through unique forms of practice, particularly experimental.'[66] Yet, the person they choose to focus on as an example is William Preece, whom they consider to be a prototype of the British late nineteenth-century electrical engineer and who disregarded the ether in favour of an energy-based ontology in his daily work. By doing so, they erase the received view of the ether as a boundary condition between old and new physics and introduce a plurality of beliefs and uses of the ether in pre-relativity times.

In a way, the arguments presented in this chapter lend support, however indirectly, to some of the suggestions made by Gooday and Mitchell. Wireless communications and the advent of radio broadcasting made it possible for the ether to remain alive both at the level of *belief* and at the level of *rhetorical use*. The easiest strategy for explaining the new and popular technology was to preserve the ether as the necessary seat for the transmission of electromagnetic waves. To be sure, it was not at the level of experimental practice that the ether was directly needed. Work with valves, capacitors and resistors was largely independent of whether or not there was an ether. But that was not the only scientific practice related to wireless: communication, explanation and justification of the new technology was, following Fleck, also part and parcel of their work. And rejecting the ether triggered more problems than solutions for those stressing the undulatory aspect of wireless radiation. Wireless waves without an ether turned out to be more mysterious than those with it, however elusive the metaphysical entity was.

As we have seen, wireless and radio broadcasting opened the door for a diversity of strategies with and about the ether, besides its role as the medium for electromagnetic waves and action at a distance. One of the most interesting outcomes of these diversified uses of the ether in the context of wireless technologies was the feedback between modernity and tradition: the ether could grant scientific robustness to wireless and radio broadcasting by linking the latter to the successful tradition of British physics, while, at the same time, it was the modernity of the radio set that granted the ether a place in the perceptions of physics of the time. Moreover, as was the case for Bragg and others, ether-based explanations of wireless became so popular that the *explanandum* turned into the *explanans* for other mysteries of modern physics. But it was the prestige of people like Oliver Lodge and Arthur Eddington as public figures of science that helped the cause of the ether in the public sphere in the 1920s. For Lodge, the ether had a strong foundational role on which the whole edifice of physics was grounded, and the success of wireless technologies was the most obvious, material and easily accessible proof of the ether's real existence. The strategy followed by Eddington was totally different: his transformed ether was less material, and more mathematical, than the one traditionally favoured in the nineteenth century; but it was, nonetheless, an ether. These two explanatory strategies complemented each other: the prestige of Lodge as a world-renowned physicist and one of the fathers of wireless communications reinforced the argument that ether was needed in order to explain the mechanisms of radio; the reputation of Eddington as the British champion of relativity was a shield against the challenges that the new, esoteric theory posed to the ether. Inventors and engineers could, and did, use both arguments in order to keep the ether as a background concept that gave sense to their activity, however unnecessary it was in their daily activity. Finally, the rise of commercial radio broadcasting triggered a surge in ether-based explanations of electromagnetic radiation, and these intensified the presence of the ether in the public sphere, even though, at times, 'ether' was only a catchword that became another term for 'radio'.

Thus, during the 1920s and well into the early 1930s, disputes on the ether transcended the restricted inner circle of specialised fundamental theoretical physics (i.e. the loci where the quantum and relativity revolutions took place) and extended into areas that the implicit history of the demise of the ether has traditionally neglected. From that historiographical point of view, the rise and popularity of wireless and radio broadcasting in those years created an irrelevant disturbance in an otherwise rather linear path towards the abandonment of the ether. On the contrary, from a more pluralist view of science, we should take the influence of wireless more seriously, as a legitimate element in the recurring debates on the ether. Furthermore, at the time, the ether was not universally viewed as a vestige from the past but was partly seen as an element of modern physics. Its identification with wireless and radio, a most innovative means of communication, gave the ether a flare of modernity.

ACKNOWLEDGEMENTS

Research for this chapter was made possible thanks to the Marconi Fellowship I was awarded by the Bodleian Libraries, Oxford, in 2013 and to the project HAR2015–67831-P of the Ministerio de Economía y Competitividad, Spanish Government. I want to thank Massimiliano Badino, Imogen Clarke, Shaul Katzir, Richard Noakes and Richard Staley for comments on earlier drafts of this manuscript, as well as all the participants in the workshop 'Ether and Modernity', which was held in San Sebastian on 28–30 March 2017.

NOTES

This is an abridged version of a longer paper under the same title and published in *Historical Studies in the Natural Sciences* 46 (2016): 460–89.

1. *The Star*, 18 December 1933; *The Northern Whig and Belfast Post*, 14 December 1933.
2. In 1920, in Leiden, Einstein gave a conference which directly addressed the subject of the ether and from which advocates of the ether often quoted: the text is published in Albert Einstein, *Äther und Relativitätstheorie* (Berlin: Springer, 1920). On the context of this conference, see Jeroen van Dongen, 'Mistaken Identity and Mirror Images: Albert and Carl Einstein, Leiden and Berlin, Relativity and Revolution', *Physics in Perspective* 14 (2012): 126–77.
3. A note on terminology: in this paper, I shall use the spelling 'ether', except in quotes, where the original spelling is in the form 'aether'.
4. This case study looks at the British case only. Initial research on similar sources in the United States, for instance, shows a different story. The term 'ether' is significantly absent from publications such as *QST*, the periodical produced by the American Radio Relay League, and the *Bell Laboratories Record: A Monthly Magazine of Information for the Members of Bell Telephone Laboratories*. One reason for this absence may well be the different explanatory strategies followed in these media, as these stress the notion of *signal* between sender and emitter rather than the

concept of *wave*. This change in explanatory strategy—using signals instead of waves—may have largely determined the needlessness for an ether in the American as compared to the British milieu.

5. See Clinton B. DeSoto, *Two Hundred Meters and Down: The Story of Amateur Radio* (West Hartford, CT: American Radio Relay League, 1936); Gordon Bussey, *Wireless: The Crucial Decade: History of the British Wireless Industry 1924–34* (London: Peter Peregrinus Ltd, 1990); Hugh G. J. Aitken, *The Continuous Wave: Technology and American Radio, 1900–1932* (Princeton: Princeton University Press, 1985); and Susan J. Douglas, *Listening In. Radio and the American Imagination* (London: University of Minnesota Press, 2004).
6. The British Broadcasting Corporation was formally founded on 15 December 1922; see William J. Baker, 'The Formation of the B.B.C.', in *A History of the Marconi Company* (London: Methuen & Co., 1970), 191–7.
7. T. Mary Lockyer and Winifred L. Lockyer, *Life and Work of Sir Norman Lockyer* (London: Macmillan & Co., 1928), p. 185.
8. Walter Hibbert, *Popular Electricity* (London: Cassell and Co., 1909), p. 225. However, as Gillian Beer pointed out, the term 'wireless' had a more positive meaning in stressing the freedom that the new technology represented; see Gillian Beer, '"Wireless": Popular Physics, Radio and Modernism', in Francis Spufford and Jenny Uglow, eds., *Cultural Babbage: Technology, Time and Invention* (London: Faber and Faber, 1996), 149–66.
9. Marconi in the *London Magazine*, March 1922, quoted in Thomas C. Bridges, *The Book of Invention* (London: George G. Harap & Co., 1925), p. 248.
10. Archibald M. Low, *Science in Industry* (London: Humphrey Milford, 1939), p. 102.
11. We find the expression 'wizard of the ether' applied to Marconi in a few news items; see e.g. 'The Ether Wizard. The Marchese Marconi, Who Is Engaged on Secret Experiments with Ultra Short Waves, Photographed in London Yesterday', *Daily Sketch*, 8 April 1931, and Arthur J. Edmonds, 'Marconi. Wizard of the Ether', *Skyrack Express*, 30 July 1937.
12. George E. Turnbull, 'From G. E. Turnbull, Assistant General Manager', *The Aerial*, 1 July 1919; my emphasis.
13. 'Wireless Experimenters' Achievements', Internal Memorandum of the Marconi Company, 23 October 1925, Marconi Archives, box 242, 1.
14. 'Crowding-Out the Ether', *Evening News*, 24 April 1926.
15. 'The Amateur's Guide to the Ether', *The Wireless World* 8 (1920): 97–8.
16. 'Who Is Who in the Ether', *The Wireless World* 24 (1929): 149.
17. Philip J. Risdon, *Wireless* (London: Ward, Lock & Co. Ltd, 1924), p. 34.
18. Ridson, *Wireless*, p. 29.
19. Laurence B. Turner, *Wireless* (Cambridge: Cambridge University Press, 1931), footnote on p. 1.
20. Richard Staley, 'On the Co-Creation of Classical and Modern Physics', *Isis* 96 (2005): 530–58.
21. Philip R. Coursey, 'Aether. The Substratum of the Universe', *The Wireless World* 8 (1920): 40.
22. *The Year Book of Wireless Telegraphy and Telephony*, 1913 (London: Marconi Press Agency, 1913), p. 415.
23. *The Year Book of Wireless Telegraphy and Telephony*, 1915 (London: Marconi Press Agency, 1915), p. 665.
24. *The Year Book of Wireless Telegraphy and Telephony*, 1924 (London: Marconi Press Agency, 1924), p. 609.
25. The entries analysed are from the following books and glossaries: *The Wireless World*, up to 1939; *The Year Book of Wireless Telegraphy and Telephony*, from 1913 to 1925; *Pitman's Radio Year*

Book, 1923–28 (London: Sir Isaac Pitman & Sons, 1923–8); Stephen O. Pearson, *Dictionary of Wireless Technical Terms. Definitions of Terms and Expressions Commonly Used in Wireless Telegraphy and Telephony* (London: Iliffe & Sons Ltd, 1926; 2nd edn, 1932); Samuel R. Roget, *Dictionary of Electrical Terms. Including Telegraphy, Telephony and Wireless* (London: I. Pitman, 1924; 2nd edn, 1931; 3rd edn, 1938); and Ralph Stranger, *Dictionary of Wireless Terms* (London: George Newnes, 1933).

26. Roget, *Dictionary of Electrical Terms*, 3rd edn, p. 125.
27. Roget, *Dictionary of Electrical Terms*, 3rd edn, p. 207.
28. John A. Fleming, *Waves and Ripples in Water, Air, and Aether* (London: Society for Promoting Christian Knowledge, 1923), p. 186.
29. Fleming, *Waves and Ripples*, p. 192.
30. Peter P. Eckersley, *All about Your Wireless Set* (London: Hodder and Stoughton, 1925), p. 94.
31. Eckersley, *All about Your Wireless Set*, pp. 95–6: 'Now, by knowing the velocity of a wave in a medium, we know the density of that medium. The denser, the faster the wave... We find then that the medium through which light and radiant heat (also a wave motion) arrive must be incredibly dense and steel must be but a gossamer structure beside this medium. Thus we say that there is an all-pervading dense medium between us and the sun and the stars which fills up the awful cold spaces between the furthest stars and ourselves as easily as it fills the little cracks and crannies between the molecules, atoms, and electrons of matter. Does not light travel through certain solids and does not Wireless, which is only a light wave of longer wavelength, titivate the carefully concealed indoor aerial of a license-avoiding listener?

    This medium—all-pervading—in which the worlds, the moons, the planets, the stars, the universe, the universe beyond the universe, float, we call the aether of space, the intangible aether which we can only apprehend as a mental picture and with which we can never come into contact with our senses. We cannot weigh, not smell, nor taste, nor see, nor hear the aether, we can only say with a certain degree of confidence that something must be there or else no light (as a wave motion) no heat (again a wave) and, worse, no Wireless, would come to let us see, to live, and to be amused!'
32. William H. Bragg, *The Universe of Light* (London: G Bell and Sons, 1933), p. 3.
33. James Rice, *Introduction to Physical Science* (London: Ernest Benn Limited, 1928), p. 76. The previous year, he had made a similar point in a book on relativity: 'We all know that the antenna of a broadcasting station radiates "waves" which can travel at an enormous speed and can excite oscillations of electricity in an aerial wire. These oscillations we "receive" and amplify for our own purposes by means of "valves, coils and condensers". Now, the electromagnetic waves which constitute light have to be radiated from something and "received" by something... Each atom is a minute "broadcasting station", emitting electromagnetic waves, each type of atom having its own characteristic wave-lengths or frequencies' (James Rice, *Relativity. An Exposition without Mathematics* (London: Ernest Benn Limited, 1927), p. 49).
34. Peter J. Bowler, *Science for All. The Popularization of Science in Early Twentieth-Century Britain* (Chicago: Chicago University Press, 2009), p. 94.
35. Charles R. Gibson, *Wireless Telegraphy and Telephony without Wires. A Popular Account on the Past and Present of Wireless Telegraphy and Telephony Which Assumes No Previous Knowledge of the Subject on the Part of the Reader* (London: Seeley, Service and Co. Ltd, 1914), pp. 18–19.
36. Charles R. Gibson, *The Mysterious Ocean of Aether* (London: Blackie & Sons, 1925), p. 75.
37. Archibald M. Low, *Tendencies of Modern Science, or, Science and Modern Life* (London: Elkin Matthews & Marrot, 1930), p. 131.

38. 'Policing the Ether', *The Wireless World* 28 (1931): 365–6.
39. A press release for *The Western Mail and Wales* (Cardiff) from 18 October 1932: 'Wireless in Aid of Industry', 1, Marconi Archives, box 242.
40. '"Congested Ether". Marconi's Experiments with Wave-Length Jumble', *Sunday Dispatch*, 13 October 1929.
41. Milena Wazeck, *Einstein's Opponents. The Public Controversy about the Theory of Relativity in the 1920s* (Cambridge: Cambridge University Press, 2014), pp. 136–47.
42. Erich Ruckhaber, *Die Äetherdynamik des Sonnensystems* (Berlin: AGV-Verlag, 1941), 6; quoted in Wazeck, *Einstein's Opponents*, p. 141.
43. Oliver Lodge, 'Address to the Wireless Society of London', *The Wireless World* 13 (1922), p. 408.
44. J. F. Corrigan, 'Radio and Relativity. How the Einstein Theory Affects Our Conception of the Ether', in *Pitman's Radio Year Book*, 1928 (London: Pitman & Sons, 1928), 100–4, p. 101.
45. Corrigan, 'Radio and Relativity', p. 104.
46. Arthur S. Eddington, *Report on the Relativity Theory of Gravitation* (London: Fleetway Press, 1918), pp. 79–80: 'This space-time may be materialised as the aether, and the aether-theory does in fact attribute electromagnetic phenomena to strains in this supposed absolute medium.'
47. The vortex-ring theory started with a paper by Hermann von Helmholtz on the stability of vortex rings in a perfect fluid and was very popular among British physicists in the second half of the nineteenth century; see Helge Kragh, 'The Vortex Atom: A Victorian Theory of Everything', *Centaurus* 44 (2002): 32–126.
48. Arthur S. Eddington, *Space, Time and Gravitation. An Outline of the General Relativity Theory* (Cambridge: Cambridge University Press, 1920), p. 39.
49. Eddington, *Space, Time and Gravitation*, p. 38.
50. Eddington, *Space, Time and Gravitation*, pp. 38–9.
51. Eddington, *Space, Time and Gravitation*, p. 39.
52. Herbert Dingle, *Relativity for All* (London: Methuen & Co., 1922), pp. 65–6.
53. Dingle, *Relativity for All*, p. 66.
54. Francis M. Denton, 'The Ether', *Pitman's Radio Year Book* (London: Pitman & Sons, 1924), pp. 25–6.
55. George G. Blake, 'The Elementary Principles of Radio Telephony', *The Wireless World* 11 (1922–23): 726–9.
56. Joseph J. Thomson, *Structure of Light. The Fison Memorial Lecture* (Cambridge: Cambridge University Press, 1925), p. 15; see also Bruce R. Wheaton, *The Tiger and the Shark. Empirical Roots of Wave–Particle Duality* (Cambridge: Cambridge University Press, 1983).
57. James Jeans, *The Mysterious Universe* (Cambridge: Cambridge University Press, 1930), p. 131.
58. James Jeans, *The Universe around Us* (Cambridge: Cambridge University Press, 1929), p. 329.
59. Beer, 'Wireless'.
60. Edward N. da C. Andrade, *The Mechanism of Nature. Being a Simple Approach to Modern Views on the Structure of Matter and Radiation* (London: G. Bell & Sons, 1930), p. 75. In his address as president of the British Association for the Advancement of Science in 1894, Lord Salisbury famously stated that 'the aether is little more than a nominative case of the verb "to undulate"' (from 'The British Association', *Nature* 50 (1894): 339–59, p. 341).
61. Andrade, *The Mechanism of Nature*, p. 79.
62. Andrade, *The Mechanism of Nature*, p. 106.
63. Hasok Chang, 'The Persistence of Epistemic Objects through Scientific Change', *Erkenntnis* 75 (2011): 413–29; Hasok Chang, *Is Water H_20? Evidence, Realism and Pluralism* (Dordrecht: Springer,

2012); Hasok Chang, 'We Have Never Been "Whiggish" (about Phlogiston)', *Centaurus* 51 (2009): 239–64.

64. I use the term 'publics of science' in the sense used in Agustí Nieto-Galan, *Science in the Public Sphere: A History of Lay Knowledge and Expertise* (London: Routledge, 2016). From this point of view, which is one of Fleckian infleunce, there is no clear distinction between esoteric and exoteric circles in science and, thus, scientific discourses like the ones taking place around the world of wireless are legitimately part of science.
65. AndAndrew Warwick, *Masters of Theory: Cambridge and the Rise of Mathematical Physics* (Chicago: Chicago University Press, 2003).
66. Graeme Gooday and Daniel J. Mitchell, 'Rethinking "Classical Physics"', in Jed Z. Buchwald and Robert Fox, eds., *The Oxford Handbook of the History of Physics* (Oxford: Oxford University Press, 2013), 721–64, p. 732.

9

Hunting for the Luminiferous Ether

The American Revival of the Michelson–Morley Experiment in the 1920s

Roberto Lalli

9.1 INTRODUCTION

Among physicists, the view is widespread that the luminiferous, or electromagnetic, ether became an outdated concept in the early twentieth century and that Einstein's special relativity killed it. A second common narrative, seen in many physics textbooks, is that the null result of the 1887 Michelson–Morley ether-drift experiment was the element of crisis, in Kuhnian terms, of Maxwell's electrodynamics and, in a few years, would lead to Einstein's theory of special relativity and the demise of the ether.[1] On the basis of these two simplified narratives, it has become part of the physicists' 'imagined past'[2] that the Michelson–Morley experiment provided the key empirical evidence that decreed the end of the ether.[3]

Science historians and philosophers have shown that this conclusion is, at least, inaccurate. Yet, most authoritative studies have only focused on the second narrative by questioning the role of the Michelson–Morley experiment in the complex interrelations of empirical tests and theoretical developments that resulted in the genesis of special relativity.[4] Only rarely have scholars addressed the first problem and questioned whether and to what extent the demise of the ether was related to the development of Einstein's theory and its reception. There are several reasons why the post-special-relativity ether attracted such scant attention from historians of science. The first reason is that, when they began to focus on the development of modern physics, the luminiferous ether was no longer mentioned in the scientific discourse. The second reason is that, in his 1905 paper, Einstein explicitly declared the ether to be 'superfluous',[5] and it was assumed that this made the concept illegitimate, although Einstein himself later expressed somewhat different views with respect to the concept of ether implicit in his theory of general relativity.[6] Another factor was that opposition to Einstein's theory and the related call for an

Lalli, R., 'Hunting for the Luminiferous Ether: The American Revival of the Michelson–Morley Experiment in the 1920s' in *Ether and Modernity*, edited by J. Navarro. © Oxford University Press 2018.
DOI: 10.1093/oso/9780198797258.003.0009

ethereal theory from the 1920s onwards was permeated by anti-Semitic tones, particularly, but not only, in Germany.[7] This lack of historical studies has contributed to the persistence of the idealised dominant narrative in which the ether is a superfluous entity that completely disappeared from the physics discourse as soon as special relativity was accepted.

Using a scientometric analysis of the American physics literature, the present paper argues that the first part of the idealised narrative is misleading and that the two parts of this narrative are deeply intertwined as they both had their historical roots in the reception of Einstein's relativity theories, particularly in the United States. In this perspective, the well-known episode of Dayton C. Miller's repetition of the Michelson–Morley experiment in the 1920s appears in a new light.

The main elements of the story are fairly familiar.[8] After four years of observations with an enlarged version of the Michelson–Morley interferometer at the top of Mount Wilson near Pasadena in California, Miller announced in 1925 that the analysis of his data showed robust evidence of an ether drift of about 10 km/s. Miller's announcement sparked a host of new experimental activities and theoretical analyses, which ended around 1930 with widespread consensus that Miller's results were due to a systematic error. None of the replications of the experiment—performed in the United States and in Europe—observed any evidence of the same kind of effect Miller had found. From the theoretical perspective, no ether-based mechanical explanation could be provided to fit Miller's data convincingly and without introducing a number of complicated assumptions about the relative motion between earth and ether. Although the scientific community at large became convinced of the non-validity of Miller's findings, he continued to claim that what he had observed was a true ether wind and that, consequently, special relativity and, with it, general relativity had been disproved. [9] In 1933, Miller was even able to publicise his views in the form of a review of the ether-drift experiments in the recently founded *Reviews of Modern Physics*, which was attempting to establish itself as a major venue of certified knowledge in physics.[10]

From the perspective of the idealised narrative described at the beginning of the present paper, these events are difficult to interpret. Indeed, scientists and experts in science studies as well as philosophers and historians of science have shown a great deal of interest in Miller's case, using it as an example in support of opposing views on how scientific knowledge evolves. The fact that the result was discarded without any reasonable alternative explanation for the effect Miller had found has, for instance, been viewed as a case of irrational behaviour in the scientific endeavour.[11] Conversely, Popper stressed that the number of experimental replications sparked by Miller's claim is a confirmation of the sound methodological basis that allows scientists to come to a conclusion as to whether or not a theory has been falsified by an empirical finding.[12] Miller's case has also been used to argue that the *experimentum crucis* does not exist in the actual practice of science: some experiments are a posteriori seen this way after a theory—special relativity in this case—has already been accepted and become dominant.[13] More recently, the case has been considered so significant that it also made its way into the 'science wars' debates.[14]

In this paper, partially based on my previous essays on the subject, I show that these views rely on a simplified view of the historical dynamics.[15] A closer analysis sheds light on the place the concept of ether had in the physics literature at the time as well as on its final disappearance from it. This makes it possible to better identify the elements of modernity that were essential in the *closure* of the experimental controversy.

9.2 AMERICAN LUMINIFEROUS ETHER BETWEEN CLASSICAL AND MODERN PHYSICS

In the last decade of the nineteenth century, American physicists did not have a common view on the ether. Judging by a survey of papers and reports published in American journals, it appears that the luminiferous ether was taken for granted as the physical carrier of electromagnetic waves, but there was no consensus on how it actually worked. Its properties either were a matter of disagreement or differed considerably according to the different perspectives held by the authors. By and large, the proposed views were an echo of the advances and discussions in the UK at the time, including attempts made by British engineer and physicist Samuel T. Preston to revitalise the century-old LeSage ether theory of gravitation or the elastic solid theory developed by Lord Kelvin.[16]

When explicitly discussing the properties of the ether in 1891, the vice-president of the physics section of the American Association for the Advancement of Science (AAAS), the American physicist Francis E. Nipher, explicitly acknowledged that the concept of mechanical luminiferous ether was fraught with difficulties.[17] The recent ether-drift experiment performed by the physicist Albert A. Michelson and the chemist Edward W. Morley was considered to be a key element in these difficulties. With their null result, Michelson and Morley had shown that there was no apparent relative motion of earth and ether on the surface of the earth. The experiment, which would later become the most famous null result in the history of physics, was well known by American physicists. However, it did not lead to any new theoretical approach to the problem of the ether. In the United States, the result was simply taken into consideration to recognise that there was a problem, but it was not a starting point for producing a new theory to solve it. The American physicists, engineers and inventors who addressed research questions on electromagnetism and optics in explicit ethereal terms before 1905 made no mention of the FitzGerald–Lorentz length-contraction hypothesis, or even of the problematic result of the 1887 Michelson–Morley experiment itself.[18]

In many cases, the ether continued to be considered an unproblematic basis for the electricity theory. One good example of this is the 1895 review of the electromagnetic theory of the Serbian-American mathematical physicist Mihajlo I. Pupin. Without any mention of the Michelson–Morley null result, Pupin maintained that the luminiferous ether was one of the 'two foundation pillars which support the splendid edifice of the modern science of electricity', and closed his eulogy by expressing the hope that everything could be reduced to the properties of the ether.[19] This naïve approach to the ether concept is far from surprising, given the status of theoretical physics in the United States in the late nineteenth century. Physics was still in a phase of transition towards professionalisation and was intellectually dependent on the British community and, less directly, on theoretical advances in Germany.[20]

The question is then whether and to what extent this state of affairs concerning the views on the ether changed after the advent of the special relativity theory and quantum physics, which gradually led to a historical transition from classical to modern physics being identified.[21] As previous historical studies had clearly shown, the reception of special relativistic ideas depended largely on different national contexts and followed specific patterns determined by the structure of the national scientific communities and their stratification in various disciplinary subcommunities of physicists, astronomers and mathematicians.[22] In the United States, the theory of special relativity itself did not enter the scientific debate until 1908.[23] It was only with the work on relativistic mechanics of two young American physical chemists, Gilbert N. Lewis and Richard C. Tolman, that special relativity—and with it Einstein's dispensation with the ether concept—became a matter of explicit debate within the American physics community.[24] Through open criticisms of the views of relativity converts, many authoritative American physicists rejected both Einstein's theory and the dismissal of the ether, on the grounds that the views of space and time offered by Einstein were contrary to *common sense*. For most American physicists, the ether *must* be present, although its nature and properties remained difficult to assess.

The appeal to common sense is exemplified by the presidential address to Section B of the AAAS by the Princeton professor of physics and founder of the American Physical Society (APS) William F. Magie: 'in my opinion the abandonment of the hypothesis of an ether at the present time is a great and serious retrograde step in the development of speculative physics. The principle of relativity accounts for the negative result of the experiment of Michelson and Morley, but without an ether how do we account for the interference phenomena which made that experiment possible?'[25]

Magie's view contains two of the tropes of crucial importance for understanding the history of the repetitions of the Michelson–Morley experiments in their national context. One is the common-sense view that the luminiferous ether was essential for making sense of optical and electromagnetic phenomena. Precise empirical work in optics and electromagnetism was a field in which American physicists had gained international acclaim and authority, symbolised by Michelson's Nobel Prize in Physics in 1907. Their experimental expertise in the field was connected to a set of beliefs about the physical mechanism underlying the phenomena they were knowledgeable about, and the new theoretical world views undermined this conceptual apparatus. The second point is the reference to the Michelson–Morley experiment as the empirical fact used as grounds for dismissing the ether. It was illogical for many American physicists to dismiss a concept on the basis of an experiment which, in their view, could not work without that concept.

9.3 THE ETHER IN THE AMERICAN PHYSICS LITERATURE UP UNTIL 1950

To explore the changing use of the concept of ether in the physics literature in the United States, I employed a statistical analysis based on papers published in *The Physical Review* (*PR*) from its establishment up until 1950. The picture resulting from this scientometric study is somewhat surprising. Although the number of contributions that explicitly referred to the ether concept remained low all through this period, the temporal distribution resembles a Gaussian curve with a peak in the early 1920s, after which the number of papers slowly diminishes during the 1930s and almost disappears in the 1940s (see Fig. 9.1).

The analysis seems to confirm that the dismissal of the ether was not directly dependent on the widespread acceptance of Einstein's special theory of relativity, and this result is confirmed by a closer examination of the topics addressed by these papers. Most of the physicists who discussed the properties of the ether and its role in electromagnetic and gravitational phenomena in the early twentieth century did so without any reference to the ongoing debate about the existence of the ether, a debate sparked by the reception of Einstein's special relativity.[26] Those who mentioned Einstein and his theory normally tried to individuate those properties of the ether that would be consistent with the special theory of relativity, in a vein similar to what had occurred in the early reception of special relativity in England.[27] A few leading physicists of the younger generation who

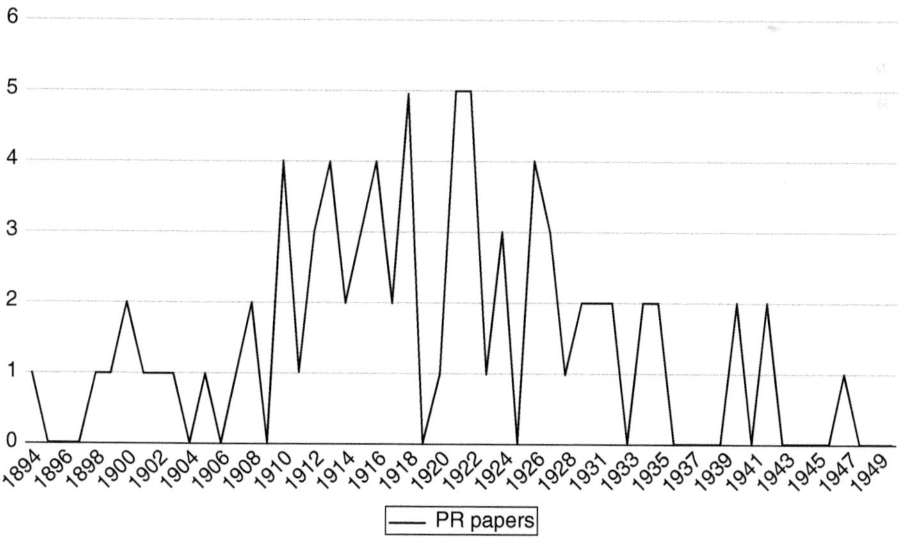

Fig. 9.1 *PR* papers containing the term 'ether' or derivatives of it, per year, between 1894 and 1950. (A normalised analysis was also performed and the results are similar: as one might expect, the main difference is that the rate of the decrease after the early 1920s is greater.)

were, in principle, more familiar with the theory of special relativity and quantum physics continued to explicitly mention the ether as late as the early 1930s, while discussing the quantum properties of electromagnetic radiation.[28]

It seems plausible that in this period the ether concept was mostly in the background of a variety of research endeavours that relied on its existence but, in most cases, it was not really part of the scientific inquiry and so the authors felt no need to mention it explicitly. Yet, this does not explain the fact that the number of papers referencing the ether continuously increased during the 1910s, peaking in the early 1920s. This growth depended on two main factors. First, those who were still referring to radiation phenomena as ethereal vibrations did not find any contradiction between this view and the relativity and quantum theories. Second, from the 1910s to the mid-1920s, the advent of the special relativity theory made the ether part of the explicit debate, and its presence in the physics literature in the United States, in fact, increased as a reaction to the attempts to dismiss it. Judging by this survey, it appears that the reception of Einstein's theory did not have any strong or evident impact on the demise of the term in specialised scientific literature. Quite the contrary, discussions on Einstein's theory increased the open scientific reflections on the ether and its properties.

After the mid-1920s, there was a major shift as the number of publications explicitly referring to the ether became strongly connected to ether-drift experiments. Figure 9.2 shows that, if we remove papers referring to ether-drift experiments from the analysis, the drop in the number of papers explicitly mentioning the ether after the mid-1920s becomes much more dramatic. It is evident that, after the early 1920s, and more prominently after

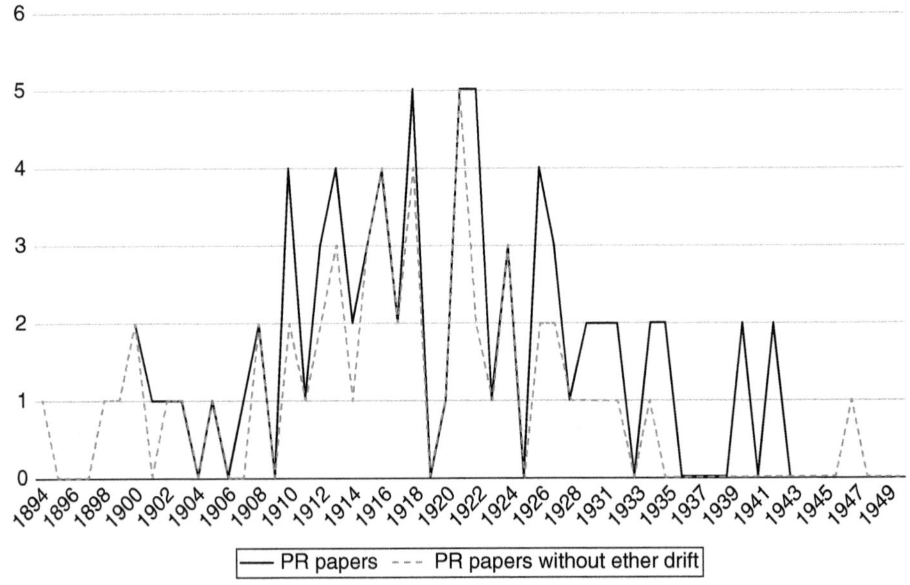

Fig. 9.2 Comparison between the annual number of *PR* papers mentioning the ether (—) and the annual number of *PR* papers not related to ether-drift experiments (- - -).

1925, the vast majority of papers referring to the ether in *PR* were related to studies of ether drift, and most of them were in response to Miller's activities in the 1920s.

9.4 MILLER'S EARLY WORK ON THE MICHELSON–MORLEY EXPERIMENT

The impact of Miller's work in the 1920s on the fate of the concept in the American physics literature invites us to explore the reasons why Miller first started performing the experiment, by placing this endeavour in the context of his scientific biography. After having earned a doctorate in astronomy in 1890, Miller came back to his native state, Ohio, in 1893 to direct the physics department of the Case School of Applied Science in Cleveland as the successor of Michelson, who had conducted there a number of important precision optical experiments on the velocity of light, with these experiments making him the first American scientist to earn the Nobel Prize.[29] These included two experiments that Michelson performed with Morley on the motion of the ether. The first one, performed in 1886, was a repetition of the Fizeau experiment on the dragging of the ether in moving water, which confirmed Hippolyte Fizeau's conclusion that the speed of light depends on the relative velocity between water and the observer.[30] The second Michelson–Morley experiment was an improvement of an interferometric experiment Michelson had designed and already performed in 1880–81 in Europe. The aim of this experiment was to measure the difference between the velocities of light in perpendicular directions, as doing so would allow the observation of an ether drift if the ether were stationary; in other words, it would allow the detection of the relative velocity between the earth and the stationary ether. However, Michelson and Morley did not observe the interferometric shift they would have expected for ether drift. They therefore concluded that the velocity between earth and ether was definitely less than one-quarter and probably less than one-sixth of the earth's orbital motion, a conclusion which led them to rule out the stationary ether hypothesis.[31]

Combined, the results of the two Michelson–Morley experiments were puzzling. The first supported the hypothesis of Fresnel-like stationary ether partially dragged within transparent materials. Yet, according to the result of the second experiment, the ether should have been completely dragged into the vicinity of massive opaque bodies, such as our planet, a result which would have been in agreement with the hypothesis proposed by Irish physicist George G. Stokes in 1845.[32]

After Miller arrived at Case School of Applied Science in the mid-1890s, his interest was completely absorbed by the major topic of the period: the recent discovery of X-rays by Wilhelm Röntgen.[33] In 1902, however, he switched to a different subject: his long-lasting friendly relationship with Morley was instrumental in this shift and the two began planning a repetition of the Michelson–Morley experiment. The explicit aim was to test the FitzGerald–Lorentz contraction hypothesis, namely, the hypothesis that the dimension of bodies contract in the direction of their motion through the ether; this hypothesis was an essential element of the new electromagnetic theory that the Dutch theoretical physicist Hendrik A. Lorentz had been developing since the early 1890s.[34]

Urged by Lord Kelvin at the 1900 International Congress of Physics in Paris, Miller and Morley decided to test the FitzGerald–Lorentz contraction hypothesis by repeating the Michelson–Morley experiment with increasing precision and, more importantly, with different materials and, later, at different altitudes. In designing their experimental setting, Miller and Morley had no doubt whatsoever about the existence of a material ether and were totally embedded in the American culture of precision experimentation. This led them to believe that the ether necessarily had to undergo some form of dynamics which would eventually be possible to detect through long sets of precise measurements, as well as by changing the experimental conditions.

These experiments with Morley were extremely important for Miller's own trajectory, as they shaped his entire approach to the search for the ether drift. One major example was the conviction that the dimension of the optical path had to be extended in order to increase the precision of the observation. In fact, the enlarged experimental apparatus built by Miller and Morley was essentially the same one that Miller would use in his 1920s experiments. Also, the experimental strategy he would use was based on the same theoretical assumptions, such as the possibility that the dragging of ether surrounding the earth changed with altitude—a hypothesis Michelson himself had tested in 1897, obtaining a negative result[35]—or that the ether contained between massive walls was dragged with the laboratory.

In order to take these possibilities into consideration, some of the observations were performed at the top of a nearby hill, and the interferometer was covered only with transparent glass. In two brief notes published in 1905 and 1907, the authors reported that the null result was confirmed: the expected ether drift was not observed in any of the different settings.[36]

Once the result of these experiments was declared null, Miller completely changed his field of research. He became one of the greatest American experts in acoustics and became famous for inventing a sound recording device called the phonodeik, which converted acoustic waves into visual images recorded on a photographic plate. When the debate on the implication of Einstein's theory of special relativity arose in the United States, Miller simply did not participate in it. He was completely absent from any discussion concerning the relevance of the ether as a fundamental piece of the world view he shared with many fellow physicists.

9.5 GENERAL RELATIVITY AND THE GRAND REVIVAL OF THE ETHER-DRIFT EXPERIMENT IN THE 1920S

While Miller had been busy with other scientific activities, his belief in the existence of the ether had not been shaken by the results of his own experiments. In most of his papers on the topic, Miller would begin by announcing his commitment to the ether as a necessary element of the wave theory of light: 'The general acceptance of the theory that light consists of wave motion in a luminiferous ether made it necessary to determine the essential properties of the ether which will enable it to transmit the waves of light and to account for optical phenomena in general.'[37] With this attitude, he shared the common-sense belief of many of his fellow countrymen that light-wave phenomena necessarily require the existence of the ether, which was defended during the debate about Einstein's special relativity around 1910. Irrespective of the result of the Michelson–Morley experiment, the very fact that it worked required a support for the light wave to travel along.

Miller's work in acoustics, we might assume, even strengthened this perspective, as it was natural for him to regard light-wave phenomena as a consequence of the movement of a medium in the same fashion as sound is an effect of the vibrations of air. His expertise in the practice of sound recording and visualisation of acoustical harmonics could only reinforce his common-sense view about the connection between the waves depicted and the motion of a medium (see Fig. 9.3). Just as the sinusoidal drawings were a faithful representation of air vibrations, the interference pattern was a faithful representation of the undulatory movement of a luminiferous ether. But, again, like many of his peers, Miller simply took the ether for granted without any deeper reflection on its nature for years, until 1920.

Why 1920? The popular campaign connected with the British expedition's result confirming one of the few predictions of Einstein's general theory of relativity had

Fig. 9.3 Miller's 'Henrici Harmonic Analyzer', from Dayton C. Miller, *The Science of Musical Sounds* (New York: The Macmillan Company, 2nd edn, 1926), p. 96. (Courtesy of the library of the Max Planck Institute for the History of Science, Berlin, Germany)

made Einstein and his theory famous overnight in a way that seemed to 'overthrow' well-established scientific views.[38] These announcements openly challenged widespread belief systems and triggered immediate reactions. Miller's response was to question the null result of the Michelson–Morley experiment he had helped establish. His reaction was specifically related to the way in which newspapers began discussing Einstein's relativity and its success in the United States, with virtually no reference to the differences between the special and the general theories.[39] Interviewed by a correspondent from the *New York Times* about what becomes of the ether, Einstein replied in early 1920: 'There is no ether, as hitherto conceived by science, which is proved by the well-known experiment of the celebrated American savant Michelson, showing that no influence by the motion of the earth on the ether is perceptible through change in velocity of light, such as ought to be produced if the old conception were true.'[40] Furthermore, in the few American textbooks addressing the topic, Einstein's special theory of relativity was described as an inductive extrapolation from the null result of the Michelson–Morley experiments; and, by extension, general relativity, too, could be perceived as being based on that result.[41]

The popular account of the new theory as exposed in American newspapers was the major reason behind Miller's decision to reconsider his previous work. He began to perceive his results, as well as those of the original Michelson–Morley experiment, as null only in that they were inferior to what was expected with respect to the accepted value of the earth's cosmic motion. In the framework of the post-1919 debate on relativity theories, the result of the Michelson–Morley experiments acquired a completely new and unexpected meaning. In the new context, any positive result, however small, had the potential to disprove a theoretical edifice that Miller neither liked nor had any possibility of understanding properly. If we trust what Miller later wrote about his motivations, these were the events that led him to resume the ether-drift experiments 'since the Theory of Relativity postulates an exact null effect from the ether-drift experiment which had never been obtained in fact'.[42] It is worth emphasising that here, and everywhere else, Miller does not make any distinction between the special and the general theory of relativity.

Initially, members of the American scientific community did not show a strong interest in Miller's proposal to repeat the Michelson–Morley experiment since, for many, the matter had been closed by Michelson himself. Miller had to strive hard to obtain funds, and Michelson was not enthusiastic about his proposal, either.[43] However, Miller could count on his high status within the American physics community. He was an esteemed experimental physicist, known for being extremely meticulous and precise. In 1921—the same year he started his repetition of the Michelson–Morley experiment—he became the secretary of the APS and was elected a member of the National Academy of Sciences (NAS), which, according to science historian John Lankford, was 'the ne plus ultra of US recognition for outstanding scientists'.[44] Because of his prestige, Miller was granted permission from the influential George E. Hale, then director of the Mount Wilson

Observatory (MWO), to use the favourable location of the observatory to repeat the Michelson–Morley experiment at high altitude, about 1,700 metres above sea level.[45]

Miller wanted to repeat the experiment at a higher altitude because he thought that a tiny signal observed fifteen years earlier could have depended on the change of that particular experimental condition. Once he was allowed to use the facilities of the MWO, Miller brought the same instrument used in his 1907 experiment with Morley to the top of Mount Wilson with the goal of repeating the observations in different seasons. Whereas previous Michelson–Morley experiments had attempted to observe the ether wind on the basis of the established values of the earth's motion in space, Miller gradually took a completely agnostic stance towards the expected value of the ether wind. The strategy he began following was to perform as many observations as possible at different times of the day and year. In his view, this would have allowed him to inductively calculate the relative motion of the earth with respect to the ether, or the absolute motion of the earth, as he would subsequently call it.

Initially, he made his observations in isolation from the American scientific community. The only real support he found was in the theoretical physicist Ludwik Silberstein. A Polish immigrant, Silberstein was the author of one of the first English monographs on special relativity in 1914.[46] He had arrived in the United States in 1920 to pursue a career as a researcher at the Eastman Kodak Company in Rochester, which also left him some time for independently pursuing more theoretically intriguing projects. While Silberstein was an expert in the theory of general relativity and able to understand its mathematical machinery, he disliked some of its physical implications.[47] A few months after the announcement that British astronomers had confirmed the gravitational light bending predicted by Einstein, Silberstein published a paper in which he deduced the same effect in the framework of the Stokes–Planck ether theory, a little-known version of the Stokes ether proposed by Max Planck in the late 1890s.[48] In connection with this theory, Silberstein urged Michelson to perform a different kind of interferometric ether-drift experiment in order to observe the relative motion of the earth with respect to the ether, as a result of the rotation of the earth (the original Michelson–Morley experiment was designed to observe the translational motion).[49]

While he was working with Michelson on the planning of this experiment, Silberstein showed a certain interest in Miller's activities, too. After the analysis of his first set of observations in 1921, Miller was cautiously optimistic that he had found a small but persistent effect of the kind expected, although there were other, unexplained effects that cast doubts on the reliability of the results.[50] When the data were still under analysis, Silberstein found a way to publicise Miller's preliminary result. In the spring of 1921, Einstein was touring the United States to collect funds for the Hebrew University of Jerusalem in support of the Zionist movement. After a lecture at Princeton University, Einstein discussed Silberstein's proposed new ether-drift experiment with him, and during the conversation, Miller's preliminary positive result was announced.[51] In the immediate future, this episode resulted only in inspiring Einstein's famous sentence that has been inscribed in the frame of a fireplace at Princeton's Institute for Advanced Studies since the early 1930s: 'Subtle is the Lord, but He is not malicious.'[52] A few weeks later, however, these events also led to a meeting between Einstein and Miller in Cleveland in which they discussed the issue in person, which left Miller with a positive impression of Albert Einstein.[53] Ultimately, these preliminary results were never published, for Miller's new data analysis did not seem to confirm the validity of the effect. Confusion surrounded the reception of the results as Hale spread the news that the result was null, whereas Miller reported that he was observing an effect which might turn out to be an ether drift.[54]

In this period of discomfort, Miller was supported by new allies who hoped for the validity of the effect Miller had reported. These were astronomer Charles L. Poor and engineer Arvid Reuterdahl, both actively campaigning in scientific and non-scientific literature against Einstein's theory, which they declared to be nonsense. In 1921, Reuterdahl had also founded a society, the Academy of Nations, which had the explicit aim of challenging Einstein's theories by coordinating the actions of many anti-relativity campaigners in the United States and Europe. All these activities had more than a hint of an anti-Semitic tone, although anti-Semitism did not seem to have been the main rationale behind them.[55] Miller corresponded with Reuterdahl and shared the urgent call for a return to the Newtonian views exposed by Poor in his book *Gravitation versus Relativity*.[56] Nonetheless, Miller never became a member of the Academy of Nations, nor did he share the anti-Semitic views of his most vocal supporters. His connection with this group remained only at the level of conceptual agreement about the fallacies of Einstein's relativity theories.

Supported by the interest of the anti-relativity campaigners, Miller painstakingly continued his observations and data analysis for four years before he came to a final conclusion which he would stand by until the end of his life. On 28 April 1925, he announced to the multidisciplinary audience of the NAS that the four-year observations at the MWO had led him to the conclusion that 'there is a relative motion of the earth and ether . . . of approximately nine kilometers per second'.[57]

9.6 THE CONTROVERSY AND ITS CLOSURE

While Miller was performing his experiments, the ether continued to be mentioned in *PR* articles in the ways previously mentioned, that is, as synonymous with light waves or as support of electromagnetic waves.[58] After Miller's 1925 announcement, however, the vast majority of the papers were direct responses to this announcement. The unequivocal conclusion of an authoritative physicist after four years of careful testing had enormous repercussions both on the theoretical debates and on experimental programmes, as well as in the popular press. Various groups immediately began planning a repetition of the experiment, and theoretical controversy soon raged in scientific journals. In April 1925, Miller had framed an explanation of the apparent contradiction between his finding and the apparent null results of the previous Michelson–Morley experiments by assuming that the ether drag diminished with altitude. This explanation shaped both the design of the experimental replications and the theoretical dispute. In the pages of *Nature*, Arthur S. Eddington immediately demonstrated that an ether drag that depended on altitude would invalidate well-established astronomical observations made by observatories located at different altitudes.[59] Those who used Miller's result as evidence in favour of the ether, such as Silberstein, did not propose any clear theoretical insight as to the exact dragging mechanism that might explain Miller's data.[60]

It was also difficult to combine the properties of the ether resulting from Miller's discovery with other empirical findings. Ironically, at the same meeting in which Miller announced his surprising positive result, two other reports presented were in favour of Einstein's theory of general relativity. Michelson reported the result of the Michelson–Gale experiment, namely, the experiment proposed by Silberstein back in 1921. The observations of the Michelson–Gale experiment were in complete agreement with Einstein's theory: no effect depending on the possible relative rotational movement between the earth and the ether was observed.[61] Hale also announced that the American astronomer Walter S. Adams—who had succeeded him as the director of the MWO in 1923—had completed the analysis of his observations of the spectrum of Syrius B, concluding that the data confirmed the third, and most observationally troubling, small effect predicted by Einstein's general relativity: the gravitational red shift.[62]

'Confusion' is the word that best captures the reactions to the different announcements, as well as the responses from theoreticians and experimenters. This confusion was increased by the press campaign that saw in the controversy between Miller's result and Einstein's theory an interesting scientific subject. News began to spread that Miller's finding disproved relativity and confirmed the existence of the ether. In these reports, Silberstein's assertion that 'relativity theory [was] dead' figured prominently.[63] At the same time, newspapers also reported widely on the various confirmations of Einstein's theories, such as the results of Adams's observations and of the Michelson–Gale experiment.[64] The scientists interviewed by journalists on the subject were not necessarily experts on the theory. Often, they were not even physicists, since the announcement was made at a gathering of the NAS. In fact, journalists relied on the authority of the scientists within the American scientific community at large without much consideration of their specific field of expertise. Einstein, called upon to comment on this finding, declared: 'If Dr. Miller's results should be confirmed . . . then the special relativity theory, and with [it] the general theory in its present form, falls.'[65]

The perceived importance of the experiments as a decisive test for Einstein's theories, sustained by the press campaign, created a deep rift between theoretical research and experimental activities. By the end of 1925, given the rumour surrounding Miller's result, the Austrian theoretical physicist Hans Thirring decided to publish a very careful and detailed analysis of Miller's observations in which he convincingly demonstrated that no movement of the ether could explain Miller's data. If one also considered the well-proven phenomenon of stellar aberration and the recent result of the Michelson–Gale experiment, Thirring's argument could hardly be refuted. Nevertheless, Thirring's detailed analysis, which he also made available in English by publishing a shorter version in *Nature*, was never cited at the time.[66]

Confusion increased as Miller confirmed the positive result in his address as the outgoing president of the APS at the joint meeting of the APS and the AAAS in late December 1925. There was a relevant difference, though. This time, Miller maintained that he had analysed all the ether-drift experiments performed in Cleveland and Mount Wilson, starting with the early tests by Michelson and Morley. The result was that all of them agreed with the recent observations. This implied that the effect had nothing to do with the altitude, which was one of the theoretical assumptions early critics had focused on. This claim also meant that some of the early experimental plans, based on this assumption, should be revisited. Even though the December 1925 report added more confusion than clarity to the possible explanation of Miller's observations in terms of a physical theory of the ether, it was rewarded with a $1,000 prize for the best report at the 1925 meeting of the AAAS.[67] The press campaign again gave prominence to this finding, sometimes even on the same page where it was reported that Michelson had confirmed Einstein's theory.[68]

In this state of confusion, the experimenters who were involved in new repetitions of the Michelson–Morley experiments stopped referring to specific theoretical assumptions. They started designing the experiments as tests of Miller's result and began focusing on one particular aspect of Miller's methodology: the repetition of the experiment at different times of the day and year. Some other features characterising Miller's experiment were instead completely discarded by those who replicated it. Miller's apparatus had a very large optical path which was not covered by massive opaque shielding, under the assumption that any shielding of this type could have dragged the ether with the optical path. The young experimenter Roy J. Kennedy designed an apparatus with an optical path that was much shorter—4 metres instead of 32 metres—and was enclosed in a sealed metal case. What Miller considered necessary elements related to defined, albeit not theoretically precise, assumptions on the behaviour of the ether were, for Kennedy, possible sources of error: the larger the path, the greater the effect of pressure and temperature gradients. After observations at different altitudes above sea level, in September 1926, Kennedy concluded that his experiment did not confirm Miller's ether drift.[69]

Since Miller continued to maintain that he had found a real effect, MWO Director Adams explicitly asked Michelson to close the matter, writing: 'I understand that Kennedy got zero effects, but have not heard the details. But what the scientific world wants is *your* final word on the subject.'[70] Michelson initially resisted this request. He was convinced that Miller's result was flawed and that Kennedy had closed the matter with a good experimental design.[71] Later, however, Michelson surrendered to Adams' pressure and performed a new Michelson–Morley experiment, together with the MWO astronomer Francis G. Pease, and Fred Pearson, Michelson's assistant at the University of Chicago. The three of them experienced some problems in constructing the apparatus as initially designed and Michelson was not very interested in pursuing this project.[72] Ultimately, Michelson, Pease and Pearson published their results in two very short notes in 1929 without any details and without even mentioning Miller.[73]

Deeply dissatisfied and after renouncing to receive a more detailed report from Michelson, Adams was helped by the MWO astronomer Charles E. St. John, who was, at the time, involved in a campaign against anti-relativity supporters to persuade scientists that astronomical observations had confirmed general relativity.[74] St. John presented a report on the observational bases of relativity theories at the NAS meeting in April 1930. In this, he summarised the results of the Michelson–Pease–Pearson experiment, stating that they completely contradicted those found by Miller.[75] He was also in the position of being able to employ the result of a new repetition of the Michelson–Morley experiment just performed in Jena by the German physicist Georg Joos.[76] Joos's experimental result was considered extremely reliable, particularly due to the technological superiority of Joos's experimental design: for example, Joos had used a motor-driven apparatus, whereas Miller had to walk with the interferometer; Joos took photos of the fringe shifts, whereas Miller had to read them while walking; and, finally, like Kennedy, Joos had enclosed the optical path in a metal shielding, while the path of Miller's interferometer was covered only by transparent glass.

In his written report, St. John stressed all the technical advances put forward by Joos and, at the same time, used the results of Michelson, Pease and Pearson because of the high level of credibility Michelson enjoyed in American scientific circles. This strategy of confronting Miller in public debates, backed by an American authority plus persuasive arguments concerning experimental practices, virtually closed the controversy in the United States. According to St. John, after the meeting ended, the 'men who counted' were convinced by his arguments. St. John reported that the experts who agreed with him left the meeting believing that Miller was 'in a tragic position . . . the only thing for him to do is to reconstruct his instrument and find out for himself that he has been wrong'.[77] That the controversy on the experimental result closed in 1930 is confirmed by the fact that the pursuit of experimental replication of the Michelson–Morley experiment stopped in 1930, although Miller continued to argue in print and at conferences that his result was real and that none of the other replications had ever tested the hypothesis that the ether was dragged by opaque and massive materials.[78]

In the confrontation between Joos's mechanised experiment and Miller's methodology, which was heavily based on the ability and skills of the experimenters and enriched by moral values such as patience and self-dedication, one can see the opposition of two different styles of experimentation. Joos's instrument and methods represented, also at a symbolic level, the new and the modern against the style of the one-man operation, which united in the same figure the instrument builder and the experimenter. Compared to new advances in experimentation, Miller's approach was beginning to be perceived as obsolete. By contrast, Joos's experiment could be seen as the embodiment of the epistemic value of mechanical objectivity, and, thus, it was persuasive enough to close the controversy.[79]

Yet, Miller continued to promote his views and published a very long review article in *Reviews of Modern Physics*, with this article slightly resurrecting the interest in his finding, from the theoretical perspective. However, no new repetition of the Michelson–Morley experiment was planned to test Miller's claim again. In 1935, only one experiment was performed with the aim of explicitly testing Miller's hypothesis that the ether is dragged within a massive enclosure. Apparently, the result, which was contrary to Miller's claim, did not have any impact at all on the discussion because no one had ever given much credit to this hypothesis.[80]

9.7 CONCLUSION: THE AFTERLIFE OF MILLER'S EXPERIMENT

The closure of the controversy in around 1930 also had effects on the persistence of the ether concept in the physics literature. Papers mentioning the ether in *PR* radically dropped off in the 1930s. Still, they did not disappear altogether. Interestingly, this post-controversy presence of the ether was to be found in works by a few general relativity experts who saw Miller's experimental result as an inspiration that could guide the search for an alternative theory of gravitation. Most notably, this attempt was pursued by the American physicist and former assistant of Albert Einstein, Nathan Rosen. In 1939–40, Rosen developed a theory of gravitation in flat space, arguing that the main advantage of this was to 'explain the principal results obtained by D. C. Miller in his 'ether-drift' experiment'.[81] Rosen's work had an impact on some serious attempts to formulate alternative gravitational theories in the post-World War II period, when various research groups began to pursue different research programmes on relativistic theories of gravitation more actively. Although some of them relied on Miller's finding, these attempts no longer had any connection with the concept of ether. None of them used the term 'ether', which had apparently gone beyond the limits of what had become acceptable scientific discourse.[82]

The historical discussion concerning Miller's repetition of the Michelson–Morley experiment is normally framed in terms of the contradiction between Miller's results and the acceptance of relativity theories. It is quite natural to see the case from this perspective since Miller himself, not to mention Einstein, understood and portrayed his ether-drift experiments this way. Disproving the theory was certainly Miller's main motivation for performing the experiment. The contrast between Miller's result and Einstein's theories remained the central focus of the debate all through the 1920s. This view also motivated the persistence with which Miller continued to claim that he had found a positive result and that, therefore, relativity theory, both special and general, had to be abandoned or, at least, revised. This was also the way in which most scientists perceived the events at the time and how most philosophers and sociologists discussed it afterwards.[83]

If we do not remain at the level of the dichotomy between ether drift and relativity, however, the case illustrates a set of interesting mechanisms concerning both the use of the ether concept and the disciplinary divides in the United States in the first half of the twentieth century. As I have attempted to show, the ether was being used by a few scientists in the scientific literature, along with relativistic or quantum concepts. The ether survived in the physics literature as a useful concept, and its dismissal does not seem to be unequivocally connected with the passage between classical and modern physics, however complex and multifaceted. In fact, the debate on special relativity, and even more importantly, the survival of the concept of the ether after the 1920s were strongly connected to the reception of general relativity in the United States and, more specifically, to the case this essay has focused on. The ether drift became a major subject of discussion in both experimental and theoretical papers after Miller announced his positive result in 1925. On the other hand, the end of the controversy about Miller's result and the formation of the consensus that the ether drift he found was spurious seemed to coincide with the sudden decline of the ether in the literature. In this sense, it is striking that those very few who continued to use the ether after the controversy ended were general relativity experts. This confirms that the dismissal of the ether from the literature was not a consequence of the success of Einstein's relativity theories in any straightforward way.

By the time Miller announced his positive result, the enthusiasm about the theory of general relativity, which five years previously had motivated Miller to perform the experiment at the MWO in the first place, was already declining. The mid-1920s corresponds to the beginning of what science historian Jean Eisenstaedt has called the 'low-water mark of general relativity', which lasted at least up until the mid-1950s.[84] As I, along with Alexander Blum and Jürgen Renn, have argued elsewhere, one of the most important elements characterising this period was that theorists did not trust the extreme physical predictions of the theory.[85] The majority of those few theoretical and mathematical physicists who conducted research related to the theory did so with the aim of substituting it with a more encompassing theory.

In this situation, Miller's results were seen by some as a welcome inspiration to guide theoretical lucubration towards a new framework surpassing Einstein's theory of gravitation. In fact, shortly after the first 1925 announcement and without waiting for confirmation of the result, theorists, particularly in the United States, attempted to exploit Miller's result to clarify the links between special and general relativity theory or to produce new advances in the theory. Contrary to Einstein's own belief, these theorists did not believe that Miller's ether drift had necessarily disproved both theories.[86] Ultimately, whether or not Miller's result was seen as an inspiration depended on the theoretical preferences of the scientists involved. In these endeavours, however, there was always a strong separation between experimental activities and theoretical constructs. Theorists did not seem to follow the experimental outcomes and, in turn, their work did not spark new experiments, in contrast to what happened in the early 1920s when Silberstein collaborated with Michelson in producing a new kind of ether-drift experiment to test the Planck–Stokes ether theory against the predictions of general relativity.

The idealised narrative I presented in the introduction—which sees the dismissal of the ether as a result of the special relativity theory being accepted, which was, in turn, a consequence of the null result of the Michelson–Morley experiment—played a fundamental role in the case presented here. It had already become a widespread narrative in the popular press and in textbooks, at least in the United States, back in the 1910s. The specific national and historical context, with all the emphasis on Michelson's international prestige and the role of the ether in the common-sense view of electromagnetic phenomena, consolidated this narrative. From the events described in this paper, the self-perpetuating effect on actual research of the 'imagined past' of physicists is evident, even when this imagined past was strongly connected to specific contexts. This narrative widespread in the United States was in fact one of the main reasons Miller decided to repeat the Michelson–Morley experiment in the 1920s.

What can be seen from the study presented here instead is that the definitive disappearance of the ether concept from the physics literature did not occur until *after* the controversy on Miller's result was over. This fate was not related to the success of general relativity, which was actually in a moment of decline. Rather, the demise of the ether was probably related to the controversy on Miller's experiment, since it polarised the discussion about the ether in very specific terms in connection with the debate on the existence or not of the ether drift. The strength of the controversy as well as of the beliefs held by Miller, who did not change his mind even after persuasive arguments were presented, had perhaps a more significant role in making the ether concept an illegitimate one: when compared with more 'modern' endeavours, powerfully represented by Joos's mechanised experiment, the ether was now being seen as inextricably linked to an old-fashioned, and less objective, style of doing physics.

NOTES

1. Allan Franklin, 'Physics Textbooks Don't Always Tell the Truth', *Physics in Perspective* 18 (2016): 3–57.
2. For the notion of imagined past see Alexander Blum, 'The Literature Review as Imagined Past', *Isis* 108 (2017): 827–9.
3. For evidence of the widespread presence of this narrative, see the Wikipedia page http://en.wikipedia.org/wiki/Luminiferous_aether.
4. For the debates on this point, see Gerald Holton, 'Einstein, Michelson, and the "Crucial" Experiment', *Isis* 60 (1969): 133–97; John Stachel, 'Einstein and Michelson: The Context of Discovery and the Context of Justification', *Astronomische Nachrichten* 303 (1982): 47–53; Hans Reichenbach, *The Philosophy of Space & Time* (New York: Dover Publications, 1958); Adolf Grünbaum, *Philosophical Problems of Space and Time* (New York: Knopf, 1963); and Jeroen van Dongen, 'On the Role of the Michelson–Morley Experiment: Einstein in Chicago', *Archive for History of Exact Sciences* 63 (2009): 655–63.
5. Albert Einstein, *'Zur Elektrodynamik Bewegter Körper'*, *Annalen Der Physik* 322 (1905): 891–921, p. 892.
6. Albert Einstein, *Ether and the Theory of Relativity* (London: Methuen & Co. Ltd, 1922); Ludwik Kostro, *Einstein and the Ether* (Montreal: Apeiron, 2000); see also Chapter 10.

7. Jeroen van Dongen, 'Reactionaries and Einstein's Fame: "German Scientists for the Preservation of Pure Science", Relativity, and the Bad Nauheim Meeting', *Physics in Perspective* 9 (2007): 212–30; Roberto Lalli, 'Anti-Relativity in Action: The Scientific Activity of Herbert E. Ives between 1937 and 1953', *Historical Studies in the Natural Sciences* 43 (2013): 41–104.
8. Loyd S. Swenson, *The Ethereal Aether: A History of the Michelson–Morley–Miller Aether-Drift Experiments, 1880–1930* (Austin: University of Texas, 1972).
9. Interview with Robert S. Shankland, 'Memories of Dayton Miller', 15 December 1981, Case Western Research University Archives, Robert Shankland Papers, 3hs55, folder 3; see also 'Interview of Robert Shankland by Loyd S. Swenson, Jr. on 1974 August 20', Niels Bohr Library & Archives, American Institute of Physics, College Park, MD, https://www.aip.org/history-programs/niels-bohr-library/oral-histories/4886-1, accessed 7 September 2017.
10. Dayton C. Miller, 'The Ether-Drift Experiment and the Determination of the Absolute Motion of the Earth', *Reviews of Modern Physics* 5 (1933): 203–42. For a discussion on the status of *Reviews of Modern Physics*, see Roberto Lalli, 'A New Scientific Journal Takes the Scene: The Birth of Reviews of Modern Physics', *Annalen Der Physik* 526 (2014): A83–7.
11. Michael Polanyi, *Personal Knowledge: Towards a Post-Critical Philosophy* (1974); Paul Feyerabend, *Against Method: Outline of an Anarchistic Theory of Knowledge* (London: Routledge, 1978). The first paper to propose an alternative explanation for the systematic effect observed by Miller was published only in 1955: R. S. Shankland, S. W. McCuskey, F. C. Leone and G. Kuerti, 'New Analysis of the Interferometer Observations of Dayton C. Miller', *Reviews of Modern Physics* 27 (1955): 167–78.
12. Karl R. Popper, *Logik der Forschung* (Wien: Springer, 1934).
13. Imre Lakatos, 'Falsification and the Methodology of Scientific Research Programme', in Imre Lakatos and Alan Musgrave, eds., *Criticism and the Growth of Knowledge* (Cambridge: Cambridge University Press, 1970), 91–195.
14. Harry M. Collins and Trevor Pinch, *The Golem: What You Should Know about Science* (Cambridge: Cambridge University Press, 1998).
15. Roberto Lalli, 'The Reception of Miller's Ether-Drift Experiments in the USA: The History of a Controversy in Relativity Revolution', *Annals of Science* 69 (2012): 153–214; Roberto Lalli, 'The Interplay of Theoretical Assumptions and Experimental Practice in the History of 20th-Century Ether-Drift Experiments', in Lucio Fregonese and Ivana Gambaro, eds., *Società Italiana degli Storici della Fisica e dell'Astronomia. Atti del XXXIII Convegno Annuale* (Pavia: Pavia University Press, 2016), 343–60. The approach of the first paper is partially inspired by Peter Galison, *How Experiments End* (Chicago: Chicago University Press, 1987).
16. To give a few examples: Frank H. Bigelow, 'The Status of the Solar Magnetic Problem', *Science* 2 (1895): 509–13; Reginald A. Fessenden, 'A Determination of the Nature and Velocity of Gravitation', *Science* 12 (1900): 740–5; and A. E. Dolbear, 'Does the Ether Absorb Light?', *Science* 21 (1893): 150. For the classical work on the history of ether and electromagnetism in the nineteenth century, see Edmund T. Whittaker, *History of the Theories of Ether and Electricity: From the Age of Descartes to the Close of the Nineteenth Century* (London: Longman, 1910); for a thorough discussions of the development in Britain, see Babara Giusti Doran, 'Origins and Consolidation of Field Theory in Nineteenth-Century Britain: From the Mechanical to the Electromagnetic View of Nature', *Historical Studies in the Physical Sciences* 6 (1975): 133–260.
17. Francis E. Nipher, 'The Ether', *Science* 18 (1891): 119–22.
18. See e.g. W. S. Franklin, 'The Electrical Theory of Gravitation', *Science* 12 (1900): 887–90; Reginald A. Fessenden, 'The Electrical Theory of Gravitation', *Science* 13 (1901): 28–31; and C. A. Chant, 'The Physical Basis of Color', *Science* 17 (1903): 864–8.

19. Mihajlo I. Pupin, 'Tendencies of Modern Electrical Research', *Science* 2 (1895): 861–80, p. 861.
20. See Daniel J. Kevles, *The Physicists: The History of a Scientific Community in Modern America* (Cambridge, MA: Harvard University Press, 1987).
21. Richard Staley, 'On the Co-Creation of Classical and Modern Physics', *Isis* 96 (2005): 530–58; Graeme Gooday and Daniel Mitchell, 'Rethinking Classical Physics', in Jed Z. Buchwald and Robert Fox, eds., *The Oxford Handbook of the History of Physics* (Oxford: Oxford University Press, 2013), 721–64.
22. Thomas F. Glick, ed., *The Comparative Reception of Relativity* (Dordrecht: Springer, 1987); Stephen G. Brush, 'Why Was Relativity Accepted?', *Physics in Perspective* 1 (1999): 184–214.
23. Stanley Goldberg, *Understanding Relativity: Origin and Impact of a Scientific Revolution* (Boston: Birkhäuser, 1984). A paper on the existence of the ether made its way into the scientific literature in 1907, but the author did not have Einstein's theory in mind, nor the related progress made in electrodynamics by Lorentz and Poincaré: Daniel F. Comstock, 'Reasons for Believing in an Ether', *Science* 25 (1907): 432–3.
24. Gilbert N. Lewis and Richard C. Tolman, 'The Principle of Relativity, and Non-Newtonian Mechanics', *Philosophical Magazine* 18 (1909): 510–23.
25. William Francis Magie, 'The Primary Concepts of Physics', *Science* 35 (1912): 281–93, p. 290. See also Alfred D. Cole, 'Section B: Physics', *Science* 28 (1908): 254–6; Louis T. More, 'On the Recent Theories of Electricity', *Philosophical Magazine* 21 (1911): 196–218; and D. L. Webster, 'On the Existence and Properties of the Ether', *Proceedings of the American Academy of Arts and Sciences* 48 (1912): 511–27.
26. See e.g. Reginald A. Fessenden, 'Transformation of Gravitational Waves into Ether Vortices', *Science* 38 (1913): 553–4; Francis E. Nipher, 'Momentum Effects in Electric Discharge', *Science* 29 (1909): 237–8; S. J. Barnett, 'On the Question of the Motion of the Ether in a Steady Electromagnetic Field', *Physical Review* 31 (1910): 662–5; and Louis Weisberg, 'On the Partition of Energy between Matter and Ether', *Physical Review* 6 (1915): 126–37.
27. See especially Leigh Page, 'The Distribution of Energy in the Normal Radiation Spectrum', *Physical Review* 7 (1916): 229–40, and Leigh Page, 'Is a Moving Mass Retarded by the Reaction of Its Own Radiation?', *Physical Review* 11 (1918): 376–400. For the reception of special relativity at Cambridge University, see Andrew Warwick, 'Cambridge Mathematics and Cavendish Physics: Cunningham, Campbell and Einstein's Relativity 1905–1911 Part I: The Uses of Theory', *Studies in History and Philosophy of Science* 23 (1992): 625–56, and Andrew Warwick, 'Cambridge Mathematics and Cavendish Physics: Cunningham, Campbell and Einstein's Relativity 1905–1911 Part II: Comparing Traditions in Cambridge Physics', *Studies in History and Philosophy of Science* 24 (1993): 1–25.
28. Robert A. Millikan and G. Harvey Cameron, 'A More Accurate and More Extended Cosmic-Ray Ionization-Depth Curve, and the Present Evidence for Atom-Building', *Physical Review* 37 (1931): 235–52.
29. Richard Staley, *Einstein's Generation: The Origins of the Relativity Revolution* (Chicago: Chicago University Press, 2008).
30. Albert A. Michelson and Edward W. Morley, 'Influence of Motion of the Medium on the Velocity of Light', *American Journal of Science* 31 (1886): 377–85.
31. Albert A. Michelson and Edward W. Morley, 'On the Relative Motion of the Earth and the Luminiferous Ether', *American Journal of Science* 34 (1887): 333–45.
32. For careful analyses of the problems related to the motion of the ether, see Michel Janssen and John Stachel, 'The Optics and the Electrodynamics of Moving Bodies', *Max-Planck-Institut für*

Wissenschaftsgeschichte, Preprint 265 (2004): 1–36, and Tetu Hirosige, 'The Ether Problem, the Mechanistic Worldview, and the Origins of the Theory of Relativity', *Historical Studies in the Physical Sciences* 7 (1976): 3–82.

33. William Fickinger, *Miller's Waves: An Informal Scientific Biography* (Cleveland: Xlibris Corporation, 2011).
34. Hendrik A. Lorentz, *Versuch einer Theorie der electrischen und optischen Erscheinungen in bewegten Körpern* (Leiden: Brill, 1895); Hendrik A. Lorentz, 'Simplified Theory of Electrical and Optical Phenomena in Moving Systems', *Koninklijke Nederlandsche Akademie van Wetenschappen Proceedings* 1 (1899): 427–42.
35. Albert A. Michelson, 'The Relative Motion of the Earth and Ether', *American Journal of Science* 18 (1897): 475–8.
36. Edward W. Morley and Dayton C. Miller, 'Report of an Experiment to Detect the FitzGerald–Lorentz Effect', *Philosophical Magazine* 9 (1905): 680–5; Edward W. Morley and Dayton C. Miller, 'Final Report on Ether-Drift Experiments', *Science* 25 (1907): 525.
37. Dayton C. Miller, 'Significance of the Ether-Drift Experiments of 1925 at Mount Wilson', *Science* 63 (1926): 433–43, p. 433; the same quotation also appears in Miller, 'The Ether-Drift Experiment and the Determination', p. 204.
38. Alistair Sponsel, 'Constructing a "Revolution in Science": The Campaign to Promote a Favourable Reception for the 1919 Solar Eclipse Experiments', *The British Journal for the History of Science* 35 (2002): 439–67; Marshall Missner, 'Why Einstein Became Famous in America', *Social Studies of Science* 15 (1985): 267–91.
39. 'Given the Speed Time is Naught', *New York Times*, 7 December 1919.
40. 'Einstein Expounds his New Theory', *New York Times*, 3 December 1919.
41. Robert Daniel Carmichael, *The Theory of Relativity* (New York: Wiley, 1913; 2nd edn, 1920).
42. Miller, 'The Ether-Drift Experiment and the Determination', p. 217.
43. George E. Hale to Miller, 19 July 1920; Miller to Hale, 11 November 1920; and Miller to Hale, 19 January 1921, Niels Bohr Library and Archives of the American Institute of Physics, George E. Hale Papers (GHP), roll 93.
44. John Lankford, *American Astronomy: Community, Careers and Power, 1859–1940* (Chicago: Chicago University Press, 1997), p. 250.
45. Three years later, a young experimenter, Roy J. Kennedy, asked the new director of the MWO, Walter S. Adams, to perform another kind of ether-drift experiment there, and his request was refused: see Kennedy to Adams, 25 July 1924, and Adams to Kennedy, 31 July 1924, Huntington Library, Walter Adams Papers (WAP), folder 38.650.
46. Ludwik Silberstein, *The Theory of Relativity* (London: Macmillan, 1914).
47. Ronny Desmet, 'Whitehead and the British Reception of Einstein's Relativity: An Addendum to Victor Lowe's Whitehead Biography', *Process Studies Supplements* 11 (2007): 1–44; Piotr Flin and Hilmar Duerbeck, 'Silberstein, General Relativity and Cosmology', in Jean-Michel Alimi and André Füzfa, eds., *Albert Einstein Century International Conference, AIP Proceedings Conference* 861 (Melvile: AIP Press, 2006), 1087–94; Giulio Peruzzi, and Matteo Realdi, 'The Quest for the Size of the Universe in Early Relativistic Cosmology (1917–1930)', *Archive for History of Exact Sciences* 65 (2011): 659–89.
48. Ludwik Silberstein, 'The Recent Eclipse Results and Stokes–Planck's Æther', *Philosophical Magazine* 39 (1920): 161–70. This little-known theory was made public in Hendrik A. Lorentz, 'Stokes's Theory of Aberration in the Supposition of a Variable Density of the Aether', *Koninklijke Nederlandsche Akademie van Wetenschappen Proceedings* 1 (1899): 443–8.

49. Ludwik Silberstein, 'The Propagation of Light in Rotating Systems', *Journal of the Optical Society of America* 5 (1921): 291–307. This experiment, corresponding to the Sagnac effect due to the earth's rotational motion, had actually already been proposed by Michelson himself in Albert A. Michelson, 'Relative Motion of Earth and Æther', *Philosophical Magazine* 8 (1904): 716–19.
50. Telegram, Miller to Hale, 22 April 1921, GHP, roll 92.
51. 'Hope Soon to Prove if Einstein is Right', *New York Times*, 13 May 1921, reprinted in József Illy, *Albert Meets America: How Journalists Treated Genius during Einstein's 1921 Travels* (Baltimore: Johns Hopkins University Press, 2006), 199–202. The evidence that it was Silberstein to mention Miller's result is circumstantial. However, as I explain in Lalli, 'The Reception', it is difficult to imagine that others knew about Miller's results at that time. The episode is quoted in several of Einstein's biographies, e.g. Albrecht Fölsing, *Albert Einstein: A Biography* (London: Penguin, 1997), p. 503.
52. Banesh Hoffmann and Helen Dukas, *Albert Einstein, Creator And Rebel* (London: Hart-Davis, MacGibbon, 1973), p. 146. This sentence became famous when Abraham Pais used part of it as the title of his scientific biography of Albert Einstein: Abraham Pais, *'Subtle is the Lord': The Science and the Life of Albert Einstein* (Oxford: Oxford University Press, 1982).
53. Miller wrote that Einstein 'was exceedingly pleasant and not at all insistent about the theory of relativity'; Miller to Thomas C. Mendenhall, 2 June 1921, Worcester Polytechnic Institute, University Archives and Special Collections, Thomas Corwin Mendenhall Papers, box 10, folder 8; quoted also in Holton 'Einstein, Michelson', p. 367. Miller repeated this opinion in Miller to Joseph Larmor, 9 June 1921, Royal Society Archives, Joseph Larmor Papers.
54. Hale wrote: 'Professor Dayton Miller . . . has just repeated the Michelson–Morley experiment on Mount Wilson with negative result . . . displacements of the fringes . . . were finally proved to be of instrumental origin and almost wholly eliminated', in Hale to Larmor, 21 December 1921, GHP, roll 22; Dayton C. Miller, 'Ether-Drift Experiments at Mount Wilson Solar Observatory', *Physical Review* 19 (1922): 407–8.
55. Milena Wazeck, *Einsteins Gegner* (Frankfurt: Campus Verlag, 2009); Milena Wazeck, 'Marginalization Processes in Science: The Controversy about the Theory of Relativity in the 1920s', *Social Studies of Science* 43 (2013): 163–90.
56. Charles L. Poor, *Gravitation Versus Relativity: A Non-Technical Explanation of the Fundamental Principles of Gravitational Astronomy and a Critical Examination of the Astronomical Evidence Cited as a Proof of the Generalized Theory of Relativity* (New York: Putnam's Sons, 1922); Miller to Reuterdahl, 13 December 1923, Arvid Reuterdahl Papers, box 4, folder 33, University of St. Thomas Special Collection, O'Shaughnessy-Frey Library, University of St. Thomas, St. Paul, MN. Elsewhere, Miller wrote: 'I agree with everything that [Poor] says and feel that he has stated the situation regarding the Relativity Theory better than any one else', in Miller to Dunn, 6 December 1922, WAP, box 8, folder 46.802.
57. Dayton C. Miller, 'Ether-Drift Experiments at Mount Wilson', *Proceedings of the National Academy of Sciences* 11 (1925): 306–14, p. 314.
58. See e.g. John H. Van Vleck, 'The Absorption of Radiation by Multiply Periodic Orbits, and Its Relation to the Correspondence Principle and the Rayleigh–Jeans Law. Part I. Some Extensions of the Correspondence Principle', *Physical Review* 24 (1924): 330–46.
59. Arthur S. Eddington, 'Ether-Drift and the Relativity Theory', *Nature* 115 (1925): 870.
60. Ludvik Silberstein, 'Ether Drift and the Relativity Theory', *Nature* 116 (1925): 98.

61. Albert Abraham Michelson and Henry G. Gale, 'The Effect of the Earth's Rotation on the Velocity of Light, II', *The Astrophysical Journal* 61 (1925): 140–5.
62. N. S. Hetherington, 'Sirius B and the Gravitational Redshift: An Historical Review', *Quarterly Journal of the Royal Astronomical Society* 21 (1980): 246–52.
63. 'Scientists Debate Recent Tests Made of Einstein Theory', *Washington Post*, 29 April 1925. See also 'Strikes A Blow Against Relativity', *New York Times*, 29 April 1925; 'Clevelander Bombs Einstein Theory', *Cleveland Plain Dealer*, 29 April 1925; 'Local Man Proves Ether Drifts, Refuting Einstein', *Cleveland Times*, 6 May 1925.
64. 'Michelson Proves Einstein Theory', *New York Times*, 9 January 1925; 'New Tests Support Einstein's Theory', *New York Times*, 21 July 1925.
65. Edwin E. Slosson, 'The Relativity Theory and the Ether Drift', *Science News* 62 (1925): viii; quoted in Lalli, 'The Reception', p. 175.
66. Hans Thirring, *'Kritische Bemerkungen zur Wiederholung des Michelson-Versuchs auf dem Mt. Wilson'*, *Zeitschrift für Physik* 35 (1926): 723–31; Hans Thirring, 'Prof. Miller's Ether Drift Experiments', *Nature* 118 (1926): 81–2.
67. Swenson, *Ethereal Aether*, p. 212.
68. See e.g. 'Michelson Proves Einstein Theory', *New York Times*, 9 January 1925, and 'Ether-Drift Is Confirmed', *New York Times*, 9 January 1925; both appear on p. 2 of that issue.
69. Roy J. Kennedy, 'A Refinement of the Michelson–Morley Experiment', *Proceedings of the National Academy of Science* 12 (1926): 621–9.
70. Adams to Michelson, 24 November 1926, Nimitz Library Special Collection and Archives Division, Albert A. Michelson Papers (AMP), box 1, US Naval Academy, Annapolis, MD; emphasis in the original.
71. Albert A. Michelson et al., 'Conference on the Michelson–Morley Experiment Held at Mount Wilson, February, 1927', *The Astrophysical Journal* 68 (1928): 341–402; Michelson, undated handwritten note in WAP, box 46, folder Michelson 1925.
72. As Adams noted a few years later, there is no doubt that Michelson 'did not have his heart very much in the investigation'; Adams to Gale, 16 January 1930, AMP, box 1. Gale had stated that Michelson's measurements 'were pretty sketchy and hurried' and that a more elaborate set of observations was necessary since he doubted that Miller could 'have deceived himself so systematically'; Gale to Adams, 9 January 1930, AMP, box 1.
73. Albert A. Michelson, Francis G. Pease and Fred Pearson, 'Repetition of the Michelson–Morley Experiment', *Journal of the Optical Society of America* 18 (1929): 181–2; Albert A. Michelson, Francis G. Pease and Fred Pearson, 'Repetition of the Michelson–Morley Experiment', *Nature* 123 (1929): 88.
74. Jeffrey Crelinsten, *Einstein's Jury: The Race to Test Relativity* (Princeton: Princeton University Press, 2006); Klaus Hentschel, 'The Conversion of St. John: A Case Study on the Interplay of Theory and Experiment', *Science in Context* 6 (1993): 137–94.
75. He published the report two years later: Charles E. St. John, 'Observational Basis of General Relativity', *Publications of the Astronomical Society of the Pacific* 44 (1932): 277–95.
76. Georg Joos, 'Die Jenaer Wiederholung Des Michelsonversuchs', *Annalen Der Physik* 399 (1930): 385–407. For an accurate study of Joos's instrument, see Christian Sichau, 'Hard Times: The Difficult Lives of Three Instruments in the Museum', in Peter J. T. Morris P. and Klaus Staubermann, eds., *Illuminating Instruments* (Washington, DC: Smithsonian Institution Scholarly Press, 2009), 73–94.

77. St. John to Campbell, 2 May 1930, quoted in Crelinsten, *Einstein's Jury*, p. 311. A similar sentence is in St. John to Adams, 6 May 1930, WAP, folder 59.132.
78. Miller, 'The Ether-Drift Experiment and the Determination'; Dayton C. Miller, 'Comments on Dr. Georg Joos's Criticism on the Ether-Drift Experiment', Nature 45 (1934): 114. The following repetition of the Michelson–Morley experiment would occur only in the mid-1950s using short radio waves: Louis Essen, 'A New Aether-Drift Experiment', *Nature* 175 (1955): 793–4.
79. For the concept of mechanical objectivity as an epistemic value, see Lorraine Daston and Peter Galison, *Objectivity* (New York: Zone Books, 2007).
80. The experimenter measured the velocity of light in two perpendicular directions, where one of the interferometer arms was shielded while the other was not, and did not find any difference between the two: G. W. Hammar, 'The Velocity of Light within a Massive Enclosure', *Physical Review* 48 (1935): 462–3.
81. Nathan Rosen, 'General Relativity and Flat Space. II', *Physical Review* 57 (1940): 150–3, p. 150; Nathan Rosen, 'Note on Ether-Drift Experiments', *Physical Review* 57 (1940): 154–5.
82. Marvin G. Moore, 'A Quasi-Relativistic Theory of Gravitation', *Physical Review* 89 (1953): 587–90; George Szekeres, 'New Formulation of the General Theory of Relativity', *Physical Review* 97 (1955): 212–23. See also John L. Synge, 'Gardner's Hypothesis and the Michelson–Morley Experiment', *Nature* 170 (1952): 243–4.
83. Charles G. Darwin, 'Logic and Probability in Physics', *Philosophy of Science* 6 (1939): 48–64; see also the literature mentioned in Section 9.1.
84. Jean Eisenstaedt, '*La relativité générale à l'étiage*: 1925–1955', *Archive for History of Exact Sciences* 35 (1986): 115–85; Jean Eisenstaedt, 'The Low-Water Mark of General Relativity, 1925-1955', in Don Howard and John Stachel, eds., *Einstein and the History of General Relativity* (Basel: Springer, 1989), 277–92.
85. Alexander Blum, Roberto Lalli and Jürgen Renn, 'The Reinvention of General Relativity: A Historiographical Framework for Assessing One Hundred Years of Curved Space-Time', *Isis* 106 (2015): 598–620.
86. See e.g. William Francis Gray Swann, 'The Relation of the Restricted to the General Theory of Relativity and the Significance of the Michelson–Morley Experiment', *Science* 62 (1925): 145–8, and Leigh Page and C. M. Sparrow, 'Relativity and Miller's Repetition of the Michelson–Morley Experiment', *Physical Review* 28 (1926): 384–91.

10

Ether and Aesthetics in the Dialogue between Relativists and Their Critics in the Late Nineteenth and Early Twentieth Centuries

Richard Staley

10.1 'THE INTRODUCTION OF A "LUMINIFEROUS ETHER" WILL PROVE TO BE SUPERFLUOUS'

If there is something illicit about the ether now, this is largely due to the work of Albert Einstein. It is well known that, in 1905, Einstein declared that his approach to the electrodynamics of moving bodies would render the luminiferous ether 'superfluous'.[1] Identified as a defining achievement of the emerging theory of relativity, Einstein's denial of the ether has also been regarded as one of the principal reasons special relativity was slow to be accepted outside Germany (although historians have shown that the practices engaged in different approaches are as important as adherence to particular principles or beliefs in shaping physicists' responses to new work).[2] It is less well known that in 1920, as a result of dialogues with both critics and supporters, Einstein argued that 'according to the general theory of relativity, space without ether is unthinkable'.[3] At first sight this appears a remarkable about face, and it might seem surprising that, with the exception of a valuable study from Ludwik Kostro, Einstein's later views have received little popular and scholarly attention.[4] Typically, however, Einstein was conceptually exact. First killing and then resuscitating what he also called 'the sick man of theoretical physics', he spelled out exactly what he meant.[5] As we shall see, together with significant differences between the luminiferous and gravitational ethers he discussed, Einstein's theoretical perspective showed a strong consistency that will help explain why his 1920 admission never outweighed the step he took in 1905.

Whether pointing to the irrelevance or integrity of the ether, Einstein was a deft and eloquent advocate. One aim of this chapter will be to explore what discussions of a

Staley, R., 'Ether and Aesthetics in the Dialogue between Relativists and Their Critics in the Late Nineteenth and Early Twentieth Centuries' in *Ether and Modernity*, edited by J. Navarro. © Oxford University Press 2018.
DOI: 10.1093/oso/9780198797258.003.0010

gravitational ether reveal about the values informing theoretical development, by examining the earlier work of the Austrian physicist Ernst Mach as well as that of Einstein. Mach was widely recognised as the principal advocate of the need to replace Newtonian concepts of absolute time and space with concepts based on relative measures, and for his antimetaphysical stance, but his major 1883 account of mechanics also discussed the possibility of a gravitational medium.[6] Mach's work on mechanics had been preceded by physiological and psychological research on the aesthetics of perception. This, too, had often concerned gravitation, especially as a result of the asymmetric nature of the horizontal and vertical dimensions of vision. A second aim of this chapter will therefore be to identify hitherto unexplored relations between treatments of theory and the aesthetics of visual perception, centred on forms of symmetry. Mach's and Einstein's work proved controversial. My third concern will be to build on their dialogues with critics to examine social and aesthetic dimensions in the environment of political debate, commercial publishing and disciplinary discussion that marked the emergence of general relativity in post-war Germany. For some, like the anti-Semitic Philipp Lenard, these questions turned on the images integral to different forms of physics.

10.2 THE AESTHETICS OF SCIENTIFIC PRINCIPLES? MACH VERSUS NEWTON, AND THE GRAVITATIONAL MEDIUM

The preface to Mach's 1883 *Die Mechanik in ihrer Entwickelung historisch-kritisch dargestellt* promised it would be no textbook imparting theorems and offered instead a 'clarificatory', 'or, to say it still more clearly, antimetaphysical' tendency.[7] Yet, Mach's principal arguments against metaphysical assumptions arrive only after more than 200 pages. Just as mechanical science rested on much instinctive knowledge and extensive experience that had only gradually been brought to conscious understanding, Mach needed to convey considerable conceptual experience before engaging his sharpest critiques. Treating Newton, he was especially careful to explain the character and significance of his accomplishments before showing weaknesses in their foundations. Discussing the terms in which Mach praised and critiqued Newton will help develop an understanding of the values informing Mach's approach. Recognising their range and variety will, in turn, show us why Mach countenanced a gravitational medium.

For Mach, Newton's main performance lay in two primary feats. The first was extending the realm of mechanical physics through the discovery of general gravitation, which required unparalleled imagination, or *Phantasie*. Perhaps guided by the principle of continuity, it took imaginative boldness to recognise that the accelerating force determining the curved path of the moon and planets was the gravitational force recognised on earth.[8] But Mach also praised Newton's restraint from speculating about hidden causes.[9] Historians and philosophers of science have emphasised this more often than Mach's references to imagination, tying it to Mach's abjuration of metaphysics and his argument that the principal aim of science is the economic description of facts. Introducing the book, Mach described the 'economy of communication and conception' as belonging to the essence of science, arguing also that in this lies its 'soothing, enlightening and aesthetic moment'.[10] While observing that Mach explicitly described an aesthetic import to economy, I want to register a methodological caution. Some of the values informing theoretical choice have regularly been treated as aesthetic, such as simplicity or—especially pertinent to this study—symmetry.[11] Rather than offering a comprehensive analysis of aesthetic properties in theory or describing some features as inherently aesthetic, here I will observe the ways that Mach and others described theory, as a preliminary basis for developing a clear historical perspective on how such views emerged; and note to begin with that physicists rarely used the term 'aesthetic'.

For Mach, Newton's second primary feat concerned bringing the expression of the principles of mechanics to completion in the form currently accepted—yet now one needed to go further.[12] Newton had clearly recognised the distinction between mass and weight, but his definition of mass in terms of the quantity of matter was 'unfortunate' in its circularity with his definition of density.[13] In contrast, Newton's clear and general identification of the equality of action and reaction was perhaps his most important service in articulating mechanical principles.[14] Mach thought the initial impulse towards this surely rested on instinctive experience, like the resistance experienced when one tries to set a body in motion—also closely related to our experience of mass. Mach argued that concepts of action and reaction and mass are so intimately bound they could not be separated, and offered his own definition of mass. It will be helpful to outline this at some length, for the concept of symmetry invoked will help me establish a link with Mach's understanding of visual aesthetics.

Mach began by noting that, according to the principle of symmetry, when two bodies similar to each other in all respects are placed opposite each other, they will produce equal and opposite accelerations in the direction of their line of junction. The principle cannot offer any guidance if there is some difference in form or chemical nature but, if mechanical experience clearly shows the existence of a property of bodies that determines accelerations, then, Mach wrote, we can arbitrarily state that 'we call bodies that produce equal and opposite accelerations when acting mutually on each other, bodies of equal mass'.[15] This could be generalised by defining a unit of mass; and that the accelerations are always of opposite sign, and there are solely positive masses given this definition, are facts taught only by experience. As Mach put it, in his concept of mass 'there lies no theory . . . it contains solely the exact fixation, designation and naming of a fact'.[16]

In Mach's scientific ideal, then, concepts express empirical relations. Mach had first expressed his approach to mass (without referring to symmetry) in a short article submitted to the prestigious *Annalen der Physik* in 1867. It was returned unpublished over a year later, so he sent it to *Carls Reportorium* and then reprinted it (with caustic comments on the typical contents of the *Annalen*) in the footnotes to a lecture on the conservation of work.[17] This early rebuff clearly confirmed Mach in his self-understanding as a critic among his physicist peers.

The next section of *Die Mechanik in ihrer Entwickelung* took up Newton's views of time, space and motion. Mach thought Newton's concept of absolute time smacked of medieval philosophy, and described his stance as 'metaphysical' because it was defined in contrast to any perceptible change.[18] Similarly, the concept of absolute space left the realm of facts. In contrast, all the basic theorems of mechanics expressed experiences of relative locations and motions of bodies—and they should not be extended beyond the bounds of experience. Mach first developed his fundamental point in abstract terms, as follows. When two bodies K and K′ affect each other's motion, it is impossible to recognise this unless there are other bodies, A, B, C, and so on, present which allow the motion to be determined. There are then two errors in thinking one could disregard the other bodies and speak of the motion of K in absolute space. First, we cannot know how K would behave in the absence of A, B, C, and so on, since our laws of mechanics have been formulated in the presence of other bodies (including, as Mach would go on to argue, the law of inertia, with the centrifugal effects evident in Newton's rotating bucket experiment). Second, without such bodies, we would have no way of evaluating the behaviour of K and proving our statements.[19]

This denial of absolute space was the context in which Mach introduced the possibility of a gravitational 'medium', writing:

> It could well be possible that the isolated bodies A, B, C . . . only play an accidental role in the determination of the motion of K, and that the motion is determined by the medium in which K finds itself. Then this must be put in the place of Newton's absolute space.[20]

Mach described some of the essential requirements of such a hypothetical medium, suggesting a potential model in a recent hydrodynamic proof that, in a frictionless fluid, a rigid body only experienced resistance with a change in velocity—noting the proof had assumed the current law of inertia but raising the possibility that this could instead be regarded as the first fact from which one could proceed. And then he went on to discuss Newton's bucket experiment in concrete detail, arguing that Newton had neglected to consider the existence of the earth and the stars, and it was therefore a mistake to attribute the centrifugal effects described to absolute space. Finally, in order to go beyond the problem that Newton's statement of the law of inertia did not specify the bodies or coordinate system with respect to which a given body remains at rest or in uniform motion, Mach offered an expression for the law in terms of the average acceleration in relation to distant masses. Arguing his expression was as appropriate as Newton's, Mach admitted it also suffered from a difficulty—the impossibility of including all the masses involved in completing the summation required.[21]

Thus, Mach's gravitational medium was one of several responses to the inadequacy of absolute space and of Newton's formulation of the law of inertia. Several elements of these responses were clearly hypothetical and provisional; yet, on the basis of reading a later edition of *The Science of Mechanics*, John Norton has noted ambiguities in Mach's formulations and argued that his strictures against metaphysics and against typical understandings of causation show Mach can only have meant to redescribe Newton's views in ways that removed any reference to absolutes.[22] Surveying briefly just some of Mach's stances has shown that his antimetaphysics was accompanied by considerable intellectual flexibility. I would argue that, rather than pursuing the conservative goal Norton suggests, in offering a specific, if as yet inadequate, model of a gravitational medium, as well as an alternative conceptual formulation of inertia, Mach's considerations reflected important features of his search for economy. He had begun the dual tasks of communicating and conceptualising a new physics, recognising that the principal problem lay in finding empirical facts that would allow one to go further than Newton's laws (which he wrote might come in extending physical and astronomical knowledge to a region in the heavens where more massive and complex motions occurred).[23]

10.3 SYMMETRY AND THE AESTHETICS OF VISION

I will later consider briefly some of the varied senses in which Einstein drew on Mach (among others) in developing special and general relativity. Before doing so, it will be helpful to indicate two ways in which Mach's mechanics can be related to his extensive studies of perception, through the concept of symmetry and the multifaceted significance of gravitation. Historians and philosophers of science have shown that considerations about harmony, equilibrium and symmetry have long played a significant role in scientific thought, but also argued that uses of the concept of symmetry in a modern sense first emerged in the geometrical work of Legendre in 1794 and crystallographic studies from the early 1800s.[24] Pierre Curie's 1894 study of symmetry in physical phenomena is usually regarded as the first explicit formulation of symmetry arguments as such, while Hon and Goldstein have shown that Einstein's 1905 work entered a tradition of the discussion of symmetries in electric and magnetic theory by Rowland, Heaviside and Hertz from the 1880s; however, we currently lack a comprehensive history of symmetry in this formative period.[25] Mach's work offers a novel perspective because his treatments of symmetry move between aesthetics and physical laws, as well as between physics and physiology; it presents important precedents to a line of inquiry Arthur I. Miller has developed linking the visual thought of Poincaré and Einstein to art and later experimental aesthetics.[26]

Discussing Mach's analysis of action and reaction, I noted his reference to instinctive, bodily grounds for the principle. Earlier in his book, Mach had pointed to the relevance of instinctive experience in a particularly revealing way. There, Mach described the first assumption Archimedes invoked when discussing the law of the lever, that magnitudes of equal weight at an equal distance from their point of support are in equilibrium. Mach drew a simple diagram (see Fig. 10.1) to show the possibility of analysing it further, writing: 'One could think that independent of all experience (according to the principle

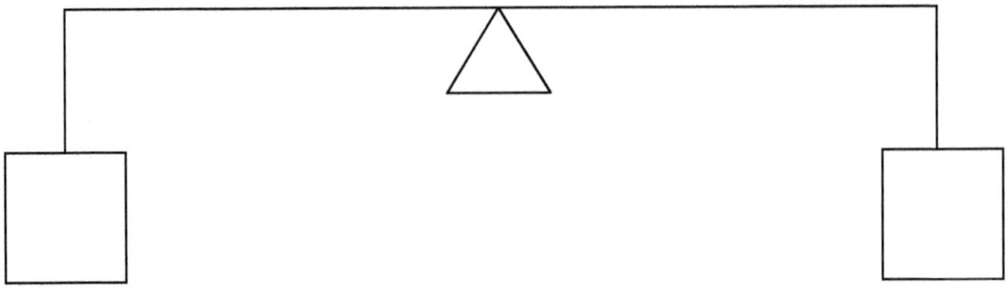

Fig. 10.1 Mach's illustration of Archimedes' law of the lever, after his *Mechanik*, Fig. 2.

of sufficient reason) it is self-evident that the symmetry of the entire arrangement gives no ground for the bar to rotate in one rather than another direction.'[27]

Introducing us thus to what he later calls the principle of symmetry (without ever defining it precisely), Mach's clarificatory point concerns the considerable negative and positive experiences hidden in the assumption. Negatively, for example, we typically forget to note that factors like different colours on the lever's arms or the position of the observer have no influence.[28] Revealingly, Mach develops his point further by writing that, if the observer 'places themselves in the plane of symmetry of the arrangement, then proposition 1 is manifestly a highly persuasive *instinctive* insight, which is determined by the symmetry of our own body'.[29]

As this shows, Mach focuses on instinctive knowledge so often partly because of the extent to which our own experience guides our conceptions, for we are part of the mechanical world.[30] As a result, many mechanical phenomena could—or should—be investigated physiologically as well as physically; in significant cases, such investigations reveal subtle distinctions. Mach's earlier research on visual perception had often turned on questions of symmetry and difference shaped by the character of our eyes and life in a gravitational environment. Begun in 1861, his studies led Mach to articulate important distinctions between what he referred to as physiological, geometrical and physical spaces from 1866. He also gave lectures on why we have two eyes and symmetry in 1866 and 1871, but I can illustrate his views by examining his 1886 book, *Beiträge zur Analyse der Empfindungen*.[31] There, Mach begins his analysis of what he termed 'space sensations' by asking the reader to examine two letters of the same size and shape but different colours (see Fig. 10.2). That we recognise them as the same form at one glance shows that sight perceptions contain the same sensation components; these, then, are space sensations.

Mach then showed that our perception of lines and forms is physiologically conditioned by our sense organs in quite specific ways, by taking his readers on a visual journey through geometric forms in different orientations, and ink spots repeated in different positions and proportions. They demonstrate a point he also made about learning to read. As Mach noted, the whole apparatus of the eye, and especially its motor apparatus, is symmetrical about the median plane of the head. This means that symmetrical movements of looking will determine the same or similar space sensations, and he wrote, 'Children constantly confuse the letters b and d, as also p and q', for these letters excite the same or similar sensations.[32] In strong contrast, looking upwards and downwards is

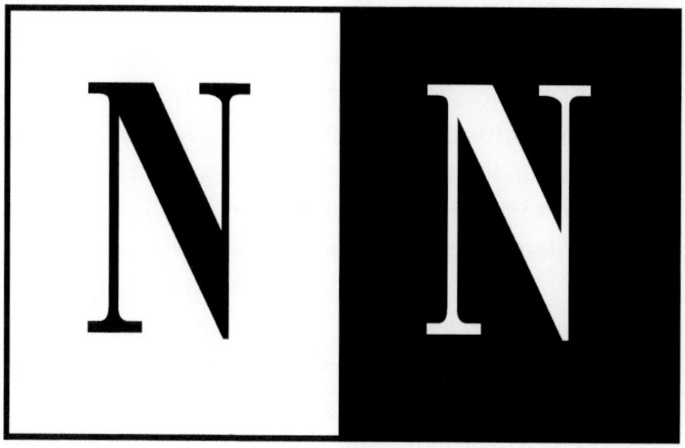

Fig. 10.2 After Mach, *Analyse der Empfindungen*, Fig. 2.

connected with fundamentally different space sensations, reflecting the asymmetrical nature of the motor apparatus of the eye in relation to the horizontal plane. Mach wrote, 'The direction of gravity is so determinative and important for the ordinary motor apparatus of the body that this circumstance must also find its expression in the eyes which serve it.' Among the consequences, he pointed to the fact that 'the letters b and p, further d and q are not confused by children'.[33] Further, although beauty is not the principal concern of writing, in the capital letters of the Latin alphabet, only the six letters F, G, L, P, Q and R do not show a form of symmetry; the rest of the letters have either vertical symmetry, like A; horizontal symmetry, like B; or centric symmetry, like N.

Mach regarded symmetries of this nature as a partial guide to aesthetic appreciation. He argued the pleasing effect of symmetry is largely a result of the repetition of sensations, and the effect in question takes place in symmetrical figures only where there is a repetition of sensation. Mach paid particular attention to the aesthetic advantages of straight lines, and especially vertical and horizontal lines, which he related to the structure of the retina, but did not investigate the nature and limits of the physiological basis for visual aesthetics at great length.[34]

In taking up this brief study of visual aesthetics, I have pursued two aims some distance from our primary subject in the ether. My immediate aim has been to establish significant continuities between different aspects of Mach's work. His research was distinctive in crossing the terrains of physics and psychology in this way, and in 1871 he described its ambitious scope, writing:

> The motions of the moon and the other heavenly bodies are accurately known. Our knowledge of the motions of our own body is by far not so complete. The mountains and natural divisions of the moon have been accurately outlined on maps, but physiologists are just beginning to find their way in the geography of the brain. The chemical constitution of many fixed stars has already been investigated. The chemical processes of the animal body are questions of much greater difficulty and complexity. We have our *Mécanique celeste*. But a *Mécanique sociale* or a *Mécanique morale* of equal trustworthiness remains to be written.[35]

At their broadest, Mach's studies concerned this rather open and potentially problematic attempt to determine the basis for psychological or social and moral phenomena. My second aim has been to prepare grounds for an examination of the terms in which relativists and their opponents met after World War I; these will suggest that aesthetic dimensions were engaged in the social and moral terms in which relativity and the ether were debated, related to what were soon described as 'Jewish' versus 'German' physics.

But Mach was also distinctive in pursuing these disciplinary relationships without insisting on physical or physiological foundations. As Alexandra Hui has shown, Mach came to understand significant elements of hearing (and, in particular, the phenomena of accommodation) to be a matter of cultural development and learning—he could find no physiological basis for them.[36] In his lecture on why we have two eyes, Mach discussed differences in the artistic depictions of ancient Egypt and China, arguing these were as true to physiological perception as the advanced use of perspective in Western painting. Without defining eyes by physiology, then, Mach wrote: 'Change man's eye and you change his conception of the world. We have observed the truth of this fact among our nearest kin, the Egyptians, the Chinese, the lake-dweller.'[37] In the final and significantly extended edition of *Beiträge zur Analyse der Empfindungen*, which was published in the early twentieth century, Mach addressed the question of education and evolution somewhat soberly, writing:

> To deny the influence of pedigree on psychical dispositions would be as unreasonable as to reduce everything to it, as is done, whether from narrow-mindedness or dishonesty, by modern fanatics on the question of race. Surely everyone knows from his own experience what rich psychical acquisitions he owes to his cultural environment, to the influence of long vanished generations, and to his contemporaries. The factors of development do not suddenly become inoperative in post-embryonic life.[38]

Finally, I would note that in his discussions of different forms of symmetry, Mach relates the aesthetics of perception to physical phenomena, but does not explicitly treat the latter (or the physical principle of symmetry) as a matter of aesthetic choice.

10.4 MACH AS CRITIC AND PROPHET

Mach was intensely aware of the novelty of his stances, and the success of *Die Mechanik in ihrer Entwickelung* surprised him. In seven editions by 1912, with English, French, Italian and Russian translations, Mach responded to objections and support. His prefaces, appendices and blossoming additions gradually chart his transformation from singular critic to one among several kindred spirits, and later a burgeoning group of 'relativists'. Without discussing this process in detail, we should note that in 1897 he devoted considerable attention to Hertz's book on mechanics, highlighting an increasing focus on the possibility of expressing gravitation (and perhaps all forces) in terms of the consequences of motion in a medium, in the same way that electric and magnetic forces had been treated.[39] Along these lines, in 1901 and 1904 Mach mentioned Paul Gerber's use of the perihelion of mercury to provide a theory of gravitation in which gravitational effects propagated with the velocity of light, in a discussion that was later important for Einstein's critics. Mach wrote this spoke for the ether as the medium for gravity—still more directly connecting the possible gravitational or inertial 'medium' he described in 1883 with the luminiferous 'ether'.[40]

By 1908, Mach identified an international group of 'relativists' in his discussion of inertia, listing eight people whose contributions ranged from closely argued technical papers on coordinate systems and rotation, to overviews of the physical sciences, including J. B. Stallo, James Thomson and Ludwig Lange. He thought there would soon be no more significant opponents to his view.[41] Several features of this moment are worth noting. First, Mach clearly used the term 'relativist' broadly, referring to relational understandings of time, space and motion. Second, he did so before learning of Einstein's work (like many others) through Minkowski's famous 'Space and Time' lecture in September 1908. Mach then asked Philipp Frank to explain the theory to him and subsequently noted respects in which his earlier views on matter, space and time were consistent with the direction being taken by Lorentz, Einstein and Minkowski.[42] Yet if the 70-year-old Mach was far from being able to participate actively in electrodynamic research, earlier discussion of his perspective is also likely to have helped prepare more general grounds for the appreciation of Einstein's approach.[43] Finally, note that, although this is unlikely to have been in response to Mach, as we shall see, Einstein also adopted the label 'relativist'.

Later, Einstein noted that the works of Hume and Mach were central to his ability to approach absolute time and the concept of simultaneity critically in 1905.[44] He drew generally on Mach's attack on absolutes without referring to the diverse forms of space and time Mach discussed, although the Olympia Academy read Mach's accounts of mechanics and sensations (and Einstein had also read his 1896 textbook on the theory of heat). Einstein's paper also offered a novel sense for symmetry in theory, with its opening sentence noting that the customary conception of electrodynamics led to asymmetries in the theoretical treatment of moving bodies that were not inherent in the phenomena—which depended only upon the relative motion of a magnet and conductor.[45] This has commonly been regarded as expressing aesthetic concerns, or even predominantly aesthetic concerns, as stated in the influential work by Gerald Holton.[46] Hon and Goldstein have countered this with a persuasive argument that Einstein's work ended a debate on the symmetry of electric and magnetic theory initiated by Rowland, Heaviside and Hertz by relating this instead to a newly unified electromagnetic theory.[47] Nevertheless, while it is important to note that Einstein does not use aesthetic language in describing his point, Hon and Goldstein may be incorrect in thinking there were no aesthetic grounds simply because there were physical grounds. Einstein turned to the ether in his second paragraph, describing it as 'superfluous' to the extent that his theory would neither require a space at absolute rest nor assign a velocity vector to a point in empty space in which electromagnetic processes took place.[48]

While the customary conception of electrodynamics was principally due to Lorentz's work and commonly attracted attention for electron theory and its implications as an electromagnetic view of nature, Einstein's approach led to an increasing focus on the principle of relativity. Among German-speaking physicists, it was gradually perceived to be central to a new theory of relativity, and by 1911 Arnold Sommerfeld counted this a secure possession.[49] As his approach won ground, Einstein drew more or less consciously on several elements of Mach's thought in going beyond the initial limitation of the principle of relativity to treating uniform motion and inertial frames of reference. Later, Einstein described the origins of general relativity as based on a 1907 thought experiment on the experience of a man falling from a rooftop—which had led him to recognise that the gravitational field was relative to the observer and had stimulated his first attempts to link mathematical treatments of accelerated motion and gravitation; Miller interprets Einstein's thought experiment as an example of visual thinking abstracted from the world of phenomena, working with different senses of the German term for intuition, *Anschauung*.[50] Although he never noted any link to Mach's work in this respect, Einstein's thought experiment can be regarded as a variation on a thought experiment on a load on a falling table that Mach treated in discussing action and reaction (a thought experiment, in turn, based on his experiments on the physiology of falling and our sense of orientation).[51] A second, more conscious respect in which Einstein drew on Mach concerned his articulation of what he called 'Mach's principle' to define the view that inertia depends upon the other bodies of the universe, a principle that played a key role in shaping his approach but that he ultimately discarded.[52] Finally, as we shall now see, Einstein also returned to Mach's consideration of an inertial or gravitational medium, but this came later and was prompted by both a friend and critics.

10.5 EINSTEIN IN DIALOGUE: THE RELATIVIST MEETS HIS CRITICS

Following Einstein's research closely and developing a supportive relationship with him, the Dutch theoretician and 1902 Nobel laureate H. A. Lorentz accepted the conceptual simplicity and consistency of his approach, but took a different stance. Lecturing in Göttingen in 1910, Lorentz allowed that both Einstein and Minkowski denied the ether and true time, but stated he preferred to see space and time as separate and distinct, upholding both an ether and a conception of true time.[53] In June 1916, Lorentz wrote to Einstein to discuss different ways of describing the effect of the earth's rotation on electromagnetic waves passing through a circuit encircling the earth. It seemed obvious to think of an ether in which the standing waves are located, and Lorentz described a Machian approach in which the influence of distant bodies was less manifest but which differed little from an ether-based interpretation. Considerations like this left relativity and Einstein's theory of gravitation entirely valid, but less pressing as the only possible explanations.[54] Einstein responded that he had often considered similar examples and agreed that general relativity was closer to the ether hypothesis than the special theory was. This new ether theory, however, would not violate the principle of relativity, because the state of the metric tensor '$g_{\mu\nu}$ –ether' would not be that of a rigid body in an independent state of motion; instead, every state of motion would be a function of position determined by material processes.[55]

Einstein's nod came in correspondence and he did not address the ether publicly until he had formulated a response to wide-ranging attacks. These came especially from Ernst Gehrcke, an experimentalist in the optical department of the Imperial Institute for Technical Physics in Berlin, and the Heidelberg experimentalist and 1905 Nobel laureate Philipp Lenard.[56] Gehrcke's first article on the limitations of relativity was published in 1911, while in 1912 and 1913 he addressed the sense of absolute motion in relativity but also described its success as the product of mass suggestion. In 1916 Gehrcke accused Einstein of plagiarising Gerber's account of a gravitational force propagating at the speed of light; Gehrcke he said Einstein must have known about Gerber's idea because it was discussed in Mach's mechanics (Einstein's recent obituary of Mach had shown that he knew Mach's work).[57] In 1918, both Gehrcke and Lenard introduced their own ether theories (see Chapter 7 for a discussion of Lenard's work). This is the context in which Einstein described his views on the ether in two revealing publications.

The first addressed the general audience of *Die Naturwissenschaften*, the German equivalent to *Nature*. Artfully, Einstein used the literary device of a dialogue between a 'Critic' and a 'Relativist' (noting this referred to the physical theory and not philosophical relativism).[58] The Critic states he has come personally so that the Relativist will not back out as before, and promises to stay until all his criticisms have been answered, a formulation revealing some ambivalence even as it affirms Einstein's commitment to debate. At different points, Einstein considers positions taken by both Gehrcke and Lenard, allowing the Critic to suggest that a colleague's accusations of plagiarism smack of impropriety (without naming Gehrcke).[59] Lenard's recent discussion of the way a jolt in a moving railway carriage might be treated as a result of the train braking, or of the environment moving, prompts the Relativist to distinguish between convenient and conceivable coordinate systems, noting that relativity offers two equivalent (in principle) ways of describing the same factual phenomena, rather than two different hypotheses about its cause.[60] The dialogue ends with the Critic admitting that refuting the Relativist's point of view was more difficult than he had thought but, before leaving, he asks how 'the sick man of theoretical physics fares, the ether, that many of you have declared to be definitely dead?'[61]

Einstein's Relativist offers a concession. Given twists and turns of fortune, he admits one cannot pronounce the ether's demise. Without any privileged state of motion, special relativity had indeed denied the ether, but since, in general, relativity empty space has physical characteristics, this introduces an important difference:

> empty space in this sense has according to the general theory of relativity physical qualities which are mathematically characterized by the components of the gravitational potential, that determine the metrical behavior of this part of space as well as its gravitational field. One can quite well construe this circumstance in such a way that one speaks of an Ether, whose state of being is different from point to point. Only one must take care not to attribute to this 'Ether' properties similar to properties of matter (for example every point a certain velocity).[62]

Two years later, Einstein gave the ether a still more extensive history, delivering an inaugural lecture in Leiden at Lorentz's invitation. There he distinguished the 'Machian ether' of general relativity from Lorentzian and other forms of ether. His summary made the ether seem inevitable:

> According to the general theory of relativity space without ether is unthinkable; for in such space there not only would be no propagation of light, but also no possibility of existence for standards of space and time (measuring-rods and clocks), nor therefore any space-time intervals in the physical sense.[63]

But this admission, with caveats insisting ether could not be endowed with qualities characteristic of ponderable media, or motion, was greeted variously by Einstein's opponents. Johannes Riem described Einstein as being forced to acknowledge the ether in a roundabout way, while Gehrcke recognised how little Einstein had actually conceded, writing 'Einstein's concept of ether = not ether, but rather non-Euclidean space'.[64]

10.6 TYPEFACING RELATIVITY AND ITS CHALLENGERS

Einstein's readiness to define a general relativistic ether was a significant concession. Along with the work of others like Hermann Weyl, the full mathematical characterisation Einstein gave it suggests an ether might have continued to form a part of physicists' approach to gravitational theory and unified theories. That it did not reflects two features. First, it was surely significant that critics like Lenard defined their alternative approaches in terms of different concepts of ether, so the ether became a mark of distinct alternatives to Einstein's approach. Second, Einstein's developmental path had moved through concepts of space and measurement, and specific forms of non-Euclidean geometry; so the conceptual and technical heart of his theory had been presented independently of ether, and this continued despite his 1920 discussion. Einstein might *write* that space without ether would be unthinkable, but neither his research papers nor popular writings before or afterwards encouraged people to actually use that thought.

Still more importantly, for all its eloquence staging a reasoned debate on principle, Einstein's dialogue failed to convince his critics—and the fronts on which they met him only widened. We have many excellent studies of the determined opposition Einstein received from a small group of German physicists, including especially the experimentalist Johannes Stark, in addition to Gehrcke and Lenard.[65] A cascading complex of issues concerning the relations between experiment and theory, Berlin and the provinces, popular and expert opinion, common sense and abstract thought, and 'German' and 'Jewish' physics became entwined in debates these figures stirred in the volatile post-war environment. Gehrcke had already fanned discontent beyond typical bounds, while Lenard played the role of an aggrieved authority. Known for his nationalism, he began by describing Einstein's work as 'un-German' in a letter to Stark prior to an anti-relativity meeting Paul Weyland had organised in Berlin on 24 August 1920. This meeting featured anti-Semitic literature and led Einstein to make the association between anti-Semitism and Weyland's and Gehrcke's critiques public, in a newspaper article in which he also promised the possibility of a scientific debate at the Annual Meeting of German Scientists and Physicians in Bad Nauheim in September.[66] I cannot offer a comprehensive treatment here but, by way of conclusion, will use a study of the publishing forums of the German physics community to briefly show how some of the issues involved were related to images of different kinds of physics—and treated aesthetically.

Once he had moved to Berlin, Einstein usually published in the proceedings of the Prussian Academy of Science or the premier research journal *Annalen der Physik*, where Gehrcke also published his critiques. In 1917 Einstein published his popular book with the Braunschweig firm Friedrich Vieweg & Sohn, but *Äther und Relativitätstheorie* appeared in the lists of the noted Jewish publisher Julius Springer, who also published *Die Naturwissenshaften* (edited by Einstein's Jewish friend and former AEG physicist, Arnold Berliner). The several editions of Lenard's book *Über Relativitätsprinzip, Äther, Gravitation* appeared with the Leipzig press of S. Hirzel from 1918. Hirzel also published the principal informal journal for physics, *Physikalische Zeitschrift*, then edited by Peter Debye. Providing rapid publication, disciplinary news and extensive advertising, its pages offer an unparalleled overview of the German physics community.

In December 1919, Hirzel celebrated the announcement of the Nobel Prizes for 1918 and 1919 with a paired spread in *Physikalische Zeitschrift*, for they published work from both Max Planck and Johannes Stark (see Fig. 10.3).[67] Yet, the symmetry on the page belied a significant difference in perceived prestige and standing, a difference which might be traced back to Planck's increasing advocacy of the role of theoretical physics rather than experiment in celebrating the rise of the Copernican movement of special relativity.[68] This was only sharpened by the extraordinary public acclamation Einstein began to receive internationally, across fault lines left by war. When Hirzel announced a new edition of Lenard's 1918 book on relativity, ether and gravitation in early 1920, somewhat unusually their advertisement outlined the advantages Lenard saw in his perspective:

> The author handles the subjects named from the standpoint of a scientist who feels himself guided more by experience of the material world than by philosophical demands. He grounds his worldview not solely on mathematical equations but on mechanisms. The high value of the book lies in these differences from the customary handling of the matter.[69]

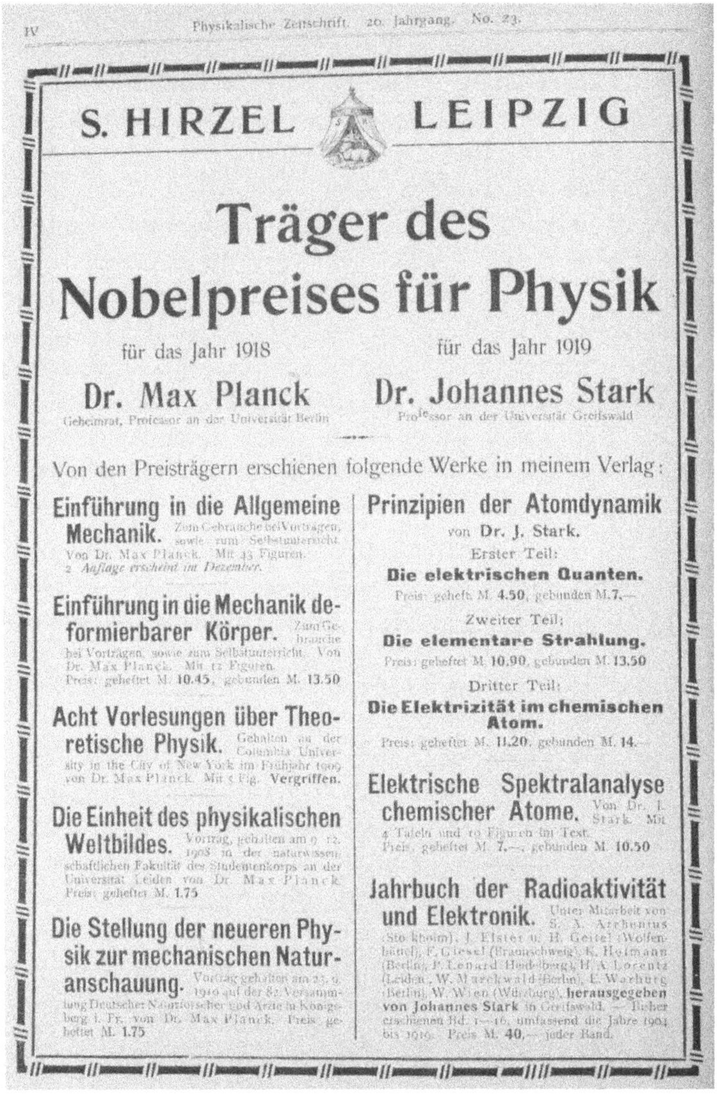

Fig. 10.3 The publisher of *Physikalische Zeitschrift*, Hirzel, paired advertisements for their two Nobel Prize-winning authors (announced in 1919).

This was a succinct statement of the characteristics of the sober experimentalist that Lenard identified with good physics, and framed oppositionally. While its text was set in a roman typeface, the title used a common art nouveau, or *Jugendstil*, typeface also featured in an ad Hirzel ran for Springer's publications on relativity in October (as well as in AEG advertisements for new staff); this typeface contrasts strongly with the *Fraktur* typeface that was used for the advertisement of Stark's Nobel lecture and which appeared on the same page (see Fig. 10.4).[70]

Fig. 10.4 Contrasting *Fraktur* and art nouveau typefaces in advertisements for Johannes Stark's book on changes in the structure and spectrum of chemical atoms, and literature on Einstein's relativity from the press of Julius Springer.

Scientific journals and books were typically typeset in a roman typeface, but *Fraktur* had been the official typeface of the German empire since its founding, and in 1910 Adolf Reinecke had offered a manifesto for its 'Germanness', in a book which featured a physiological comparison of font readability.[71] Adherents of both roman typeface and *Fraktur* argued over the aesthetic advantages of the two fonts and, in April 1918, a new *Bund für deutsche Schrift* was founded, aiming to extend the use of *Fraktur* beyond its traditionally *völkisch* associations.[72] In previous issues, Hirzel's ads had used the font only rarely, for some personnel ads and notes on the journal price, as well as for three war memoirs they published from von Hindenburg, von Bernardi and Vollerthun. Before the war, Lenard and Stark had often published in roman type; but both Lenard's 1918 book and now Stark's pamphlet were printed in *Fraktur*. The first ad for Stark's lecture appeared on 1 September, shortly before the Bad Nauheim meeting and, from December 1920, Hirzel's ads for the third edition of Lenard's book now appeared in *Fraktur* as well, stating that it included his response to the debate in Bad Nauheim (see Fig. 10.5).[73]

The Bad Nauheim discussion realised publicly the dialogue Einstein had scripted in 1918 and, indeed, at the meeting, Lenard brought up his understanding of the braking train. He also linked his treatment of different forms of physics to two different kinds of pictures, associating Einstein's mathematical thought with the first and asserting that a reference to processes in space is characteristic of the second, for which the ether is indispensable. When he pressed Einstein to address the gap between images of the first kind and the 'intuitive' (or 'comprehensible': *anschaulichen*) pictures of the second kind, Einstein argued that 'what people regard as *anschaulich*, and what not, has changed. Views on *Anschaulichkeit* are to some extent a function of time. I think physics is conceptual, not

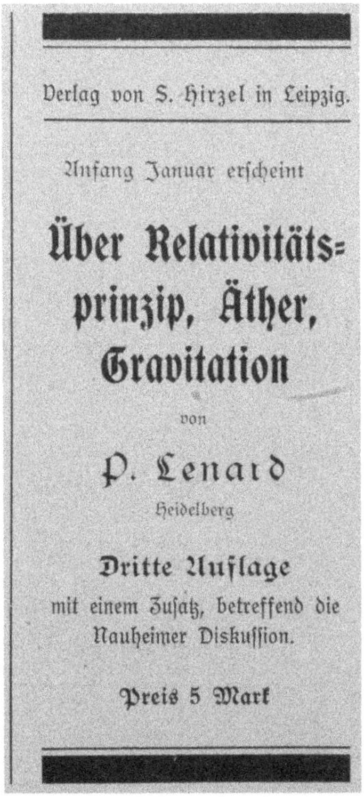

Fig. 10.5 The *Fraktur* font chosen to advertise the third edition of Lenard's *Über Relativitätsprinzip, Äther, Gravitation*.

anschaulich.'[74] Many of those present thought that the account of the meeting published in *Physikalische Zeitschrift* was flawed, and Berliner blamed this on Hirzel putting pressure on space, rather than on Debye's transcription.[75] Lenard was so unhappy with the German Physical Society and the management of the meeting that, at the end of the meeting, he resigned from the society, finding his next real hopes for German physics in the emergence of Hitler and the Nazi party as early as 1924. Who and what kind of physics counted as German was critical for Lenard; in 1936 his *Deutsche Physik* was published in *Fraktur* by the Munich firm J. F. Lehmann.

It would be inappropriate to draw too strong a conclusion from the patterns that I have picked out here. German newsprint of the day was most often printed in *Fraktur*, publishers conserved fonts and images set for specific advertisements over many issues, and the most striking feature of *Physikalische Zeitschrift* advertisements is their typographic and imagistic variety. But these font choices do suggest a further aesthetic aspect to the struggles fought over relativity and ether, featuring visual differences that some observers may have linked to what Lenard and others sought to define as 'Jewish' versus 'German' physics. Social distinctions were being pressed in a kind of physics associated with ether, and in publishing fonts.

Ernst Mach might not have known what to make of the popularity of *Fraktur*, with its broken lines shattering the symmetry of individual letters, but he would have recognised the fanaticism of Gehrcke and Lenard. Yet, his son Ludwig did not. As Geroen Wolters has shown, it was Ludwig Mach's desire to modulate the reputation of his father that led him to fraudulently insert a preface to the posthumous 1921 publication of Ernst Mach's unfinished text *Die Prinzipien der physikalischen Optik* that was purportedly written in 1913 and repudiated relativity. In 1923 Ludwig Mach wrote to Lenard to establish contact, and Lenard suggested that Ludwig's father would not have appreciated the mathematical direction relativity had taken, also conjecturing that the Mach family was of northern blood.[76] Such overtures would have meant nothing to Ernst Mach, but the rise of anti-Semitism changed Germany so thoroughly that Einstein was fortunate to leave it in 1933.

For Mach, ether was always only a hypothetical possibility, initially worth considering as an alternative approach to gravitation given the unsatisfactory nature of Newton's account of absolute space and the law of inertia, and gradually gaining significance with the rise of electromagnetic approaches. In contrast, Einstein's approach to electrodynamics and mechanics led him to declare the luminiferous ether superfluous in 1905, but the role of the metric tensor determining the properties of spacetime later led him to briefly highlight the possibility of thinking in terms of a general relativistic ether. While this was stimulated positively by correspondence with Lorentz, it also figured in his dialogue with critics; but Einstein's conceptual flexibility and the generosity of his search for common ground was not reciprocated. Finally, this chapter has paired its exploration of the ways Mach, Einstein and Lenard described the values informing theoretical and experimental work (which helps explain why they accepted diverse concepts of ether) with an exploration of Mach's studies of visual sensation, and especially his accounts of symmetry. My aim in doing so has been to bring our study of the aesthetics of theory into contact with elements of Mach's experimental research on aesthetics, thereby pointing to suggestive, if somewhat glancing, interrelations between discussions of imagination, economy and the role of images in physics; accounts of visual similarity and difference; and typefaces that may have served to associate kinds of physics with social identity in the critical period in which some German physicists argued for ether and against Einstein's relativity. These moments underline just some of the ways in which, as Einstein noted of intuition or *Anschaulichkeit*, attitudes towards both ether and aesthetics have been functions of time.

NOTES

1. Albert Einstein, '*Zur Elektrodynamik bewegter Körper*', *Annalen der Physik*, 17 (1905): 891–921, p. 892. I refer the original Einstein publications here; for electronic versions, see http://einsteinpapers.press.princeton.edu.
2. Goldberg and Glick provide examples of the importance of concepts of the ether in the reception of the theory, while Warwick argued persuasively for the need to relate such approaches to a richer understanding of their technical complement in the theoretical technologies of mathematical physics; see Stanley Goldberg, *Understanding Relativity: Origin and Impact of a Scientific Revolution* (Boston: Birkhäuser, 1984); Thomas F. Glick, ed., *The Comparative Reception of Relativity* (Dordrecht: D. Reidel, 1987); and Andrew Warwick, *Masters of Theory: Cambridge and the Rise of Mathematical Physics* (Chicago: University of Chicago Press, 2003).
3. Albert Einstein, *Äther und Relatvitätstheorie* (Berlin: Julius Springer, 1920), p. 15.
4. Ludwik Kostro, 'An Outline of the History of Einstein's Relativistic Ether Concept', in Jean Eisenstaedt and Anne J. Kox, eds., *Studies in the History of General Relativity* (Boston: Birkhäuser, 1992); Ludwik Kostro, *Einstein and the Ether* (Montreal: Apeiron, 2000).
5. Albert Einstein, '*Dialog über Einwände gegen die Relativitätstheorie*', *Die Naturwissenschaften* 6 (1918): 697–702, p. 701.
6. Ernst Mach, *Die Mechanik in ihrer Entwickelung historisch-kritisch dargestellt* (Leipzig: F.A. Brockhaus, 1883).
7. Mach, *Mechanik*, p. v.
8. Mach, *Mechanik*, pp. 175, 177.
9. Mach, *Mechanik*, p. 180.
10. Mach, *Mechanik*, p. 5.
11. For such an analysis, see James W. McAllister, *Beauty and Revolution in Science* (Ithaca, NY: Cornell University Press, 1996), pp. 41–4, 185–7 on symmetry.
12. Mach, *Mechanik*, pp. 175–6.
13. Mach, *Mechanik*, p. 181.
14. Mach, *Mechanik*, p. 185.
15. Mach, *Mechanik*, p. 203.
16. Mach, *Mechanik*, p. 204.
17. Ernst Mach, '*Über die Definition der Masse*', *Repertorium für physikalische Technik, für mathematische und astronomische Instrumentenkunde (Carl's Repertorium der Physik)* 4 (1868): 355–9; Ernst Mach, *Die Geschichte und die Wurzel des Satzes von der Erhaltung der Arbeit* (Prague: J.G. Calve, 1872), pp. 50–4.

18. Mach, *Mechanik*, pp. 208, 209.
19. Mach, *Mechanik*, p. 214.
20. Mach, *Mechanik*, p. 215.
21. Mach, *Mechanik*, pp. 215 (on the medium) and 216–22. See Julian B. Barbour, 'General Relativity as a Perfectly Machian Theory', in Julian B. Barbour and Herbert Pfister, eds., *Mach's Principle: From Newton's Bucket to Quantum Gravity* (Boston: Birkhäuser, 1995), 215–18.
22. Norton presents a strong reading that he recognises has not been shared by the majority of Mach's readers who, like Einstein, took Mach to advocate the need for a new approach to inertia and gravitation. See John D. Norton, 'Mach's Principle Before Einstein', in Barbour and Pfister, *Mach's Principle*, 9–57.
23. Mach, *Mechanik*, p. 221.
24. Giora Hon and Bernard R. Goldstein, *From Summetria to Symmetry: The Making of a Revolutionary Scientific Concept* (Dordrecht: Springer, 2008); Shaul Katzir, 'The Emergence of the Principle of Symmetry in Physics', *Historical Studies in the Physical and Biological Sciences* 35 (2004): 35–65; Katherine Brading and Elena Castellani, 'Introduction', in Katherine Brading and Elena Castellani, eds., *Symmetries in Physics: Philosophical Reflections* (Cambridge: Cambridge University Press, 2003), 9–11.
25. Giora Hon and Bernard R. Goldstein, 'Symmetry and Asymmetry in Electrodynamics from Rowland to Einstein', *Studies in History and Philosophy of Modern Physics* 37 (2006): 635–60.
26. Arthur I. Miller, *Insights of Genius: Imagery and Creativity in Science and Art* (Cambridge, MA: MIT Press, 2000).
27. Mach, *Mechanik*, p. 9.
28. Mach, *Mechanik*, p. 9.
29. Mach, *Mechanik*, p. 10. Note that Archimedes did not use a concept of symmetry but equilibrium, and the instinctive approach Mach described anachronistically reveals his own concerns, as argued in Hon and Goldstein, *Summetria*, pp. 39–44.
30. Mach, *Mechanik*, p. 282.
31. Ernst Mach, '*Über das Sehen von Lagen und Winkeln durch die Bewegung des Auges*', *Sitzungsberichte der Mathematisch-naturwissenschaftlichen Classe der Kaiserlichen Akademie der Wissenschaften Wien* 43 (1861): 215–24; Ernst Mach, '*Bemerkungen über die Entwicklung der Raumvorstellungen*', *Zeitschrift für Philosophie und philosophische Kritik* N.F. 49 (1866): 227–32; Ernst Mach, '*Über die physikalische Bedeutung der Gesetze der Symmetrie*', *Lotos. Zeitschrift für Naturwissenschaften* 21 (1871): 139–47; Ernst Mach, *Beiträge zur Analyse der Empfindungen* (Jena: Fischer, 1886); English translations of Mach's lectures were published as 'Why Has Man Two Eyes?' and 'On Symmetry' in Ernst Mach, *Popular Scientific Lectures*, trans. Thomas J. McCormack (Chicago: The Open Court Publishing Company, 1894), 66–88 and 89–107, respectively.
32. Mach, *Analyse der Empfindungen*, p. 48.
33. Mach, *Analyse der Empfindungen*, p. 50.
34. For recent scientific work on these issues, often drawing on similar materials, see Arthur P. Shimamura and Stephen E. Palmer, *Aesthetic Science: Connecting Minds, Brains, and Experience* (Oxford: Oxford University Press, 2012).
35. Mach, 'Symmetry', p. 90.
36. Alexandra Hui, 'The Aesthetics of Attention: Ernst Mach's Accomodation Experiments, His Psychophysical Musical Aesthetics, and His Friendship with Eduard Kulke', in *The Psychophysical Ear: Musical Experiments, Experimental Sounds, 1840–1910* (Cambridge, MA: MIT Press, 2013), 89–122.

37. Mach, 'Why Has Man Two Eyes?', p. 82.
38. Ernst Mach, *The Analysis of the Sensations, and the Relation of the Physical to the Psychical*, trans. C.M. Williams, from 5th German edn (Chicago: Open Court, 1914), p. 309; Hui, *The Psychophysical Ear*, p. 120.
39. Ernst Mach, *Die Mechanik in ihrer Entwickelung historisch-kritisch dargestellt* (Leipzig: F.A. Brockhaus, 3rd edn,1897), pp. 232–7, and ix–x and 253–6 on Hertz and the medium.
40. Ernst Mach, *Die Mechanik in ihrer Entwickelung historisch-kritisch dargestellt* (Leipzig: F.A. Brockhaus, 4th edn, 1901), p. 199.
41. Ernst Mach, *Die Mechanik in ihrer Entwickelung historisch-kritisch dargestellt* (Leipzig: F.A. Brockhaus, 6th edn, 1908), p. 257.
42. Gereon Wolters, 'Mach and Einstein, or, Clearing Troubled Waters in the History of Science', in Christoph Lehner, Jürgen Renn and Matthias Schemmel, eds., *Einstein and the Changing Worldviews of Physics* (Berlin: Springer, 2012), 39–57, pp. 48–9.
43. Renn's valuable study is too centred on Einstein to give a clear understanding of Mach's general contribution; see Jürgen Renn, 'The Third Way to General Relativity: Einstein and Mach in Context', in Jürgen Renn and Matthias Schemmel, eds., *The Genesis of General Relativity, Volume 3: Gravitation in the Twilight of Classical Physics* (Berlin: Springer, 2007), 21–76.
44. John D. Norton, 'How Hume and Mach Helped Einstein Find Special Relativity', in M. Dickson and M. Domski , eds., *Discourse on a New Method: Reinvigorating the Marriage of History and Philosophy of Science* (Chicago and La Salle, IL: Open Court, 2010), 359–87.
45. Einstein, *'Elektrodynamik bewegter Körper'*, p. 891.
46. Gerald Holton, *Thematic Origins of Scientific Thought: Kepler to Einstein* (Cambridge, MA: Harvard University Press, 1973), p. 366.
47. Hon and Goldstein, 'Symmetry and Asymmetry in Electrodynamics from Rowland to Einstein', pp. 650–6.
48. Einstein, *'Elektrodynamik bewegter Körper'*, pp. 891–2.
49. Richard Staley, *Einstein's Generation: The Origins of the Relativity Revolution* (Chicago: University of Chicago Press, 2008), pp. 241–9 and 294–344.
50. Albert Einstein, *'Grundgedanken und Methoden der Relativitätstheorie in ihrer Entwicklung dargestellt'*, in *The Collected Papers of Albert Einstein, Volume 7: The Berlin Years: Writings, 1918–1921*, ed. Michel Janssen, Robert Schulmann, József Illy, Christoph Lehner and Diana Kormos Buchwald (Princeton: Princeton Univ. Press, 2002), 245–81; Miller, *Insights*, pp. 44–7, 313–19.
51. Mach, *Mechanik*, p. 191; Wolters, 'Mach and Einstein', pp. 43–4; Richard Staley, 'Ernst Mach on Bodies and Buckets', *Physics Today* 66 (2013): 42–47, pp. 46–7. In his 1916 obituary of Mach, Einstein noted with some wonder how close Mach was to demanding a general theory of relativity, almost fifty years earlier: A. Einstein, 'Ernst Mach', *Physikalische Zeitschrift* 17 (1916): 101–4.
52. Julian B. Barbour, 'Einstein and Mach's Principle', in Renn and Schemmel, eds. *Gravitation*, 596–604; Barbour and Pfister, *Mach's Principle*.
53. H. A. Lorentz, *'Alte und neue Fragen der Physik'*, *Physikalische Zeitschrift* 11 (1910): 1234–57, p. 1236.
54. H. A. Lorentz to A. Einstein, 6 June 1916, in Albert Einstein, *The Collected Papers of Albert Einstein, Volume 8: The Berlin Years: Correspondence, 1914–1918. Part A: 1914–1917*, ed. A.J. Kox, Martin J. Klein, Robert Schulmann and Józef Illy (Princeton: Princeton University Press, 1998), 295–9. For a much more detailed study, see Kostro, *Einstein and the Ether*.
55. A. Einstein to H. A. Lorentz, 16 June 1916, in Einstein, *Collected Papers, Vol. 8A*, 299–301, p. 300.
56. For a study of their responses to the ether, see Milena Wazeck, *Einstein's Opponents: The Public Controversy about the Theory of Relativity in the 1920s*, trans. Geoffrey S. Koby (Cambridge: Cambridge Univ. Press, 2014), pp. 138–57.

57. Ernst Gehrcke, *'Zur Kritik und Geschichte der neuren Gravitationstheorien'*, *Annalen der Physik*, 356 (1916): 119–24, p. 124, as quoted in Wazeck, *Einstein's Opponents*, pp. 276.
58. Einstein, *'Dialog'*, p. 697.
59. Einstein, *'Dialog'*, p. 698.
60. Einstein, *'Dialog'*, pp. 700–1.
61. Einstein, *'Dialog'*, p. 701.
62. Einstein, *'Dialog'*, p. 702.
63. Einstein, *Äther*, p. 15.
64. E. Gehrcke to O. Kraus, 7 December 1920, quoted in Wazeck, *Einstein's Opponents*, p. 144.
65. Klaus Hentschel, *Interpretationen und Fehlinterpretationen der speziellen und der allgemeinen Relativitätstheorie durch Zeitgenossen Albert Einsteins* (Basel: Birkhäuser, 1990); Wazeck, *Einstein's Opponents*.
66. Michel Janssen, Robert Schulmann, József Illy, Christoph Lehner and Diana Kormos Buchwald, 'Einstein's Encounters with German Anti-Relativists', in Einstein, *The Collected Papers of Albert Einstein, Volume 7*, 101–13; Jeroen Van Dongen, 'Reactionaries and Einstein's Fame: "German Scientists for the Preservation of Pure Science", Relativity, and the Bad Nauheim Meeting', *Physics in Perspective* 9 (2007): 212–30.
67. S. Hirzel Verlag, *'S. Hirzel Leipzig, Träger des Nobelpreises für Physik für das Jahr 1918, Dr. Max Planck, für das Jahr 1919, Dr. Johannes Stark'*, *Physikalische Zeitschrift* 20 (1919): iv.
68. Staley, *Einstein's Generation*, pp. 324–8.
69. S. Hirzel Verlag, *'Verlag von S. Hirzel in Leipzig soeben erschien in neuer vermehrter Auflage, Über Relativitätsprinzip, Äther, Gravitation, von P. Lenard, Preis 5 Mark'*, *Physikalische Zeitschrift* 21/13 (1920): ii.
70. S. Hirzel Verlag, *'S. Hirzel in Leipzig soeben neu erschienen: Änderungen der Strukture und des Spektrums chemischer Atome, Nobelvortrag, gehalten am 3. Juni 1919 in Stockholm, von Dr. Johannes Stark, Professor der Physik an der Universität Greifswald, Preis 2.25 Mark'*, *Physikalische Zeitschrift* 21/19 (1920): v; S. Hirzel Verlag, *'Literatur über Einsteins Relativitätstheorie aus dem Verlage von Julius Springer in Berlin W9'*, *Physikalische Zeitschrift* 21/19 (1920): v.
71. Adolf Reinecke, *Die deutsche Buchstabenschrift: Ihre Entstehung und Entwickelung, ihre zweckmässigkeit und völkische Bedeutung* (Leipzig-Borsdorf: A. Hasert u.C., 1910), pp. 264–5.
72. Silvia Hartmann, *Fraktur oder Antiqua: Der Schriftstreit von 1881 bis 1941* (Frankfurt am Main: P. Lang, 1998), p. 89.
73. S. Hirzel Verlag, *'Verlag von S. Hirzel in Leipzig. Anfang Januar erschient Über Relativitätsprinzip, Äther, Gravitation von P. Lenard, Heidelberg, Dritte Auflage mit einem Zusatz, betreffend die Nauheimer Diskussion. Preis 5 Mark'*, *Physikalische Zeitschrift* 21/23/24 (1920): v.
74. *'Allgemeine Diskussion über Relativitätstheorie'*, *Physikalische Zeitschrift*, 21 (1920): 666–8, p. 666.
75. Janssen et al., 'Einstein's Encounters', p. 109.
76. Gereon Wolters, *Mach I, Mach II, Einstein und die Relativitätstheorie: Eine Fälschung und ihre Folgen* (Berlin: Walter de Gruyter, 1987), pp. 368–9. A similar generational difference has been noted in Malachi Haim Hacohen, 'The Culture of Viennese Science and the Riddle of Austrian Liberalism', *Modern Intellectual History* 6 (2009): 369–96.

11

Umberto Boccioni's *Elasticity*, Italian Futurism and the Ether of Space

Linda Dalrymple Henderson

11.1 INTRODUCTION

For most of the last century, art historians studying modern art operated in a vacuum, with little knowledge of the scientific backdrop for the early decades of the twentieth century, when so many crucial artistic innovations were made. The 'ether physics' of the late nineteenth and early twentieth century had been largely occluded by the rise of Einstein (and relativity theory) to celebrity status after the announcement of the 1919 eclipse expedition's confirmation of one of his postulates.[1] Within art history, gaining a clearer understanding of the layperson's world view in the first two decades of the century was made even more difficult by the emergence in the 1940s of the myth of a connection between Pablo Picasso's cubism and Einstein's relativity theory—a discourse intended primarily to validate cubism in some way.[2] That phenomenon effectively closed off curiosity about early twentieth-century science as popularly known and, hence, knowledge of the vital role played by the ether of space in the cultural imagination of this period.[3] The present book, which brings together scholarship on the ether in the history of science, thus provides a critical new resource for scholars of modernism across the humanities.[4]

In a number of publications over the last twenty-plus years, I have argued for the importance for artists of the ether and various scientific discoveries and developments in the late nineteenth century, such as the X-ray, radioactivity, the electron and wireless telegraphy, as will be discussed further.[5] The present chapter focuses on Umberto Boccioni's 1913 painting *Elasticity* and explores his response to the ether in both its scientific and its occult contexts, which were often closely linked in this era.[6] Along with Boccioni, Wassily Kandinsky was the other modern artist for whom the ether was most central to the development of his theories and art.[7] But, in fact, for every modern artist—who as a painter or sculptor was necessarily engaged with the long-standing question of the nature of matter and space—the prevalence of the ether for the first two

Henderson, L.D., 'Umberto Boccioni's *Elasticity*, Italian Futurism, and the Ether of Space' in *Ether and Modernity*, edited by J. Navarro. © Oxford University Press 2018. DOI: 10.1093/oso/9780198797258.003.0011 decades of the century is highly relevant. Thus, from the invention of cubism by Picasso and Georges Braque as well as the science-inspired art of Marcel Duchamp to the emergence of the abstract styles of Kandinsky, František Kupka, and Kazimir Malevich, the ether is a critical ingredient of any history of the art of the early twentieth-century.[8]

Although I have previously suggested the centrality of the ether to Boccioni's well-known sculpture *Unique Forms of Continuity in Space* of 1913 (Fig. 11.7), this essay focuses primarily on Boccioni's later 1912 painting *Elasticity* (Fig. 11.1) and explores more fully the context for the artist's interest in the ether, including the important role of Sir Oliver Lodge's writings.[9]

Recovering the ether illuminates in important new ways both Boccioni's writings and the evolution of his artistic style during 1912–13. Although the importance of science for the futurists Boccioni and poet/founder F. T. Marinetti has been overshadowed by their adulation of technology, a closer reading of their texts reveals how central new scientific ideas, along with aspects of occultism, were in stimulating their creative imaginations.

Elasticity is a highly important, if less well-known, painting by Boccioni that reflects a critical moment in the development of his mature style and artistic theories in the latter half of 1912. Having absorbed the lessons of French cubism in fall 1911 and early 1912, the

Fig. 11.1 Umberto Boccioni, *Elasticity*, 1912. Pinacoteca Brera, Milan. (Mondadori Portfolio/Electa/Art Resource, NY)

artist now put that transparency and fluidity to new use to create a 'style of motion' focused on 'dynamic continuity'—in stark contrast to the stasis of cubism.[10] Although Boccioni himself was a rider, there was much more at stake for him in this painting than simply depicting a galloping horse in a modern technological landscape. His title *Elasticity*, which has largely eluded scholars apart from its obvious association with the necessary muscular elasticity of a horse and rider at a gallop, is an important clue. *Elasticity*, this essay argues, is a response to Boccioni's engagement with the ether and, specifically, elasticity as one of its basic characteristic of the ether, '*ce milieu élastique*' [this elastic milieu], as Emile Picard described it in his popular science book *La science moderne* of 1905.[11] This larger context for Boccioni's art complements the long-standing recognition of the artist's interest in the philosophy of Henri Bergson, who was himself a product of the late nineteenth-century era, in which ether physics was dominant.[12]

We know of Boccioni's interest in science as early as 1907, when he recorded in a December 1907 diary entry, 'How, where, when can I study all that chemistry and physics?'[13] And in the April 1910 'Futurist Painting: Technical Manifesto', Boccioni, its primary author, indicated the group's interconnected interests in science and the occult, making the first reference to X-rays by any modern artist: 'Who can still believe in the opacity of bodies since our sharpened and multiplied sensitiveness has already penetrated the obscure manifestations of the medium? Why should we forget in our creations the doubled power of our sight, capable of giving results analogous to those of the X-rays?'[14] Reiterating this theme a year later in his notes for a May 1911 lecture at the Circolo Artistico in Rome, the artist declared, 'What needs to be painted is not the visible but what has heretofore been held to be invisible, that is, what the clairvoyant painter sees.'[15]

Without fuller knowledge of Boccioni's scientific/occult contexts, however, most scholars of futurism have focused on the central iconographical elements of *Elasticity*, while also noting the fluid relationship of the horse and rider to their surroundings. Thus, for example, the pioneering futurist scholar Marianne Martin wrote of *Elasticity* in 1968: 'Boccioni has pictured the heroic new citizen, albeit in the guise of a historic knight, with horsepower (electricity) under his control. . . . Like energy in its limitless flow, he and his horse (his inspiration) are physically inseparable from their surroundings—steam, steel, electricity, the city.'[16]

In general, *Elasticity* has been viewed as an homage to the newly electrified outskirts of Milan and a sign of the futurists' enthusiastic embrace of energy, electricity and technology.[17] As Marinetti declared in 'The Founding and Manifesto of Futurism' of 1909, 'We intend to sing the love of danger, the habit of energy and fearlessness.'[18] In the February 1910 'Manifesto of the Futurist Painters', Boccioni likewise celebrated the 'iron network of speedy communications which envelopes the earth.'[19] And Marinetti in a text of 1911 imagined a future world of electrical bliss, asserting, 'With surprising ease, scientists govern masses of docile electrons. The earth, which we have long known to be entirely composed of tiny electrified particles, is now regulated like an enormous Ruhmkorff generator. Eyes and other human organs are no longer simple sensory receptors, but true accumulators of electric charge.'[20]

Fig. 11.2 Umberto Boccioni, *The City Rises*, 1910. The Museum of Modern Art, New York. (© The Museum of Modern Art/Licensed by SCALA/Art Resource, NY)

As a prelude to examining *Elasticity* further, it is useful to consider Boccioni's development of his initial futurist painting style as embodied in the triptych *States of Mind* (see Fig. 11.4 for the panel entitled *The Farewells*), a centrepiece of the group's exhibition in Paris in February 1912. Before that exhibition, Boccioni and the other futurists' painting techniques had been transformed by their encounter with cubism on a visit to Paris in October–November 1911.

Previously, his style had been characterised by works such as *The City Rises* of 1910 (Fig. 11.2), in which active brushstrokes suggest the whirling, frenzied energy of a Milanese construction site, centred on a large, almost Pegasus-like workhorse.[21] Here Boccioni employed the discrete strokes of pure colour he had adopted from Italian divisionist painters to achieve his goal of rendering 'dynamic sensation'.[22] While that paint application would remain in future works, his overall style was about to change radically.

As noted above, in the April 1910 'Futurist Painting: Technical Manifesto', Boccioni had referred specifically to X-rays, although he had no clear sense yet of their artistic implications. However, beginning in 1909, the cubist painters Picasso and Braque in Paris had been responding to the new conceptions of reality suggested by X-rays and other aspects of the early twentieth-century science—as well as the possible existence of a fourth dimension of space.[23]

As demonstrated in Picasso's *The Poet* of summer–fall 1911 (Fig. 11.3), their new mode of painting rejected modelling with visible light and shade and replaced three-dimensional perspective with multiple viewpoints. Metonymical signs identify parts of the

Fig. 11.3 Pablo Picasso, *The Poet*, 1911. The Solomon R. Guggenheim Foundation, Peggy Guggenheim Collection, Venice. (Cameraphoto Arte, Venice / Art Resource, NY)

figure and are overlaid in transparent planes assembled on an underlying grid structure. Having adopted a monochromatic palette, the cubists nonetheless activated their surfaces by means of brick-like divisionist strokes (sans colour), their own response to later stages of French neo-impressionism.

When Boccioni saw cubist paintings by Picasso, Braque and others on the futurists' fall 1911 visit to Paris, he realised the power of X-ray-like transparency and overlay as a means of combining multiple viewpoints and perceptual experiences.[24] Upon his return to Milan he transformed the three paintings of his *States of Mind* triptych he had been working on, including *States of Mind: The Farewells* (Fig. 11.4).[25]

Here the subject matter is typical of the futurist embrace of modern life: on a railway station platform with an engine belching steam, travellers, in the various panels of the triptych, are saying good-bye, departing or remaining. With his strong roots in the symbolist aesthetic that suggested specific linear configurations could convey specific meanings, Boccioni signifies the moment of farewell by the intertwining lines he had used in his earlier drawings for the project. Now, however, the swirling lines have become both smoke and curvilinear greenish passages with images of embracing couples. The impact of cubism is evident in the painting in the multiple views of the locomotive—a profile view with its stencilled number and a frontal view with its headlamp—and in the grid lines so characteristic of cubist painting. Closer examination of the painting also reveals two passenger cars (shown in outline from a higher vantage point) that extend obliquely towards the upper left and right corners of the painting, and several of the embracing green couples have one member inside the train cars' windows. Boccioni, however, endows the cubist grid with energy by conceiving of these marks as 'force lines', evoking the 'lines of force' of Michael Faraday and James Clerk Maxwell.[26]

At the time of their February 1912 Paris exhibition, Boccioni and his colleagues presented a new theoretical text, 'The Exhibitors to the Public', which stands as the first

Fig. 11.4 Umberto Boccioni, *States of Mind: The Farewells*, 1911. The Museum of Modern Art, New York. (© The Museum of Modern Art/Licensed by SCALA/Art Resource, NY)

full articulation of their theories—and further illuminates paintings like *States of Mind: The Farewells* as well as aspects of *Elasticity.* The goal of futurist painting is now described as the creation of a 'synthesis of *what one remembers* and of *what one sees*' or the 'the simultaneousness of states of mind'—exactly the effect Boccioni was now able to convey in *The Farewells.*[27] Balancing the emotional and intuitive emphasis of 'The Exhibitors to the Public' are Boccioni's overt and more subtle references to aspects of science, including 'force lines'. According to the artist, 'Every object reveals by its lines how it would resolve itself were it to follow the tendencies of its forces.'[28] He named this focus on invisible forces and energies 'physical transcendentalism' and declared, 'You must render the invisible, which stirs and lives between intervening obstacles.'[29]

Clearly, Boccioni's interest in science was not a passing fancy, but something integral to his thinking. Even before his reference to X-rays in the futurists' 'Technical Manifesto', the initial 'Manifesto of the Futurist Painters' of February 1910 had declared the group's commitment to painting 'today's life, incessantly and tumultuously transformed by victorious science'.[30] As suggested earlier, active interest in the new science was not limited to the futurists, but was, in fact, an international phenomenon among laypersons and artists in the early twentieth century.

11.2 THE SCIENTIFIC AND PHILOSOPHICAL BACKGROUNDS TO CUBISM AND FUTURISM

The interest of the general public had been drawn to contemporary science by a series of scientific discoveries and developments beginning in the 1890s: Wilhelm Röntgen's discovery of the X-ray in 1895, J. J. Thomson's identification of the electron in 1897, Marie and Pierre Curie's isolation of radioactive elements in 1898, and Ernest Rutherford's subsequent work on radioactivity and the structure of the atom. In addition, the emergence of wireless telegraphy in the late 1890s (of particular interest in Italy because of Guglielmo Marconi's activities) focused attention on the electromagnetic waves Heinrich Hertz had identified in 1888 and, in particular, on the space-filling ether, the medium in which they were understood to vibrate.[31]

All of these phenomena pointed to the existence of an invisible reality just beyond the reach of sense perception, one in which distinctions between matter and space were no longer clear cut. On the model of radioactivity (which was regularly speculated to be a universal property), matter was no longer seen as stable but as fluctuant and thought to be in the process of dematerialising back into the ether.[32] An article on radioactivity was, in fact, found in Boccioni's studio after his death in 1916.[33] The theme of the dematerialisation of matter was widely promulgated, for example, in the bestselling books of French science writer Gustave Le Bon, *L'Evolution de la matière* of 1905 and *L'Evolution des forces* of 1907.[34] At the same time, matter was discussed as possibly cohering from the ether, as in the 'electric theory of matter' propounded by Oliver Lodge and Joseph

Larmor—an idea Boccioni mentions specifically in his treatise *Pittura scultura futuriste*, which was published in 1914.[35]

'Without the slightest flaw in its infinite continuity', as Maxwell had declared, the ether also suggested a new image of space as filled with visible and invisible waves of varying frequencies coursing through it.[36] Boccioni refers directly to Hertzian waves in his writings, and Marinetti's poetic paradigm of 'wireless imagination' or the use of words without the connecting 'wires' of syntax drew its inspiration from wireless telegraphy.[37] As the essays in this volume document, the ether was far from banished by the publication of Einstein's special theory of relativity in 1905, as the narrative usually goes. It remained central to the world view of most laypersons until 1919, and a good number of scientists, including Lodge, continued to advocate for its importance. Thus, for example, Sir J. J. Thomson declared in his presidential address before the British Association for the Advancement of Science in 1909, 'The ether is not a fantastic creation of the speculative philosopher; it is as essential to us as the air we breathe.... The study of this all-pervading substance is perhaps the most fascinating duty of the physicist.'[38]

As previously discussed, cubists had been the first painters to respond directly to the new realities beyond sense perception suggested by contemporary science. And Boccioni had found in works like *The Poet* (see Fig. 11.3) a model for his transparent overlay of multiple perceptual experiences to create the 'simultaneousness of states of mind' in *States of Mind: The Farewells*. At this stage, however, his forms (the locomotive, in particular) were not as fluid as those of Picasso and Braque, where the boundaries of objects truly disappear. With their divisionist brushstrokes and interpenetrating space and form, cubist paintings effectively suggested the dematerialisation of matter, as discussed by Le Bon.[39] Here the radioactivity-like effect is akin to what anyone could readily observe in the popular parlour toy invented by Crookes, the spinthariscope, as emissions from a speck of radium struck the fluorescent screen within, producing flashing scintillations.

Beyond the extensive discussion of X-rays, radioactivity and ether vibrations in this period, the writings of Henri Bergson, focused on reality as duration, flux and continuity, supported the theories of both cubist and futurist painters.[40] In his 1896 *Matière et mémoire*, for example, Bergson had not only cited the works of Faraday, Maxwell and Lord Kelvin but had also made such provocative statements as 'Matter thus resolves itself into numberless vibrations, all linked together in uninterrupted continuity, all bound up with each other, and traveling in every direction like shivers through an immense body.'[41] Given Bergson's own grounding in ether physics, as well as his friendship with Le Bon, it is not surprising that his ideas meshed so well with Boccioni's interest in the newest popularised conceptions of matter and ether. Indeed, the ether stood as the primary embodiment in science of the theme of continuity so central to Bergson's philosophy.[42]

As in the case of Bergson, whose *Matière et mémoire* Boccioni read in the French edition, many French-language popular scientific and occult sources were accessible to the artist, along with the texts readily available in Italy. Among the latter were the popular *Manuali Hoepli* 'pocket books', covering topics from science to the occult, as well as writings geared to a lay audience by the prominent Italian physicist Augusto Righi.[43] There was also considerable traffic between Milan and Paris, with both books and futurist artists and writers travelling between the two cities.[44] Arnaldo Ginna, an experimental cinema pioneer who would come into the futurists' milieu with his brother and fellow artist Bruno Corra, noted in 1910 that they regularly bought 'books on spiritualism and the occult from the [Parisian] publishers Dourville [*sic*] and Chacornac', including theosophical texts by 'Blavatsky and Steiner, Besant, . . . Leadbeater, Edoard [*sic*] Schuré'.[45] Against this backdrop of individuals, and ideas flowing between France and Italy, we may return to Boccioni's *Elasticity*, first looking briefly at his painting *Matter* (of summer 1912), which preceded it (Fig. 11.5).

11.3 BOCCIONI'S STYLE AND THEORIES FROM MID-1912 ONWARDS: *MATTER* AND *ELASTICITY*

Both *Matter* and *Elasticity* represent a distinct move beyond the *States of Mind* triptych and the surface grid lines that Boccioni had adopted from Picasso and Braque. Here and, especially in *Elasticity*, space itself seems to bend or fold to create Boccioni's 'force lines'. The two paintings also reflect the scientific concerns Boccioni set forth in a rarely discussed passage near the end of *Pittura scultura futuriste*, which he had finished writing by spring 1913:

> Don't forget that life resides in the unity of energy, that we are centers that receive and transmit, in such a way that we are indissolubly bound to everything. . . .
>
> Why does science have the courage to formulate hypotheses transcending the experimental, and art, which is intuition itself, still remains at the stage of making experimental copies of reality and little sentimental and nostalgic games? Why be terrified of breaking with traditional representation? The electric theory of matter, according to which matter is only energy or, in other words condensed electricity, and exists only as *force*, is a hypothesis confirming my intuition. . . .
>
> The latest scientific hypotheses, the innumerable possibilities offered by chemistry, physics, biology and all the discoveries of science, the life of the infinitesimally small, the fundamental unity of the energy that gives us life, all push us to create through [our] plastic sensibility analogies with the new and marvelous conceptions of nature.
>
> Around us pass energies that are being observed and studied; from our bodies emanate fluids of power, attraction or repulsion . . . Hertzian waves carry the feverish pulse of the races thousands of kilometers across oceans, across deserts. . . . Electrons revolve in the atom by tens of thousands, separated one from the other like the planets of the solar system and, like them, have an orbit and a speed inconceivable for us, and still the atom is invisible to our eyes and to optical instruments.[46]

Here Boccioni enumerates specifically the scientific ideas that represented 'the new and marvelous conceptions of nature' he and Marinetti embraced. He could have encountered such ideas in a variety of scientific sources, including the numerous editions of Oreste Murani's *Fisica* (in the *Manuali Hoepli* series; in its eighth edition in 1908), which included a chapter entitled 'Cathode Rays, X-Rays, and Radioactivity'. There Murani discussed the *'grandissima rapidità'* of the electrons that composed cathode rays—as well as Crookes's spinthariscope.[47] Righi, in *La moderna teoria dei fenomeni fisici: Radioattività, ioni, eletroni* (published in 1904) likewise notes the spinthariscope and discusses the probable radioactivity of all matter; there he also asserts, 'Nothing prevents us from supposing that matter, and hence all known bodies, consist of aggregations or systems of electrons.'[48]

Fig. 11.5 Umberto Boccioni, *Matter*, 1912. Private Collection. (Bridgeman-Giraudon / Art Resource, NY)

When Boccioni decided in summer 1912 to paint a portrait of his mother on her balcony and title it *Matter* [*Materia*] (Fig. 11.5), he was creating a visual manifesto of his new understanding of the nature of subatomic elements and their interrelationship with the ether.[49] Although the ether would be his primary focus in *Elasticity*, in *Matter* he nonetheless suggests its presence in the form of the glowing lines descending on his mother; these can be read as visible and invisible rays or ether vibrations—that is, what Boccioni would term the 'sensible conductor of dynamic forces'.[50] However, his central goal in *Matter* seems to have been establishing the fluidity of his mother's mass, which dissolves into its surroundings (or coheres from them), a process emphasised by the particulate light greenish-blue strokes on the surface of the canvas, often detached from any specific object. Even more so than in cubist painting, Boccioni creates an image of continuous diffusion and cohesion truly suggestive of the radioactive emissions observable in the spinthariscope. In *Matter* he realised the goal he had announced to Carlo Carrà in an April 1912 letter from Berlin, 'I'm not interested in anything but matter expressed according to myself.'[51]

If *Matter* embodied Boccioni's new understanding of matter, in *Elasticity* he turned his attention directly to the subject of the ether, or 'atmosphere', as he termed it. As of spring 1912, he had a growing interest in sculpture. 'These days I'm obsessed with sculpture,' he wrote his friend Vico Baer in March 1912, and that fall he would publish his 'Technical Manifesto of Futurist Sculpture'.[52] There he argued that 'sculpture should give life to objects by rendering their extension into space palpable, systematic, and plastic'. 'We must model the atmosphere which surrounds our objects,' he urged, rejecting 'the liquefaction of objects favored by the vision of the Impressionists' (as in Claude Monet's *Le Pont de l'Europe, Gare Saint-Lazare*, shown in Fig. 11.6).[53] It was the goal of the 'solidification of Impressionism', as he termed it, that he now took up in painting and sculpture, and, surprisingly, conceptions of the ether would be his vehicle.[54]

Elasticity, painted in fall 1912, functioned, in a way, as a laboratory for Boccioni to work out sculptural ideas about the interaction of matter and space as well as the 'style of motion' he had again declared a goal in the April 1912 sculpture 'Technical Manifesto'.[55] In the painting Boccioni depicts the muscular forms of the galloping horse flowing one into the other and also suggesting in places their interaction with the surrounding 'atmosphere' or, in reality, the ether. In many ways, the 'dynamic continuity' of *Elasticity* foretells the fluidity of Boccioni's drawings such as *Muscular Dynamism* of 1913 (The Museum of Modern Art, New York) and, ultimately, *Unique Forms of Continuity in Space* (Fig. 11.7).[56] In the concluding section of *Pittura scultura futuriste* (preceding the extensive quotation noted earlier) Boccioni talks overtly of the ether, declaring, 'We want to model the atmosphere, to denote the forces of objects, their reciprocal influences, the unique form of continuity in space,' equating these goals with the 'materialization of the fluid, of the etherial, the imponderable.'[57] Although here his obvious referent is his 1913 sculpture of that title, his comments are equally applicable to his first efforts in this direction in *Elasticity* and in his early sculptures. As we shall see, Boccioni's changing conception of atmosphere and increasing interest in the 'solidification of Impressionism', central themes of his writings of 1912–13, respond to new ideas about the ether itself in the late nineteenth and early twentieth centuries, particularly in the writings of Lodge.

In the 'Technical Manifesto of Futurist Sculpture', Boccioni had declared the 'complete abolition of finite lines and the contained statue'.[58] There, too, he speculated on a painterly solution to the problem of erasing the boundaries of a sculpture, considering 'the intuitive shading of white, grey, black'—a chiaroscuro effect of light/dark gradation—on the sculpture's edges.[59] Indeed, a kind of colourful chiaroscuro operates in *Elasticity*, creating the curvilinear volumes of the horse's and rider's component elements. Gone are the black 'force lines' conspicuously applied to the surface of *States of Mind*, and even the glowing rays of *Matter* are absent. Apart from the rectilinear electrical transmission towers interpenetrating the figures, dynamic, curved elements predominate. And both his curved and his straight edges suggest the bending or moulding of a palpable substance, far from the abstract, linear scaffolding of Picasso's cubist paintings. In the preface to his Paris sculpture exhibition of June–July 1913, Boccioni would extend his concept of 'force lines' to 'force forms', an effective description of the sense of forms being created by the elastic forces of the ether in *Elasticity.*[60]

Boccioni's 'Plastic Foundations of Futurist Sculpture and Painting' essay of March 1913 makes clear how sculptural his ideas about 'atmosphere' had become. Celebrating science in his first sentence ('Our constructive idealism has taken its laws from the new certainties given us by science'), he continues,

> When I say that sculpture must try and model the atmosphere, I mean that I want to suppress... all the traditional and sentimental values concerning atmosphere, the recent naturalism which veils objects, making them diaphanous or distant like a dream, etc., etc. For me atmosphere is like a materiality that exists between objects, distorting plastic values. Instead of making it float overhead like a puff of air (because culture taught me that atmosphere is intangible or made of gas, etc.), I feel it, seek it, seize hold of it and emphasize it by using all the various effects which light, shadows and streams of energy have upon it. In this way I create the atmosphere! ...
>
> Now this *tangible measuring* of what formerly appeared to be empty space, this clear superimposition of new strata on what we call real objects and the shapes that determine them—this new aspect of reality is one of the foundations of our painting and sculpture. It should now be clear, then, why endless lines and currents emanate from our objects, making them live in the environment which has been created by their vibrations.
>
> The *distances* between one object and another are not just empty spaces, but are occupied by material continuities made up of varying intensities, continuities which we reveal with perceptible lines that do not correspond to any photographic truth. This is why in our paintings we do not have just objects and empty spaces, but only a greater and lesser intensity and solidity of space.
>
> With this, what I meant by referring to the *solidification* of Impressionism becomes abundantly clear.[61]

Examining Boccioni's larger cultural context more closely—from popular scientific texts to writings by occultists responding to the ether and physics more generally—is the critical step necessary for interpreting his developing theories and art from mid-1912 onwards, including his preoccupation with the 'solidification of Impressionism'. The occult aspect of this context is particularly important, since both spiritualism and theosophy attracted eager audiences in Italy in this period, including the futurists.[62] And because the ether so readily transcended divisions between science and occultism, it serves as a vital thread through the cultural history of this moment.

11.4 *ELASTICITY* IN ITS SCIENTIFIC AND OCCULT CONTEXTS

Boccioni may well have first encountered the 'marvelous' new science, including electrons and the ether, as well as the ideas of Lodge, in the milieu of theosophy. The painter's diary entries of 1908 show him grappling with issues of belief and rejecting 'the monopoly of one church', since humanity is, as he states, 'on the eve of universal brotherhood', one of the three stated 'Objects of the Theosophical Society'.[63] It was in a December 1907 diary entry that Boccioni had queried, 'How, where, when can I study all that chemistry and physics?', and the following passage from the 1907 *Manuali Hoepli* edition of Giuseppe Giordano's *Teosofica* suggests such a text as his stimulus:

> By now, anybody who keeps up with the modern scientific movement is no longer unaware that recent decades have seen a rapid succession of the most marvelous and surprising discoveries in the field of Chemistry and Physics; and that, thanks to a multitude of famous scientists (not at all Theosophists, for the most part), the concept that we had twenty years ago of chemical elements, of various forms of energy, and of matter in general, has been entirely transformed.
>
> The names of Faraday, Herz [*sic*], Lorentz, Zeemann [*sic*], Crookes, Thompson [*sic*], Rutherford, Lodge, Plücker, Hittorf, Roentgen and Lenard appear to furnish ample guarantee that the Theosophical views on the true makeup and subdivision of physical matter are not empty speculations, or the fantasies of a crowd of hallucinators or impostors. [64]

Theosophists Annie Besant and C. W. Leadbeater, whose works were regularly translated into French and Italian, drew extensively on ether physics. Like Giordano in his 1907 *Teosofia*, they regularly celebrated recent scientific discoveries, finding in the general acceptance of the ether in science a means to normalise, if also to differentiate, the central theosophical concepts of the 'astral body' and the 'ether body'.[65] Thus, Besant writes of ether physics in her discussion of the 'Etheric Double' in *Man and His Bodies* (published in 1896; translated into French in 1902):

> Modern physical science holds that all bodily changes, whether in the muscles, cells, or nerves, are accompanied by electric action, and the same is probably true of the chemical changes which are continually going on... Whenever electric action occurs ether must be present, so that the presence of the current is proof of the presence of the ether, which interpenetrates all, surrounds all; no particle of physical matter is in contact with any other particle, but each swings in a field of ether.[66]

Here Besant touches on themes highly relevant for Boccioni's *Elasticity*: muscles, electricity and ether. And the source for her erudition on electricity and ether, including the idea of particles 'swing[ing] in a field of ether', was certainly Lodge. She and Leadbeater quoted Lodge's 1907 text 'The Density of the Ether' in 'The Aether of Space', the appendix to *Occult Chemistry*, which was published in 1908; the appendix was later translated into Italian and published as *L'Etere dello spazio* that same year.[67]

Lodge was seen as a highly sympathetic figure by occultists, and his prolific writing brought his views of the centrality of the ether to a broad public. He was a member (and President, 1901–3) of the British Society for Psychical Research, which included a Milan branch, and he was well known among Italian spiritualists. As numerous scholars of Lodge and of psychical research have established, he was interested in telepathy and in the phenomena produced by spirit mediums, and he dealt with these topics in his popular writings and lectures.[68] In 1894 Lodge had participated in seances with the Italian medium Eusapia Palladino at the invitation of the French physiologist Charles Richet, editor of the *Revue Scientifique* from 1878 to 1902 and founder of the *Annales des Sciences Psychiques*; from this experience, he concluded that 'certain phenomena of this class may, under certain conditions, have a real and objective existence'.[69] Often cited in the international literature of spiritualism, Lodge was also in contact with the well-known Italian psychical researcher and criminologist Cesare Lombroso, who cited Lodge repeatedly in his publications. For example, in his *Ricerche sui fenomeni ipnotici e spiritici* (published in 1909), in a section entitled 'Radio-activity', he invokes Lodge's idea that spirits might possess an 'etherial body' (although Lombroso prefers the term 'radiant body'), allowing them to build up a 'material body capable of manifesting itself'.[70] Such a presence in occult literature, especially his openness to the theme of materialisation from the ether, would have made the British physicist's writings of particular interest to the futurists.[71]

Lodge's name featured regularly in contemporary Italian scientific sources—from mentions of his central contributions to wireless telegraphy (much discussed because of Marconi, as noted earlier) to the writings of Italian scientists like Righi and the Italian translations of English-language popular scientific books, such as Carl Snyder's *New Conceptions in Science* (published in 1903; Italian translation, 1907) and Charles Gibson's *Scientific Ideas of To-Day* (published in 1909; Italian translation, 1912).[72] Many of Lodge's own books were translated into French and, occasionally, Italian, and his active publishing of both scholarly and popular scientific texts (some 1,200 publications during his lifetime) made his ideas readily available for reprinting and excerpting in both scientific and occult texts.[73] From Boccioni's specific citing of the 'electric theory of matter', it is clear that he had somewhere encountered Lodge's ideas on electrons and the ether.

A few more words about the ether are necessary before proceeding. As the introduction to this book establishes (see Chapter 1), the ether had a long history, going back to Descartes and Newton.[74] Elasticity had been central to Newton's 'Aethereal Medium', which he described as 'exceedingly more rare and subtile than the Air' in his *Opticks* (first published in 1704).[75] With the acceptance of the wave theory of light in the 1820s, the 'luminiferous ether' became a central feature of nineteenth-century physics, and it was its quality as 'rare and subtile' that dominated popular understanding of the ether. French impressionist painters like Claude Monet (see Fig. 11.6) captured this identity of the ether as a diaphanous medium or, as critics described his paintings, an '*enveloppe lumineuse*' created by 'movement of the ether'.[76]

Yet, for physicists, the ether needed not only to be rarefied enough to pass through the smallest interstices of matter but also to have the properties of an elastic solid, in order to transmit wave vibrations. As Lodge and other physicists posited additional functions for the ether in the late nineteenth century, issues such as the rigidity and density of the ether became increasingly prominent in Lodge's own writings. While smoke rings served as one of the models for William Thomson (later Lord Kelvin) as he developed his 'vortex theory of the atom' in the 1860s, it was the smoke that remained in the public imagination as an analogue for the ether, not its potential solidity.[77]

Lodge had formulated an overview of the ether's new functions in his 1883 article in *Nature*, 'The Ether and Its Functions', describing 'one continuous substance filling all space: which can vibrate as light; which can be sheared into positive and negative electricity; which in whirls constitutes matter; and which transmits by continuity, and not by impact, every action and reaction of which matter is capable.' And, as he explained in his

Fig. 11.6 Claude Monet, *Le Pont de l'Europe, Gare Saint-Lazare*, 1877. Musée Marmottan-Claude Monet, Paris. (Scala / White Images / Art Resource, NY)

1909 *The Ether of Space*, 'The *luminiferous* it used to be called, because the conveyance of light was all it was then known to be capable of; but now that it is known to do a variety of other things also, the qualifying adjective may be dropped... we may distinguish the ultra-material luminiferous medium by calling it the Ether of Space.'[78] Lodge's many publications addressing the ether, including *Modern Views of Electricity* in its multiple editions (published in 1889, 1892 and 1907), his 1907 'The Density of the Ether' used by Besant and Leadbeater, and his February 1908 Royal Institution lecture, which was widely reprinted as 'The Ether of Space', would have made available his basic ideas on the ether, as he would then summarise them in his 1909 *The Ether of Space* and his 1913 *Continuity*.[79]

It was the new 'ultra-material' ether that engaged Boccioni, and he would certainly have read with interest Lodge's discussion of the ether in terms of elasticity, electricity, energy and even speed. According to Lodge in his *Modern Views of Electricity* (translated into French in 1907),

> In spite of immobility as a whole, [the ether] possesses that property of 'rigidity', or elastic resilience to 'shear', which is characteristic of what we would ordinarily call a solid; wherefore it would appear that it must be, throughout, in such a state of excessively fine-grained turbulent motion as would confer this property upon it... It is the gyrostatic kind of elasticity, discovered dynamically and applied ethereally by Lord Kelvin, whereby a perfect fluid can kinetically acquire some of the properties of a perfect solid.[80]

As Lodge explained further in his 1909 *The Ether of Space*, this gyrostatic motion or 'internal squirming circulation, to which every part of the ether is subject, must be carried on with a velocity of the same order of magnitude as the velocity of light'.[81]

On the theme of elasticity, Lodge clarified that rather than 'ordinary mechanical rigidity', the ether is characterised by 'electric rigidity, or electromotive elasticity'.[82] It also possesses magnetic inertia or density, and it is because of these two properties—'electric elasticity and magnetic density'—that it can transmit electromagnetic waves, travelling at the speed of light.[83] But, in Lodge's view, the ether also had a fundamental relationship to electricity and electric current. In the extreme case of a very powerful electric field, the ether could be 'sheared to and fro', producing positive and negative electric charges, or electrons.[84] In Lodge's 'electric theory of matter', these electrons would become the building blocks of matter as they acquired mass by their movement. As he summarises in *Modern Views of Electricity*, 'These immensely numerous mobile specks of electrified ether—through the forces they exert and the disturbances they originate—constitute the substratum of what appeals to our senses as matter.'[85]

In contrast to the purely light-oriented ether of the impressionists, Lodge's discussions of the ether suggest a structural field of great density as well as great energy and huge velocities, themes at the heart of futurism.[86] How provocative for Boccioni would Lodge's further discussion of elasticity in *The Ether of Space* have been:

> This is the theory then—this theory of elasticity as dependent on motion—which, in combination with the estimate of density, makes the internal energy of the ether so

> gigantic. For in every cubic millimeter of space we have . . . a mass equivalent to what, if it were matter, we should call a thousand tons, circulating internally . . with a velocity comparable to the velocity of light, and therefore containing . . . an amount of energy of the order of 10^{29} ergs . . . or what is otherwise expressible as equal to the energy of a million horse-power station working continuously for forty million years.[87]

Besant had actually quoted an earlier version of this statement in the appendix to *Occult Chemistry*, making it readily accessible.[88]

It is in this context that we can finally better understand both the form and the subject matter of Boccioni's *Elasticity*, which for so long had seemed to be only about electrical transmission and muscular elasticity. Ether is the unifying component here, filling all space and serving as 'the substratum of what appeals to our senses as matter'. This elastic ether, unlike the atmospheric spaces of impressionist painting, is a robust, energy-laden entity. Force lines are no longer marks added to the surface of the canvas, but are now 'force forms' expressed as folds and 'shears' of the ether.

Boccioni's discussion of the ether at the end of *Pittura scultura futuriste* applies equally well to *Elasticity* and *Unique Forms of Continuity in Space* (Fig. 11.7), which, as noted earlier, he connected in that text directly to the 'materialization of the fluid, of the etherial':

Fig. 11.7 Umberto Boccioni, *Unique Forms of Continuity in Space*. 1913. The Museum of Modern Art, New York. (© The Museum of Modern Art/Licensed by SCALA/Art Resource, NY)

> We ought to realize that if this infinite, this imponderable, this indivisible becomes ever more an object of inquiry and observations, that happens because in modern man a marvelous sense is beginning to awaken in the unknown depths of consciousness.
>
> Our Futurist audacity has already forced open the gates of an unknown world. We are going to create something analogous to what the physiologist Charles Richet calls *etherplastic* [eteroplastica] or *ideoplastic* [ideoplastica]. The biological mystery of mediumistic materialization is for us a *certainty*, a clarity in the intuition of physical transcendentalism and of plastic states of mind.[89]

Although Richet's term was 'heteroplastic' [*eteroplastica*], Boccioni could well have coined the term *etereplastica* or 'etherplastic' to signify his commitment to materialising the ether as he discovered it in both its occult and scientific contexts. Whether using chiaroscuro modelling to create dynamic 'force forms' in drawing and painting or sculpting them in clay ('From this day forward, clay, too, will be able to shape the atmosphere surrounding things,' he had asserted), Boccioni was seeking to model a new kind of sculptural atmosphere or ether.[90] In this context, his comments related to the 'solidification of Impressionism' in his 'Plastic Foundations' text, quoted earlier, take on new resonances: 'The *distances* between one object and another are not just empty spaces, but are occupied by material continuities made up of varying intensities.... This is why in our paintings we do not have just objects and empty spaces, but only a greater and lesser intensity and solidity of space.'[91]

'We are the Primitives of a new and completely transformed sensibility,' Boccioni had declared in the 1910 'Technical Manifesto', and he believed firmly in the ability of the sensitive artist to discern and give form to the new paradigm of reality.[92] By 1914 he had come to understand those new paradigms much more thoroughly and could assert, 'The new scientific hypotheses... lead us to create through [our] plastic sensibility analogies with the new and marvelous conceptions of nature.'[93] Central to those hypotheses were radioactivity, the space-filling ether and Lodge's 'electric theory of matter', which was grounded in the electron. The ether is, indeed, the critical, if long-missing, element of Boccioni's pursuit of the 'unique form that gives continuity in space' in painting as well as sculpture. And, just as reinserting the ether into the study of Boccioni transforms our understanding of his goals, so, too, a new phase of cultural history—of both modern art and modernism—awaits us, with the lost history of the ether as a tool.

ACKNOWLEDGEMENTS

A number of individuals contributed importantly to this essay, including my University of Texas colleague and historian of Victorian physics, Bruce Hunt. I am also very grateful to the futurism scholars Luciano Chessa, Lisa Hanstein and David Mather, as well as to Massimo Introvigne, Francesco Passanti and Elisa Valentini for assistance with resources and refining translations.

NOTES

1. See e.g. Helge Kragh, *Quantum Generations: A History of Physics in the Twentieth Century* (Princeton: Princeton University Press, 1999); on the delayed reception of relativity theory, see e.g. Thomas F. Glick, ed., *The Comparative Reception of Relativity* (Dordrecht: D. Reidel Publishing Co., 1987).
2. See Linda Dalrymple Henderson, 'Four-Dimensional Space or Space-Time: The Emergence of the Cubism–Relativity Myth in New York in the 1940s', in Michele Emmer, ed., *The Visual Mind II* (Cambridge: MIT Press, 2005), 349–97. On this subject, see also Part I of Linda Dalrymple Henderson, 'Editor's Introduction: I. Writing Modern Art and Science: An Overview; II. Cubism, Futurism, and Ether Physics in the Early Twentieth Century', *Science in Context* 17 (2004): 423–66.
3. The cubism–relativity myth also distracted attention from research on the artists for whom relativity theory was a crucial stimulus, such as the French surrealists, as Gavin Parkinson has established so convincingly. See Gavin Parkinson, *Surrealism, Art and Modern Science: Relativity, Quantum Mechanics, Epistemology* (New Haven: Yale University Press, 2008). See also Linda Dalrymple Henderson, 'Einstein and 20th-Century Art: A Romance of Many Dimensions', in Peter L. Galison, Gerald Holton and Silvan S. Schweber, eds., *Einstein for the 21st Century* (Princeton: Princeton University Press, 2007), 101–29.
4. Literature scholar Donald Benson was a pioneer in considering the relevance of the ether and an important stimulus for my work and that of my literature scholar collaborator Bruce Clarke; see Donald R. Benson, 'Facts and Fictions in Scientific Discourse: The Case of the Ether', *Georgia Review* 38 (1984): 825–37.
5. See e.g. Linda Dalrymple Henderson, *Duchamp in Context: Science and Technology and Related Works* (Princeton: Princeton University Press, 1998); Linda Dalrymple Henderson, 'Vibratory Modernism: Boccioni, Kupka, and the Ether of Space', in Bruce Clarke and Linda Dalrymple Henderson, eds., *From Energy to Information: Representation in Science and Technology, Art, and Literature* (Stanford: Stanford University Press, 2002), 126–49; Henderson, 'Editor's Introduction'; Linda Dalrymple Henderson, '*L'Ether de l'espace: Medium de l'art, de la science et de l'occultisme*', in Larisa Dryansky, Antonio Somaini and Riccardo Venturi, eds., *Repenser le medium* (Dijon: Les Presses du Réel, forthcoming).
6. For scholarship on this subject, on the transmission of scientific information by international occult periodicals, see Linda Dalrymple Henderson, 'The Forgotten Meta-Realities of Modernism: *Die Uebersinnliche Welt* and the International Cultures of Science and Occultism', *Glass Bead* 0 (2016): http://www.glass-bead.org/article/the-forgotten-meta-realities-of-modernism/.
7. See Henderson, 'Abstraction'. For Kandinsky and the ether, see Linda Dalrymple Henderson, 'Abstraction, the Ether, and the Fourth Dimension: Kandinsky, Mondrian, and Malevich in Context', in Marian Ackermann and Isabelle Malz, eds., *Kandinsky, Malewitsch, Mondrian: Der Weisse Abgrund Unendlichkeit/The Infinite White Abyss* (Düsseldorf: Kunstsammlung Nordrhein-Westfalen, 2014), 37–55 (German), 233–44 (English).
8. See Henderson, *Duchamp in Context*; Henderson, 'Vibratory Modernism'; Henderson, 'Editor's Introduction'; Henderson, '*L'Ether de l'espace*'; and Henderson, 'Abstraction'.
9. The earlier discussion of Boccioni occurs in Henderson, 'Vibratory Modernism', and Part II of Henderson, 'Editor's Introduction'.

10. For 'style of motion', see Umberto Boccioni, Carlo Carrà, Luigi Russolo, Giacomo Balla and Gino Severini, 'The Exhibitors to the Public [February 1912]', in Umbro Apollonio, ed., *Futurist Manifestos*, trans. Robert Brain, R. W. Flint, J. C. Higgitt and Caroline Tisdall (New York: Viking Press, 1973), 45–50, p. 46; for 'dynamic continuity', see Umberto Boccioni, 'Plastic Dynamism [December 1913]', in Apollonio, *Futurist Manifestos*, 92–5, p. 93. Because the volume edited by Apollonio is still the most readily available collection of futurist documents, I cite this source primarily. However, when there are problems of translation, and for texts that do not appear in it, I utilise Lawrence Rainey, Christine Poggi and Laura Wittman, eds., *Futurism: An Anthology* (New Haven: Yale University Press, 2009); this is now the best translated collection of these documents. Boccioni is considered to have been the primary author of the futurist manifestos / texts in which he participated.
11. Emile Picard, *La science moderne* (Paris: Ernest Flammarion, 1905), p. 138.
12. For Bergson and ether physics, see Henri Bergson, *Matter and Memory*, trans. Nancy Margaret Paul and W. Scott Palmer (New York: Zone Books, 1988), pp. 200–1, 208. On Boccioni and Bergson, see Brian Petrie, 'Boccioni and Bergson', *The Burlington Magazine* 116 (1974): 140–7; and Flavio Fergonzi, 'On the Title of the Painting *Materia*', in Laura Mattioli Rossi, *Boccioni's Materia: A Futurist Masterpiece of the Avant-Garde in Milan and Paris* (New York: The Solomon R. Guggenheim Museum, 2004), pp. 47–53. Bergson celebrated elasticity in a more general sense, talking of the elasticity of memory ('capable of expanding more and more'), and futurist founder F. T. Marinetti, responding to Bergson, celebrated the 'dynamism of our elastic consciousness' being realised in futurist free verse. See Bergson, *Matter and Memory*, p. 104; see also F. T. Marinetti, *Le Futurisme* (Paris: Sansot, 1911), p. 90.
13. Umberto Boccioni, diary entry, 21 December 1907, in Ester Coen, *Umberto Boccioni* (New York: The Metropolitan Museum of Art, 1988), p. 257.
14. Umberto Boccioni, Carlo Carrà, Luigi Russolo, Giacomo Balla and Gino Severini, 'Futurist Painting: Technical Manifesto [April 1910]', in Apollonio, *Futurist Manifestos*, 27–31, p. 28.
15. Umberto Boccioni, 'Selected Notes for a Lecture on Futurist Painting', in Coen, *Boccioni*, p. 239; appended to Umberto Boccioni, 'Lecture before the Circolo Artistico, Rome, May 29, 1911', in Coen, *Boccioni*, 231–9.
16. Marianne W. Martin, *Futurist Art and Theory 1909–1915* (Oxford: Clarendon Press, 1968), p. 153.
17. See e.g. Maria Fratelli, *Umberto Boccioni: Elasticità* (Milan: Skira, 2000).
18. F[ilippo] T[ommaso] Marinetti, 'The Founding and Manifesto of Futurism [February 1909]', in Apollonio, *Futurist Manifestos*, 19–24, p. 21.
19. Umberto Boccioni, Carlo Carrà, Luigi Russolo, Giacomo Balla and Gino Severini, 'Manifesto of the Futurist Painters [February 1910]', in Apollonio, *Futurist Manifestos*, 24–7, p. 25.
20. F. T. Marinetti, 'Electrical War', from *Le Futurisme* [1911], in Rainey et al., *Futurism*, p. 103.
21. For this basic stylistic history, see e.g. Martin, *Futurist Art and Theory*, pp. 79–137. On the social and political dimensions of futurism, including in *The City Rises*, see Christine Poggi, *Inventing Futurism: The Art and Politics of Artificial Optimism* (Princeton: Princeton University Press, 2009).
22. Italian divisionist painters, such as Gaetano Previati, responded to the neo-impressionist style first developed by Georges Seurat in France and which would also have an impact on cubism. For 'dynamic sensation', see Boccioni et al., 'Futurist Painting', p. 27; see also Boccioni et al., 'Exhibitors', p. 47.
23. See e.g. Part II of Henderson, 'Editor's Introduction'. For an overview of the stylistic development of cubism, as grounded in the artists' study of African sculpture and the painting of Paul

Cézanne, see e.g. Edward F. Fry, 'Introduction', in *Cubism* (New York: McGraw-Hill Book Co., 1966), 9–35.

24. On the Paris visit, including Boccioni's encounters with 'Salon' cubists such as Jean Metzinger and Albert Gleizes, see e.g. Martin, 'The Milanese Artists and Cubism in 1911', in *Futurist Art and Theory*, 120–37.
25. For all three *States of Mind* paintings and Boccioni's studies for them, see e.g. Coen, *Boccioni*, pp. 116–31.
26. On 'lines of force' in physics, see e.g. P. M. Harman, *Energy, Force, Matter: The Conceptual Development of Nineteenth-Century Physics* (Cambridge: Cambridge University Press, 1982), pp. 85 and 88–9. Such information would have been readily available to Boccioni in popular science books such as Oreste Murani, *Fisica* (Milan: Ulrico Hoepli, 1912), which was part of the '*Manuali Hoepli*' series discussed further in Sections 11.3 and 11.4; see e.g. pp. 453–4.
27. Boccioni et al., 'Exhibitors', in Apollonio, *Futurist Manifestos*, p. 26, p. 47.
28. Boccioni et al., 'Exhibitors', p. 48.
29. Boccioni et al., 'Exhibitors', pp. 47–8. In the catalogue for the London venue of the exhibition, Boccioni added the following comment for his 1911 painting *The Laugh*: 'The personages are studied from all sides and both the objects in front and those at the back are to be seen, all these being present in the painter's memory, so that the principle of the Roentgen rays is applied to the picture' (Boccioni, as quoted in Coen, *Boccioni*, p. 109).
30. Boccioni et al., 'Manifesto of the Futurist Painters', p. 26.
31. For these developments, see e.g. Alex Keller, *The Infancy of Atomic Physics: Hercules in His Cradle* (Oxford: Clarendon Press, 1983).
32. For examples of comments about universal radioactivity, see e.g. Sir Oliver Lodge, 'Electric Theory of Matter', *Harper's Monthly Magazine* 109 (1904), 383–9, p. 386, and Augusto Righi, *Modern Theory of Physical Phenomena: Radio-Activity, Ions, Electrons*, trans. Augustus Trowbridge (New York: The Macmillan Co., 1905), pp. 152–3.
33. For this article, '*Il misteri della radioattività*', see Luciano Chessa, *Luigi Russolo, Futurist: Noise, Visual Arts, and the Occult* (Berkeley: University of California Press, 2012), p. 235, note 16.
34. Gustave Le Bon, *L'Evolution de la matière* (Paris: Ernest Flammarion, 1905); Le Bon, *L'Evolution des forces* (Paris: Ernest Flammarion, 1907).
35. For Boccioni's reference, see Umberto Boccioni, *Pittura scultura futuriste (dinamismo plastico)* (Milan: *Edizioni Futuriste di 'Poesia'*, 1914), p. 326; see also Boccioni, *Futurist Painting Sculpture (Plastic Dynamism)*, trans. Richard Shane Agin and Maria Elena Versari (Los Angeles: Getty Research Institute, 2015), p. 155. Kandinsky likewise cited the 'electron theory—i.e., the theory of moving electricity, which is supposed completely to replace matter' in his 1911 treatise *On the Spiritual in Art*; see Wassily Kandinsky, *On the Spiritual in Art* [1911], in Kenneth C. Lindsay and Peter Vergo, eds., *Kandinsky: Complete Writings on Art* (New York: Da Capo, 1994), 114–220, p. 142.
36. Lodge quotes Maxwell in Oliver Lodge, *The Ether of Space* (New York and London: Harper & Brothers, 1909), p. 114.
37. For Boccioni's reference, see the quote on p. 202. See also F. T. Marinetti, 'Technical Manifesto of Futurist Literature [May 1912]', in Rainey et al., *Futurism*, 119–25, pp. 123–4; and Marinetti, 'Destruction of Syntax—Wireless Imagination—Words-in-Freedom [May 1913]', in Rainey et al., *Futurism*, 143–51, pp. 146–7; the latter text also appears in Apollonio, *Futurist Manifestos*, 95–106, pp. 99–100.
38. J. J. Thomson, 'Address', in *Report of the Seventy-Ninth Meeting of the British Association for the Advancement of Science* (London: John Murray, 1910), 3–29, p. 15. On Thomson's theories and

his interest in psychical research, see Jaume Navarro, *A History of the Electron: J. J. and G. P. Thomson* (Cambridge: Cambridge University Press, 2012).

39. See Le Bon, *L'Evolution de la matière*, Le Bon, *L'Evolution des forces*, and Gustave Le Bon, *'La Dématerialisation de la matière'*, *Revue Scientifique*, 5th ser., 2 (Nov. 19, 1904), 641–51; 2 (Dec. 10, 1904), 737–40. See also Gustave Le Bon, 'The Decay of Matter', *The Independent*, 26 July 1906, 183–6.
40. On cubism and Bergson, see Mark Antliff and Patricia Leighten, 'Philosophies of Time and Space', *Cubism and Culture* (New York: Thames and Hudson, 2001), 64-110. On Boccioni and Bergson, see Petrie, 'Boccioni and Bergson', and Fergonzi, 'On the Title of the Painting *Materia*'.
41. See Henri Bergson, *Matter and Memory*, trans. Nancy Margaret Paul and W. Scott Palmer (New York: Zone Books, 1988), pp. 200–1, 208.
42. See e.g. Sir Oliver Lodge, *Continuity: The Presidential Address to the British Association for 1913* (New York: G. P. Putnam's Sons, 1914).
43. A number of these books, including those by Righi, are cited in Section 11.4. Fergonzi notes, in passing, texts by Righi, Charles Gibson and Henri Poincaré as popularisations of electromagnetic theory, but moves immediately to focus on Bergson as the primary stimulus for Boccioni's *Matter* in 'On the Title of the Painting *Materia*', pp. 49–50.
44. Marinetti had been closely involved with literary circles in Paris from 1906 onward and published the founding manifesto of futurism in the Paris newspaper *Le Figaro* in 1909. Boccioni travelled to Paris for the Salon d'Automne in both 1911 and 1912 and to exhibit not only in the February 1912 futurism show also but in his own sculpture exhibition in summer 1913.
45. Arnaldo Ginna and Bruno Corra, as quoted in Germano Celant, 'Futurism and the Occult', *Artforum* 19 (1981): 36–42, p. 38.
46. Boccioni, *Pittura scultura futuriste*, pp. 327–9; Boccioni, *Futurist Painting Sculpture*, trans. Agin and Versari, pp. 155–6 (with slight variation in translation by LDH).
47. See Murani, *Fisica*, pp. 644, 651.
48. Righi, *Modern Theory of Physical Phenomena*, pp. 79, 144, 151. Lodge summed up his thinking (and earlier published statements) about electrons in Sir Oliver Lodge, *Electrons, or, The Nature and Properties of Negative Electricity* (London: George Bell, 1906), which was translated as *Sur les électrons* (Paris: Gauthier-Villars, 1906).
49. *Matter* [*Materia*] is a monumental painting, measuring 88⅝″ × 59″. While there is considerable literature on the painting, apart from a passing reference in Fergonzi, 'On the Title', other authors have not discussed it in relation to contemporary science; I first discussed the work in Henderson, 'Vibratory Modernism'. See e.g. Fergonzi, 'On the Title of the Painting *Materia*', and Laura Mattioli Rossi, 'Boccioni Between Painting and Sculpture', in Rossi, *Boccioni's Materia*, 35–45; Flavio Fergonzi, 'Umberto Boccioni, Materia, 1912, reworked 1913"' in *The Mattioli Collection: Masterpieces of the Italian Avant-Garde* (Milan: Skira Editore, 2003), 155–77; Laura Mattioli Rossi, ed., *Boccioni 1912 Materia* (Milan: Edizione Gabriele Mazzotta, 1995), and Christine Poggi, 'Dreams of Metallized Flesh: Futurism and the Masculine Body', in *Inventing Futurism*, 150–80. While Rossi has argued that *Elasticity* precedes *Matter* (see Rossi, 'Boccioni Between Painting and Sculpture' (pp. 37–38), and Fergonzi is uncertain (*Mattioli Collection*, p. 158), I follow the order adopted in Coen, *Boccioni*, pp. 156–7, with *Elasticity* following *Matter.*
50. Boccioni, *Pittura scultura futuriste*, p. 237; Boccioni, *Futurist Painting Sculpture*, trans. Agin and Versari, p. 128 (with slight variation in translation by LDH).
51. Umberto Boccioni to Carlo Carrà, mid-April 1912; quoted in Fergonzi, 'On the Title of the Painting *Materia*', p. 50.

52. Umberto Boccioni to Vico Baer, 15 March 1912; quoted in Rossi, 'Boccioni Between Painting and Sculpture', p. 35. Although the printed version of the sculpture manifesto bore the date 11 April 1912, the manifesto actually appeared only in mid-September. See Giovanni Lista, ed., *Le Futurisme: Textes et manifestes 1909–1944* (Ceyzérieu: Champ Vallon, 2015), p. 384.
53. For the first two quotations here, see Umberto Boccioni, 'Technical Manifesto of Futurist Sculpture', in Rainey et al., *Futurism*, 113–19, pp. 115, 119; for a less satisfactorily translated version, see Umberto Boccioni, 'Technical Manifesto of Futurist Sculpture', in Apollonio, *Futurist Manifestos*, 51–64, pp. 52, 65. For 'liquefaction of objects', see Boccioni et al., 'Exhibitors', p. 50, and Umberto Boccioni, Carlo Carrà, Luigi Russolo, Giacomo Balla and Gino Severini, 'The Exhibitors to the Public', in Rainey et al., *Futurism*, 105–9, p. 109.
54. Umberto Boccioni, 'Plastic Foundations of Futurist Sculpture and Painting', in Apollonio, *Futurist Manifestos*, 88–90, p. 89. 'Solidification of Impressionism' is also the title of Chapter 12 in *Pittura scultura futuriste*, 223–38.
55. Boccioni, 'Technical Manifesto of Futurist Sculpture', in Apollonio, *Futurist Manifestos*, p. 62. *Elasticity* (39⅜″ × 39⅜″) was included among the six paintings Boccioni showed at the Teatro Costanzi exhibition in Rome in February 1913, along with *Matter*; on the painting's dating, see the discussion in note 49.
56. For 'dynamic continuity', see Boccioni, 'Plastic Dynamism', in Apollonio, *Futurist Manifestos*, p. 93.
57. Boccioni, *Pittura scultura futuriste*, p. 325; see also Boccioni, *Futurist Painting Sculpture*, trans. Agin and Versari, p. 155 (with slight variation in translation by LDH).
58. Boccioni, 'Technical Manifesto of Futurist Sculpture', in Apollonio, *Futurist Manifestos*, p. 63.
59. Boccioni, 'Technical Manifesto of Futurist Sculpture', in Apollonio, *Futurist Manifestos*, p. 63. The importance of chiaroscuro modelling for the painting is clear in the *Study for 'Elasticity'* (The Museum of Modern Art, NY), illustrated in Coen, *Boccioni*, p. 157.
60. See Umberto Boccioni, 'Preface, First Exhibition of Futurist Sculpture', in Robert L. Herbert, ed., *Modern Artists on Art* (Englewood Cliffs, NJ: Prentice Hall, 1964), 47–57, p. 48; and Boccioni, 'Plastic Dynamism', in Apollonio, *Futurist Manifestos*, p. 94. Boccioni combines the two ideas as *'linee-forme-forza'* in *Pittura scultura futuriste*, p. 110.
61. Umberto Boccioni, 'Plastic Foundations of Futurist Sculpture and Painting', in Rainey et al., *Futurism*, pp. 140–1; see also Apollonio, *Futurist Manifestos*, pp. 88–9, with variation in translation.
62. See once again the comment by Arnaldo Ginna in Section 11.2. For the scholarship on this topic, see e.g. Celant, 'Futurism and the Occult'; Simona Cigliana, *Futurismo esoterico* (Naples: Liguori Editore. 2002); Chessa, *Luigi Russolo*; and Lisa Hanstein, 'Unseen Spirits? Occult Tradition in Italian Futurist Art and Theory', in Daniel Zamani, ed., 'Charming Intentions: Occultism, Magic and the History of Art', Special Issue No. 1, *Abraxis: International Journal of Esoteric Studies* (Summer 1913): 85–99.
63. Umberto Boccioni, diary entry, 22 March 1908, in Coen, *Boccioni*, p. 260. The three 'Objects of the Theosophical Society', including 'To form a nucleus of the universal brotherhood of humanity, without distinction of race, creed, sect, caste or color', were regularly printed in publications of the Theosophical Publishing Society, such as Annie Besant, *Man and His Bodies* (London: The Theosophical Publishing Society, 1900).
64. See Giuseppe Giordano, *Teosofia* (Milan: Ulrico Hoepli, 1907), p. 221. For Boccioni's 1907 diary comment, see Coen, *Boccioni*, p. 257.
65. H. P. Blavatsky had made this distinction clear in *Isis Unveiled* in 1877, identifying the ether with the anima mundi, or 'world soul', and a variety of other concepts but also claiming it to be 'the *modern* Ether; not such as is recognized by our scientists, but such as it *was* known to

the ancient philosophers'; see H. P. Blavatsky, *Isis Unveiled: A Master-Key to the Mysteries of Ancient and Modern Science and Theology*, vol. 1 (New York: J. W. Bouton, 1877), pp. 129, 134.

66. Besant, *Man and His Bodies*, p. 27.
67. See Annie Besant and C. W. Leadbeater, The Aether of Space', appendix to, see Annie Besant and C. W. Leadbeater, *Occult Chemistry: A Series of Clairvoyant Observations on the Chemical Elements* (London: Theosophical Publishing Society, 1908). in Besant and Leadbeater, *Occult Chemistry*, i–x. See also Annie Besant and C. W. Leadbeater, *L'Etere dello spazio* (Genoa: Tip. A. Ciminago, 1908). As their appendix suggests, the two authors were far more engaged with the new physics than Blavatsky had been, although they still differentiate the theosophical ether from that of science, naming it 'koilon' and interpreting matter's formation differently (pp. i–ii). Also see Sir Oliver Lodge, 'The Density of the Ether', *Philosophical Magazine* 13 (1907): 488–506.
68. On the Milan branch of the Society for Psychical Research, see Cigliana, *Futurismo esoterico*, p. 43. On Lodge and psychical research, see e.g. David B. Wilson, 'The Thought of Late Victorian Physicists: Oliver Lodge's Ethereal Body', *Victorian Studies* 15 (1971): 29–48; Janet Oppenheim, *The Other World: Spiritualism and Psychical Research in England, 1850–1914* (Cambridge: Cambridge University Press, 1985); Courtney Grean Raia, 'From Ether Theory to Ether Theology: Oliver Lodge and the Physics of Immortality', *Journal of the History of the Behavioral Sciences* 43 (2007): 18–43; Richard Noakes, 'The "World of the Infinitely Little": Connecting Physical and Psychical Realities Circa 1900', *Studies in the History and Philosophy of Science* 39 (2008): 323–34; Richard Noakes, 'Haunted Thoughts of the Careful Experimentalist: Psychical Research and the Troubles of Experimental Physics', *Studies in the History and Philosophy of Biological and Biomedical Sciences* 48 (2014): 46–56; Egil Asprem, 'Pondering Imponderables: Occultism in the Mirror of Late Classical Physics', *ARIES* 11 (2011): 129–65; and Egil Asprem, *The Problem of Disenchantment; Scientific Naturalism and Esoteric Discourse, 1900–1939* (Leiden: Brill, 2014).
69. Oliver J. Lodge, 'Experience of Unusual Psychic Phenomena Occurring in the Presence of an Entranced Person (Eusapia Paladino [*sic*]),' *Journal of the Society for Psychical Research* 6 (1894): 306–336, pp. 307–8. On this and other seances in which Lodge participated, see Raia, 'Ether Theory to Ether Theology', Noakes, 'Haunted Thoughts', and Oppenheim, *Other World*, pp. 150–1. Richet was a highly respected scientist who won the Nobel Prize in Physiology in 1913.
70. Cesare Lombroso, *After Death What? Spiritistic Phenomena and Their Interpretation* [*Ricerche sui fenomeni ipnotici e spiritici*], trans. William Sloane Kennedy (Boston: Small, Maynard & Co., 1909), pp. 187–8. Although not common in his discourse in the early years of the century, Lodge had used this language in his March 1902 presidential address to the Society for Psychical Research; see Oliver Lodge, 'Address by the President' in *Proceedings of the Society for Psychical Research*, vol. 17 (London: R. Brimley Johnson, 1901–1903), 37–58, p. 47, which was likely Lombroso's source.
71. See e.g. prominent French occultist Albert de Rochas drew on Lodge for one of the fifteen supporting 'Notes' at the conclusion of his *L'Extériorisation de la sensibilité* (Paris: Bibliothèque Chacornac, 1896). Rochas, along with numerous other French occultists, and scientists such as Le Bon and Righi figure prominently in Carlo Ballatore, '*Radioattività universale e radioattività umana*', *Ultra: Rivista Teosofica di Roma* 3 (1909): 55–72.
72. See e.g. Charles R. Gibson, *Scientific Ideas of To-Day Popularly Explained* (London: Seeley & Co., 1909), pp. 51, 84, 169 and 274. For the translations of these books, see Carl Snyder, *La nuova scienza* (Turin: Fratelli Bocca, 1907), and Charles Gibson, *Idee scientifiche d'oggi sulla natura della materia, dell'elettricità, della luce, del calore, ecc., alla portata di tutti* (Milan: L. F. Cogliati, 1912).
73. On Lodge's prolific publication of popular as well as scholarly articles and books, see Bruce J. Hunt, *The Maxwellians* (Ithaca: Cornell University Press, 1991), p. 91. The French translations included *Les théories modernes de l'électricité* (Paris: Gauthier-Villars, 1891), *Sur les électrons* (Paris:

Gauthier-Villars, 1906) and *La Vie et la matière* (Paris: Félix Alcan, 1909); the Italian books included *Vita e materia: Critica del libro del Prof. Haeckel L'enigma dell'universo* (Milan: Arts Regia, 1909).

74. On the history of ether theories, see G. N. Cantor and M. J. S. Hodge, eds., *Conceptions of Ether: Studies in the History of Ether Theories 1740–1900* (Cambridge: Cambridge University Press, 1981), including Daniel Siegel, 'Thomson, Maxwell, and the Universal Ether in Victorian Physics', 239–68. See also Hunt, *The Maxwellians*, and Bruce J. Hunt, 'Lines of Force, Swirls of Ether', in Clarke and Henderson, *From Energy to Information*, 99–113.
75. See Queries 18–24 in Sir Isaac Newton, *Opticks, or, A Treatise of the Reflections, Refractions, Inflections and Colours of Light* [The fourth edition, corrected, London, 1730] (New York: Dover Publications, 1952), pp. 349–54.
76. See Anthea Callen, *The Art of Impressionism: Painting Technique and the Making of Modernity* (London: Yale University Press, 2000), p. 187.
77. For the vortex atom, see e.g. Siegel, 'Thomson, Maxwell, and the Universal Ether in Victorian Physics', pp. 254–9.
78. Lodge, *Ether of Space*, p. 2.
79. See e.g. Lodge, 'The Density of the Ether'. For 'The Ether of Space', see e.g. Oliver Lodge, 'The Ether of Space', *The North American Review*, 187 (1908): 724–36; the lecture appeared in multiple periodicals (see *Reader's Guide to Periodical Literature* for 1905–1909 for this listing as well as the range of Lodge's popular writings, including on psychical research).
80. Lodge, *Modern Views of Electricity*, p. 319.
81. Lodge, *Ether of Space*, p. 103.
82. Lodge, *Modern Views of Electricity*, p. 320.
83. Lodge, *Modern Views of Electricity*, p. 321.
84. Lodge, *Modern Views of Electricity*, p. 322.
85. Lodge, *Modern Views of Electricity*, p. 3.
86. Lodge's statements about the ether's density must have startled his lay readers while stimulating their imaginations at the same time. Comparing ether to matter, he suggested that the ether was 'so dense that matter by comparison is like gossamer, or a filmy imperceptible mist, or a Milky Way' (Lodge, 'Ether of Space', p. 729); see also Lodge, 'The Density of the Ether'.
87. Lodge, *Ether of Space*, pp. 103 and 123.
88. Besant and Leadbeater, 'The Aether of Space', p. viii.
89. Boccioni, *Pittura scultura futuriste*, pp. 328–9; Boccioni, *Futurist Painting Sculpture*, trans. Agin and Versari, p. 156 (with slight variation in translation by LDH).
90. Boccioni, 'Technical Manifesto of Futurist Sculpture', quoted in Rossi, 'Between Painting and Sculpture', p. 40; inexplicably, this extension of the eighth 'Conclusion' Boccioni lists at the end of his manifesto appears in none of the English translations of the manifesto, except for that in Herbert, *Modern Artists on Art*, p. 56.
91. Boccioni, 'Plastic Foundations', in Rainey et al., *Futurism*, pp. 140–41; and Apollonio, *Futurist Manifestos*, p. 89, with variation in translation. See the quote associated with p. 211 note 61 for the full quotation.
92. See Umberto Boccioni, Carlo Carrà, Luigi Russolo, Giacomo Balla and Gino Severini, 'Futurist Painting: Technical Manifesto', in Rainey et al., *Futurism*, p. 67, with slight variation as in Coen, *Boccioni*, p. 231. This statement is missing in Apollonio, *Futurist Manifestos*, p. 30, which uses the later, 1912 Sackville Gallery London version of the document.
93. See again the full quotation on p. 211.

12

An Ether by Any Other Name?

Paul Dirac's Æther

Aaron Sidney Wright

12.1 INTRODUCTION

Readers of the 24 November 1951 edition of *Nature* were surely startled by the first item of the 'Letters to the Editors' column: a letter from P. A. M. Dirac under the headline 'Is There an Æther?'[1] His answer to this rhetorical question was that on his 'new theory of electrodynamics we are rather forced to have an æther'.[2] Even as the papers in this volume argue that the demise of the aether has been premature, Dirac's letter registered as a shock.[3] Paul Adrien Maurice Dirac (1902–84) was 50 years old; a Nobel laureate for his work in quantum theory (having received the prize in 1933); and the Lucasian Professor of Mathematics at Cambridge University. The word 'æther' had not appeared in his previous research.[4]

In sharp contrast to Dirac's letter, this same issue of *Nature* contained the notice that the 1951 Nobel Prize for Physics was awarded to John Cockcroft and E. T. S. Walton 'for their pioneer work on the transmutation of atomic nuclei by artificially accelerated atomic particles'. That December, Cockcroft's Nobel banquet speech made repeated analogies to the magic of genies and their bottles. He acknowledged that since 'the enormous power of nuclear forces became apparent... many people have wished perhaps that the genie and with him the scientist could be put back in the bottle and have blamed science for the troubles of our time'. Cockcroft did not enumerate these troubles, but Britain was in turmoil. Its empire was being dismantled: India had gained independence in 1947; Ireland had become a republic in 1949; uprisings were being brutally suppressed in Kenya; and decolonisation and Cold War anti-communism converged in the 'emergency' in British Malaya in 1948–60.[5] In the 'home countries', wartime food rationing continued until 1954. Cockcroft rejected blame for any 'troubles', and he rejected the possibility that the genies of science or scientists could be put back in their bottles. Instead, 'in these difficult times science can be one of the strongest shields of our Western

Wright, A.S., 'An Ether by Any Other Name? Paul Dirac's Æther' in *Ether and Modernity*, edited by J. Navarro.
 DOI: 10.1093/oso/9780198797258.003.0012

Civilisation'.[6] Cockcroft gave voice to a well-established connection between celebrations of scientific achievement and a view of history as linear and progressive.

The initial reactions to Dirac's æther letter from scientists in *Nature* can be characterised as domestication, attempting to bring Dirac's proposal into the writer's line of research, but without the æther concept. Leopold Infeld remarked that Dirac's central formulae could be integrated into Infeld's electrodynamics. Herman Bondi and Thomas Gold connected Dirac's proposal to their radical 'continuous creation' cosmological alternative to the big bang theory. Similarly, Erwin Schrödinger argued that many of the features of Dirac's new electrodynamics were found in the 1926 work of himself, Walter Gordon and Oskar Klein. Dirac made short, critical replies to each. Perhaps the most serious work on Dirac's theory was published by Banesh Hoffmann in the context of general relativity. Dirac did not endorse these commentaries, and insofar as his 'New Classical Theory of Electrons' was picked up by others, this was done without the æther.[7]

Reactions in the popular press took the opposite attitude, and heightened the anachronism of Dirac's æther: the sense of the past transported into the present. The New York *Times* 'Notes on Science' column gave a short résumé of Dirac's letter under the headline 'Ether, Abolished by Einstein, Restored'.[8] The Belgian physicist Léon Rosenfeld wrote an article for the Saturday Manchester *Guardian*, 19 January 1952: 'The Ghost of the "Aether": New Conceptions in Physics'. He gave a sweeping history of 'the concepts of matter and force', which were found in the 'culmination of an age-long dialectic process', to have a synthesis in modern physics.[9] Then, under the heading 'Professor Dirac's Aether', he wrote:

> In the usual formulation of electrodynamics any change in the state of motion of an electric charge is ascribed to a change in the 'electrotonic state' of the field (to use Faraday's old phrase). What Dirac shows, by a very beautiful analysis, is that one can identify the electrotonic state itself with the state of motion of the charges.

This was not very clear. He went on to link both Einstein's field of gravity and Dirac's field of electromagnetism with properties of space itself. But, after his grand narrative of dialectic and synthesis, Rosenfeld sputtered out. Was there an aether, or only its ghost?

> The electrotonic character of space, however, is not merely geometrical, like its gravitational curvature; it is of the nature of a velocity; and Dirac seems to think that a space which is not only bending but actually flowing deserves the more exalted name of aether.[10]

After all the heady metaphysics, a retreat to names—to nominalism—reads as a defeat, or, at least, a muddle. These popular accounts emphasised the spooky, difficult and nonlinear effect of Dirac's æther: an abolished entity restored, a ghost.

12.2 THE AETHER DURING DIRAC'S EDUCATION

In 1920, Dirac was 18 years old and one year from the completion of his first-class honours Bachelor's degree in electrical engineering at the University of Bristol. Like so many

Britons, he had been fascinated by Arthur Stanley Eddington's eclipse expedition and his energetic media campaign to popularise the results. The London *Times* famously blared on 7 November 1919, 'REVOLUTION IN SCIENCE. NEW THEORY OF THE UNIVERSE. NEWTONIAN IDEAS OVERTHROWN'.[11] In response, Dirac studied Eddington's popular book, *Space, Time, and Gravitation* and attended a course of lectures on relativity. The lectures were given by the philosopher C. D. Broad, who also published them as a book, *Scientific Thought*. In it, Broad framed the special theory of relativity in terms of the longer philosophical tradition of absolutists and relativists about space; within this frame, he thought that 'there is no important difference between the Ether and Absolute Space'. He argued that special and general relativity required modifications of both absolute and relative philosophies of space, but he maintained that there could be no final settling of the argument on empirical grounds. Broad thought that traditional views of space—represented by absolutists with a 'stagnant ether'—could always support *some* explanation for experimental results. Broad gave *aesthetic* grounds for preferring a relativist picture: the 'very complicated and artificial' absolutist theories.[12] Broad's lectures—combined with a tight job market for engineers, as soldiers returned from World War I—fuelled Dirac's interest in mathematics. Dirac stayed on at Bristol for two years, studying mathematics and living with his parents, before assembling enough scholarship funds to support a postgraduate degree at St John's College, University of Cambridge. He had hoped to study relativity under Ebenezer Cunningham, but was assigned to Ralph Fowler, a quantum theorist, instead.[13]

In their writing on the aether in the 1920s and 1930s, Cunningham and Eddington are exemplars of a specific attitude towards the aether: they were nominalists, admitting an aether just as a name for something already present in relativistic physics.[14] Eddington wrote:

> What we have here called the world [the collection of all point events] might perhaps have been legitimately called the aether; at least it is the universal substratum of things which the relativity theory gives us in place of the aether.[15]

Eddington took care to distinguish this uncertain and qualified renaming ('might perhaps have been legitimately called') of the aether from a more robust commitment to a jelly or a liquid substance in Euclidean spacetime. He dismissed those who wished to recover a pre-relativistic, Euclidean space, stating that

> it is only recently that conservative physicists, frightened by the revolutionary consequences of modern experiments, have begun to play with the idea of a pre-existing space whose properties cannot be ascertained by experiment—a metaphysical space, to which they arbitrarily assign Euclidean properties .[16]

Space, Time, and Gravitation is an exemplar of an attempt to domesticate relativity for a British audience. If a reader felt the need for an aether, Eddington offered a part of Einstein's theory which could accept the name.[17] But readers were left with no mistaking that a true rejection of relativity was relegated to frightened conservatives who played with metaphysics.[18]

Eddington was far from the only author to publish on Einstein's theory in the years after World War I.[19] Cunningham began a 1922 book review in *Nature*: 'The stream continues. Here are seven more books on relativity.' The books were in English, French and German and ranged from Herbert Dingle's popular *Relativity for All* to Jacques Becquerel's mathematical lectures at the École Polytechnique. Cunningham deployed a modesty about the philosophical conclusions which could be drawn from relativity; and he deployed this modesty to leave room for an aether in relativistic physics. He objected to Becquerel, who 'traces the history of the rise and fall and ultimate extinction (*as he seems to consider it*) of the idea of the "ether"' (emphasis added). Becquerel had claimed that 'mechanics and physics are reduced to geometry' by Einstein's theory. (On such reductions, see Chapter 4, this volume.) Cunningham commented: 'Is it not time that this statement should be examined more carefully?... When the physicist lapses into metaphysics he is apt to leave his terms undefined.'[20]

Cunningham continued this boundary work in his relativity textbook.[21] He argued that a weak version of an aether was left open as a *metaphysical* possibility if physics kept within its bounds. Like Eddington, Cunningham defined the aether in terms of its function in physical theories; even if all efforts to measure the aether 'were foredoomed to failure', these functions would remain. However, in a parallel to Broad, Cunningham maintained the prerogative for a more robust aether, even if it was out of reach of experimental confirmation or refutation. He insisted 'on *meta*physical grounds' on the ability to 'assert [the æther's] existence', but only in a very weak sense, because 'we have no means of identifying it' among an infinite array of possibilities. (This metaphysical *aether* was not quite the metaphysical *space* which Eddington decried.) Pinned between his desire for an aether and the consensus of thirty years of scientific studies of electromagnetism, optics, and astronomy, Cunningham marshalled disciplinary arguments to separate physics from *meta*physics. 'The æther, whatever its nature, may for ever remain concealed'.[22] While this concealment removed the aether from physical science, Cunningham insisted that this *un*observability made the world *safe* for an aether—for how could experiment refute its existence?[23] Cunningham's views are particularly relevant to this discussion because Dirac studied his textbook after he joined Cunningham's college at Cambridge University.[24]

Another textbook which Dirac studied was E. T. Whittaker's *Analytical Dynamics*.[25] The strategy of relabelling the aether as a part of relativity played both rhetorical and technical roles for Whittaker (1873–1956). He exhibited both uses in a 1927 lecture on general relativity to the London Mathematical Society.[26] He began by quoting Michael Faraday (1791–1867) on the connection between magnetism and gravitation. Faraday was revered for his physical intuition, and Whittaker quoted his 'strong feeling of the existence of a relation between gravity and electricity'. While Faraday's own investigations showed no effects, Whittaker claimed that the 'the true solution to Faraday's problem' was shown in the influence of gravity over electromagnetism in general relativity, such as through Eddington's eclipse observations. This was the central theme of Whittaker's rhetoric: contemporary physics was continuous with—or even vindicated—older physics:

> It is important to notice that gravity and electricity have been brought into connexion in essentially the same way as light and electricity were brought into connexion by the Maxwellian electromagnetic theory of light—namely, by postulating that the same 'æther' transmits both kinds of actions.[27]

But, in order for this æther to function within relativity, it was relabelled:

> It is true that we do not speak much of the æther nowadays, and certainly do not regard it as a quasi-material medium filling all space; but when we endow space itself ... with properties such as curvature, we are making it play the part of an æther.[28]

Whittaker retreated from by-then-untenable physical aethers. But this was a tactical retreat. It allowed him to position Faraday as a prescient master whose universal aether had survived from the 1850s through to the 1920s:

> The principle that one and the same aether ought to serve for all purposes was enunciated by Faraday himself: 'It is not at all unlikely', he said, 'that if there be an æther, it should have other uses than simply the conveyance of radiations'.[29]

In this passage, Whittaker deployed the (relabelled) aether as a rhetorical device which allowed him to cast the pronouncements of nineteenth-century men of science as correct in the modern day.[30]

Whittaker's technical results made these historical remarks concrete. He considered a curved spacetime surrounding a star, as described by Karl Schwarzschild's solution to Einstein's equations. And he supposed that an electromagnetic field existed in this curved space, outside the boundary of the star. Whittaker viewed this curved space—without material objects—as 'a medium' and so wrote down Maxwell's equations for a dielectric medium (not for empty space). Whittaker used the gravitational potentials of general relativity as the equations for this physical medium. (Einstein suggested that these potentials served as the aether in relativity.[31]) Whittaker used this formalism to study rays of light which aimed very close to the star. He was able to reproduce the small deflection of the path of a ray of light of the sort Eddington had observed in 1919. But he also found two strange possibilities for the path of light. One equation

> represents light which, coming from infinity towards the mass, is 'captured' by it, and never gets away again, but circles round it for ever; the [other] equation, on the other hand, represents luminous energy which is, and always has been, imprisoned in the immediate neighbourhood of the mass.[32]

This was a solid contribution to general relativity, which in retrospect falls in the lineage of 'black holes'.[33] Whittaker was not shy about the novelty of his findings: 'The capture and imprisonment of radiation by the intense gravitational field surrounding a point-mass is a remarkable theoretical possibility, markedly different from anything in pre-relativity physics.'[34] Whittaker domesticated the aether within relativity physics; by giving the aether a place in relativity, he was able to give fresh voice to the views of the great men of science of the nineteenth century.

12.3 THE ÆTHER IN *NATURE*

In the 1920s and 1930s, invocations of the aether were debated with a markedly different character in the pages of *Nature*. Senior physicists and astronomers such as Oliver Lodge (1851–1940), James Jeans (1877–1946) and Joseph Larmor (1857–1942) clashed over philosophical interpretations of the aether, with novel experiments or observations rarely at issue. These were often found in *Nature*'s 'Letters to the Editors' column.[35] Apart from historical discussions, a representative example from an issue of *Nature* closer to Dirac's 1951 letter was the astronomer E. A. Milne's 1943 review of Jeans's *Physics and Philosophy*, entitled 'Medicine for Philosophers'.[36]

Lodge in particular was unhappy with the contours of the aether discussion in the scientific literature (see chapters 2 and 3, this volume). Some rejected any aether at all. Others offered aethers as names for parts of relativistic physics. Cunningham's aether was avowedly metaphysical, and Jeans joined him in asserting that the aether was an ideal object (in the sense of British idealism). No, Lodge wanted a material aether, the kind of thing which could supply a mechanical explanation, like the corpuscles of the seventeenth century. In a reply to Jeans he wrote, 'The fact is I am instinctively unwilling or unable to go all the way... to think that nothing exists but mathematical abstractions. The old idea of some kind of a mechanical explanation still has its hold upon me.' But few scientists knew more than Lodge himself that no material aether proposal was consistent with the results of experiment. As a result, he looked to the future: 'I am unwilling to shut the door on future discovery.' Lodge had faith in the future of the aether. And he went beyond the simple metaphoric connection to religious belief: Lodge described his faith in the aether in terms of the Christian martyrdom in George Bernard Shaw's 1912 play *Androcles and the Lion*.[37] These statements echo the Victorian 'universalist' aether discourse. In 1890, the Irish 'Maxwellian' George Francis FitzGerald connected his aether theory to a universal theology: 'If it be true, ether, matter, gold, air, wood, brains, are but different motions.... Can we resist the conclusion that all motion is [God's] thought?'[38]

This was the character of debates about the aether in the pages of *Nature* during Dirac's rise to prominence (1924–32):[39] suffused with theology and conflicts among broad philosophical commitments to empiricism, realism and idealism. A related affair took place in the late 1930s. In 1937, Dirac published a speculative letter in *Nature* on the ratios of the physical constants of natural science and their consequences for cosmology.[40] He was quickly swept up in a philosophical-*cum*-disciplinary dispute between astronomers when Dingle accused Dirac, Eddington, Milne and others of being 'modern Aristotelians'—of clinging to metaphysics and forgetting the lessons of the great Galileo Galilei. Dirac mostly kept away from the sweeping philosophies.[41] He professed to be uncomfortable with general philosophical arguments.[42] With this framing, Dirac's choice to insert himself into the aether debate—particularly in *Nature*—is all the more strange; its arguments were orthogonal to his professed preferences.

Further, *Nature* was an unusual place for Dirac to send a publication. The magazine had never achieved the prominence among physicists as it had among other scientists, often serving for quickly published notices of new results, without full technical detail. *Nature*'s 'Letters to the Editors' column could serve as a venue for brief notices that could serve as ends in themselves for more experimental researchers such as Ernest Rutherford.[43] But Dirac made little use of such publications. Dirac's research was published most often in the *Proceedings of the Royal Society of London* (*PRSL*). For example, in 1930 Dirac published his radical 'Theory of Electrons and Protons' in the *PRSL*; ten months later, a two-page summary appeared in *Nature*.[44] However, in a prelude to 1951's 'Is There an Æther?', in 1936 Dirac sent *Nature* a short letter with a similarly provocative title: 'Does Conservation of Energy Hold in Atomic Processes?'[45]

Both the trajectory of Dirac's work and his professed aversion to philosophical debates heighten the strangeness of 'Is There an Æther?'. I suggest that Dirac's letter to *Nature* is best understood as a rhetorical manoeuvre—as purposeful anachronism.[46] Even if the aether lived in physics beyond the advent of special relativity, by the 1950s Dirac's 'Æther' came as a shock.[47] And it was not just any æther. 'Ether' and 'aether' were perfectly accepted spellings in the pages of *Nature*. Dirac's choice of the grapheme æ (ash) recalled the eighteenth-century typography of Isaac Newton's *Opticks* and the nineteenth-century typography of James Clerk Maxwell's publications in the *Philosophical Transactions of the Royal Society of London*.[48] Dirac's title accentuated the incongruity of a figurehead of modern physics—quantum theory and relativity—asking a question associated with older scientists. Further, Dirac's choice of *Nature* as venue of publication suggests he desired a broad audience of non-specialist scientists. In this sense, 'Is There an Æther?' is an example of a strongly *rhetorical* use of the aether.[49] However, Dirac's rhetoric had the opposite aim of many authors who invoke an unusual scientific term. Whereas artists, authors or inventors have used the aether as a signal of modernity (see Chapters 2, 3, 6, 8 and 11, this volume), among scientists in 1951, the aether was an anachronism. Anachronisms—whether obsolete words, or horse-drawn carriages in contemporary cities—draw attention.

12.4 AN ÆTHER FOR QUANTUM ELECTRODYNAMICS

Dirac self-fashioned as opposed to general philosophical arguments in physics. A natural inference might be that he would avoid debates over the aether. However, a reader of Dirac's papers who was unfamiliar with this self-fashioning might be surprised by this position. Throughout his career, Dirac balanced appeals to empiricist principles with appeals to formal results and aesthetic qualities. In a famous early paper, Dirac embraced the positivist attitude expressed in Werner Heisenberg's formulation of quantum mechanics.[50] According to Dirac, Heisenberg's 'very satisfactory' framework

> enables one to calculate just those quantities that are of physical importance, and gives no information about quantities . . . that one can never hope to measure experimentally.[51]

In a different mode, Dirac's 1930 'hole' theory of electrons contained a series of striking metaphysical proposals: empty space is full of an infinite sea of electrons; this sea is unobservable, that is, it does not have an infinite charge; these electrons are in strange 'negative-energy' states, where higher velocity means *lower* energy; and 'holes' in this sea behave like positively charged particles, that is, protons.[52] In an echo of Cunningham's aether, Dirac's sea did not contradict *any* measurements because it was out of reach of *all* measurements. In developing a theory of electrons *and* protons with only electrons as fundamental entities, Dirac boasted to the British Association for the Advancement of Science in 1930 that he had achieved the 'dream of the philosophers' of a world composed of a single substance.[53] The 'Science Service Washington DC Daily Science News' bulletin reported on his lecture under the headline 'New Theory of Space Fills It with Energy Lack':

> Dr. Dirac's paper was praised highly by Sir Oliver Lodge who said: 'When developed this theory will probably lead us in the direction we wish to go'.
>
> Sir Oliver also said that space is the important thing and that matter is merely incidental and accidental.[54]

By 1932, Dirac's 'hole' theory became a theory of electrons and *positrons* (particles with a positive charge and the same mass as the electron, also called anti-electrons).[55] In a series of experimental and theoretical advances, Dirac's theory was—astoundingly—matched with experiment.[56] Dirac's friend and hiking companion, the Soviet physicist Igor Tamm (1895–1971) wrote to Dirac on 5 June 1933 to congratulate him on the success of his theory. But, even after the positron was discovered, Tamm was cautious about the 'metaphysical' aspects of Dirac's work:

> One often hears people [saying] that the unobservable ne[g]ative electrification of the world, produced by the negative energy electrons, is a metaphysical notion. I personally am also inclined to think that this notion in its present form will find no place in the future structure of physics, but nevertheless I think one has to use this notion and work with it. The notion of the immovable and unobservable ether, formed by Lorentz about 1890, was also a metaphysical one and was banned from physics in some 15 years by Einstein, but the Lorentz transformation and other premises for Einstein's work would be hardly established, if one did discard [this] notion right from the beginning.[57]

Tamm's letter reflects the dominant use of the word 'metaphysical' in a pejorative sense among mid-twentieth-century physicists. In Tamm's view, such metaphysical notions as Dirac's vacuum sea and Lorentz's ether were unavoidable trappings of physical theory at their particular time. But as physics progressed, these notions 'will find no place in the future structure of physics'. Whatever Dirac's self-understanding as unphilosophical, his peers saw metaphysical entities in his work.

In the wake of Dirac's Nobel Prize-winning success (shared with Schrödinger in 1933), the 'hole' theory became a minor industry in quantum theory.[58] A circle of physicists clarified, elaborated and modified Dirac's work. Dirac's evolving picture of the vacuum was central to the 'hole' theory. What in 1930 was a perfectly smooth sea of electrons throughout all of empty space had become, by 1934, a 'polarizable' medium, with currents and concentrations of charge.[59] Because of this, the sea could no longer be unobservable. Physicists calculations became littered with senseless infinities.[60]

In the 1930s and 1940s, Dirac advanced several objections to the state of relativistic quantum electrodynamics (QED): its limited range of applicability; its questionable use of physical concepts; and its unpleasing aesthetics. Initially, Dirac's scepticism about QED was widely shared.[61] Like many physicists, Dirac was committed to the unity of nature, and concomitant unity of science (see Chapter 4, this volume).[62] The new QED worked very well for atomic physics, in which physicists studied the relationship between atomic electrons and the central nucleus (protons and, after 1932, neutrons). The natural next step for physicists was to try to extend the QED formalism to cover phenomena *within* the nucleus. They postulated fields of force to mediate the relation between protons, neutrons and the recently discovered mesons. Quickly, 'meson physics' became a dominant subject for theoretical physics, which struggled to understand the waves of interesting new results from experimental nuclear physics, cosmic ray studies and cyclotron particle accelerators.[63]

On this ground, QED faltered.[64] In a 1940 address, the American physicist J. Robert Oppenheimer was torn over 'why the theory of the mesotron, and more generally the quantum theory of fields, has failed so completely to deepen our understanding of nuclear forces and processes'. Oppenheimer's reflections were echoed by Dirac's argument in a popular article 'Quantum Mechanics and the Aether' (1954): '[QED] works only in a limited domain. The application [of QED] to mesons has met with no success at all.' Dirac's assessment of nuclear physics' successes was similar to those of other physicists. However, Dirac went farther and used the unity of science to damn all of QED:

> One is thus led to doubt the validity of the whole structure of quantum field theory with its technic for removing infinities.[65]

This strong conclusion was not generally supported by Dirac's peers. Oppenheimer's 1940 lecture moved on from the acute difficulties of mesotron (now meson) theory: 'the discovery of the mesotron... has also given us some confidence that the fundamental ideas of the theory are right; for the mesotron was a prediction... of this theory'.[66] The 'fundamental ideas' Oppenheimer upheld were precisely what Dirac sought to dislodge. As progress in QED was made, by the late 1940s, Dirac's opposition became iconoclastic.[67] I suggest that the rhetorical force of Dirac's anachronistic 'æther' was meant to draw attention to his iconoclastic critique of QED.

Unity was only part of Dirac's aesthetics of science; he famously also appealed to the mathematical beauty of successful scientific theories. After his 1933 Nobel Prize, his commitment to mathematical beauty—and its public expressions—increased in intensity.[68] (Dirac held that beauty in mathematics was like beauty in art: it was not possible to give mathematical beauty a simple definition; rather, practitioners knew it when they saw it.) By 1943, he expressed his commitment to mathematical beauty as more than a guide; beauty became a requirement: 'The [mathematical] scheme, to be acceptable, must be neat and beautiful.'[69] In 1951, Dirac applied this aesthetic standard directly to the new QED:

> Recent work by Lamb, Schwinger, Feynman and others has been very successful... but the resulting theory is an ugly and incomplete one, and cannot be considered as a satisfactory solution of the problem of the electron.[70]

Despite the practical success of post-World War II QED in finding numbers to compare with experiment, Dirac maintained his aesthetic critique of the theory.

However, Dirac's critique of QED was not wholly negative. He diagnosed as the problem the underlying non-quantum theory of electrons and electromagnetic field. From the late 1930s through the 1950s, Dirac developed proposals for new *classical* theories from which he hoped to build a QED without senseless and ugly infinities. He was not alone in this work; other theories were due to Max Born and Leopold Infeld.[71] This was the general strategy Dirac took in 1938 and again in 1951:

> The trouble of the present quantum electrodynamics should be ascribed primarily, in my opinion, not to a fault in the general principles of quantization, but to our working from a wrong classical theory. To make progress one should therefore re-examine the classical theory of electrons and try to improve on it.[72]

This re-examination was the immediate intellectual context for Dirac's 'Is There an Æther?' letter.

QED was formed from the combination of two interrelated parts: a theory of the electromagnetic field, on the Maxwellian model; and a theory of particulate electrons moving at relativistic speed, on the Lorentzian model (see Chapters 5 and 7, this volume). Lest one think that while he was rebuilding these classical theories, Dirac had abandoned his positivist commitments, in 1938 he declared:

> Our aim will be not so much to get a model of the electron as to get a simple scheme of equations which can be used to calculate all the results that can be obtained from experiment.

Further, he rejected any objection to the results of this mathematical enquiry on the basis that it might violate one's expectations about the physical world. Provided his equations described experiments—and provided they respected 'principles' such as relativity and conservation of energy—'it should not be considered an objection to the theory that it is not based on a model conforming to current physical ideas'.[73] He was aiming at something new.

Dirac's idea in 'A New Classical Theory of Electrons' (1951) was to do more with less. One group of difficulties in QED was grounded in the model of electrons from Dirac's 1928 relativistic theory: an extensionless electric charge at a point in spacetime.[74] The only shape for an electron which respects the principles of special relativity is no shape at all (see Chapter 5, this volume).[75] This leaves physicists with a singularity of charge, and mathematically manipulating this singularity leads to infinities:

> If one assumes the charge of the electron to be concentrated at a point, one gets an infinite self-energy, which is physically meaningless, and if one assumes it to be spread through a small volume, as Lorentz himself did, one gets into great complexities when one tries to treat the acceleration of an electron relativistically.

And, far from solving this problem, things seemed worse in QED. Dirac's way around this difficulty was to build an electrodynamics and electron theory without writing down a separate description of electrons *at all*. Instead, he tried to build charges out of the electromagnetic field itself.[76]

Dirac managed to use these variables to describe the motion of beams of charge, but not individual charges themselves. In his discussion of these results, Dirac expressed his view on the relationship between classical and quantum theories. His new classical theory

> does not provide a detailed description for individual electrons of the beam in the way the usual theory of electrons does. Such a detailed description is not needed when quantum phenomena are not being considered, so it *ought not to appear* in a classical theory and it is an *advantage* of the new theory not to give it.

Here 'classical' functioned as a label for physical phenomena, and for the theory that described these phenomena. It included special relativity, and was opposed to quantum phenomena, which were described by quantum theories. Dirac used this division normatively to present his limited classical theory as *superior*.[77] Dirac appealed to the classical/quantum distinction and his positivist view of theory to argue that apparent defects of his new electrodynamics were really advantages.

12.5 DIRAC'S HISTORICIST ÆTHER

Dirac's 'Is There an Æther?' began with a capsule history of twentieth-century aether theories:

> In the last century, the idea of a universal and all-pervading æther was popular as a foundation on which to build the theory of electromagnetic phenomena. The situation was profoundly influenced in 1905 by Einstein's discovery of the principle of relativity.... It was soon found that the existence of an æther could not be fitted in with relativity, and since relativity was well established, the æther was abandoned.

This capsule gives a recognisable iteration of the scientists' view of this history.[78] Dirac's letter skipped from the introduction of relativity to the development of quantum mechanics in the 1920s. He formed a historical argument that this further change in physical theory had blunted the force of earlier anti-aether arguments:[79]

> If one re-examines the question in the light of present-day knowledge, one finds that the æther is no longer ruled out by relativity, and good reasons can now be advanced for postulating an æther.[80]

Rather than Cockcroft's linear picture of genies escaped from their bottles (see Section 12.1), Dirac allowed that newer scientific theories could resuscitate cast-off scientific entities.

Dirac took as his starting point the same physical quality of aethers as Einstein had rejected in 1920: possessing a velocity. On the picture of physics in 1905, Dirac accepted that an aether with a velocity at each point in space was in conflict with relativity. But this was not the end of the story:

> This argument [against the aether] is unassailable from the 1905 point of view, but at the present time it needs modification, because we have to apply quantum mechanics to the æther. The velocity of the æther, like other physical variables, is subject to uncertainty relations.

Dirac argued that quantum mechanics could rescue the aether from relativity. He was not proposing that Einstein's interpretation of aether-drift experiments was incorrect.[81] Like his vacuum 'sea' of electrons, Dirac's æther was meant to be unobservable. This unobservability of the æther was established in the mathematical transition from classical to quantum theory. Dirac argued from an analogy to the physics of hydrogen atoms to show that, in some cases, 'a thing which cannot be symmetrical in the classical model may very well be symmetrical after quantization'.[82] Rosenfeld's characterisation of Dirac's æther as a 'ghost' had some merit (see Section 12.1). Quantum mechanics resurrected the æther, but only at the price that it be unobservable like Broad and Cunningham's metaphysical aethers.

This marked a shift in Dirac's thinking about classical and quantum theories. By 1938 he had cleared some intellectual space to propose theories 'not based on a model conforming to current physical ideas'. But he still constrained his classical theory building to mathematics which obeyed the restrictions of special relativity (see Section 12.4). The *PRSL* received 'A New Classical Theory of Electrons' on 3 July 1951. In it, a mathematical condition on the electromagnetic field shuffled excess mathematical structure into material for building electrons (a gauge-fixing condition.) This condition respected the symmetries of special relativity.[83] In his November letter to *Nature*, this condition on the electromagnetic field was made proportional to the æther velocity, thus breaking the symmetries of relativity. Nothing transpired in those four months to change Dirac's conviction in special relativity. Rather, he was allowing his classical theories to range farther from 'current physical ideas'. His goal was a quantum theory, and so long as *after quantisation* his strange theory obeyed the symmetries of relativity, he was content. For Dirac, here, 'classical' did not designate a time period but rather a range of phenomena, such as light with a limited spectrum of frequencies. The 'passage' from a classical theory to a quantum theory was a procedure which physicists applied on paper, first writing down a description of a system in Hamiltonian mechanics and then applying new quantum rules for how the symbols could be manipulated.[84] Dirac was not concerned with rehabilitating a nineteenth-century æther theory. He was proposing 'A *New* Classical Theory of Electrons' (emphasis added).

How did Dirac's æther proposal compare to Eddington's or Whittaker's? Dirac's æther was much stronger, more physical, than that of Einstein (*c.*1920), Eddington, Cunningham, or Whittaker (see Section 12.2). He was not suggesting that some part of a relativistic theory—a manifold of events or gravitational potentials—might be *called* an aether. He identified a specific formal entity—the electromagnetic potential—which represented a velocity at each point in space and would cause charges to move. In my view, the substantialness of his aether was similar to that of mid-nineteenth-century aethers.[85] Its velocity obeyed quantum probabilities; but so did the velocity of electrons. Here are the final lines of Dirac's letter to *Nature*:

> Its [the æther velocity's] physical significance in the theory is that if there is any electric charge it must flow with this velocity, and in regions where there is no charge it is the velocity with which a small charge would have to flow if it were introduced.
>
> We have now the [æther] velocity at all points of space-time, playing a fundamental part in electrodynamics. It is natural to regard it as the velocity of some real physical thing. Thus with the new theory of electrodynamics we are rather forced to have an æther.[86]

Dirac really was suggesting that something like a nineteenth-century aether could take its place in contemporary physics. If it was unobservable, it was motivated by Dirac's physics, unlike Cunningham's, and so Dirac would not consider it *meta*physical. History was a crucial element in Dirac's argument for the æther, but it was a very different manifestation than Whittaker's attempts to vindicate the British heroes of electromagnetism. In his effort to provide new foundations for QED, he argued that the development of physical ideas could reopen the conclusions of earlier science.

12.6 CONCLUSION

Through the 1950s, Dirac continued to develop his new electrodynamics and spoke at conferences on his æther theory, including the 1953 Lindau gathering of Nobel laureates.[87] But overall, this work was to be a disappointment. In the summer of 1956 he returned to Lindau. In notes for his talk, he expressed his frustration with QED's direction, and he insisted that the question of the aether was still open:

> When I was here three years ago I spoke about the possibility of reviving the aether and you may wonder what is the present position with regard to the aether. It is closely tied up with the problem of the description of the vacuum and so long as this problem remains unsolved one cannot assert anything definite about the aether.... The problem[s of QED are] so far from being solved that the question of the aether is still open.[88]

Whether or not this was an accurate assessment of QED, Dirac's bold argument failed to attract his fellow physicists to a new aetherial theory. He was not able to develop a rival to QED. John Cockcroft also spoke at the Lindau meeting, on 'Scientific and Technical Problems in the Development of Nuclear Power'. He reviewed different designs for reactors then under construction and gave optimistic estimates of their efficiency and low costs of dealing with nuclear waste, including by dumping it into the oceans. He was optimistic. He began, 'Ladies and gentlemen, we are now embarking on the very great scientific and technical adventure of making use of the fission of uranium... to produce a new source of power. In Britain, in particular, we are going to develop nuclear power on a large scale.'[89] On 17 October 1956, the first industrial–military nuclear power station was opened by Queen Elizabeth II, at Calder Hall, Cumbria, England. Since then, for better or for worse, neither the nuclear genie nor the ætherial ghost have returned from whence they came.[90]

ACKNOWLEDGEMENTS

I would like to thank Jaume Navarro and Connemara Doran for reading earlier drafts of this essay, and for their comments alongside the other participants of the 'Ether and Modernity' workshop, 29–31 March 2017, at the Donostia International Physics Centre, University of the Basque Country. I presented this material to the History of Science Working Group at Stanford University on 9 February 2017; I thank all the participants, particularly J. D. Bjorken, Francis Everett, Michael Friedman, Hannah Leblanc and Tom Ryckman. All errors are my own.

NOTES

1. It is unclear to what degree, if any, letters to the editors were reviewed before publication; see Melinda Baldwin, '"Keeping in the Race": Physics, Publication Speed and National Publishing Strategies in Nature, 1895–1939', *The British Journal for the History of Science* 47 (2014): 257–79, and Imogen Clarke, 'The Gatekeepers of Modern Physics: Periodicals and Peer Review in 1920s Britain', *Isis* 106 (2015): 70–93.
2. P. A. M. Dirac, 'Is There an Æther?', *Nature* 168 (1951): 906–7, p. 907; P. A. M. Dirac, 'A New Classical Theory of Electrons', *Proceedings of the Royal Society of London A* 209 (1951): 291–6.
3. Helge Kragh, *Dirac: A Scientific Biography* (Cambridge: Cambridge University Press, 1990), p. 200.
4. Dirac's æther work has been discussed by both technical and popular biographies; see Helge Kragh, 'Electrons and Ether', in Kragh, *Dirac*, 189–204, and Graham Farmelo, *The Strangest Man: The Hidden Life of Paul Dirac, Mystic of the Atom* (New York: Basic Books, 2009), 344.
5. These high-profile examples were widely known. I make no assertions about Cockcroft's personal views. He was a high-level administrator of Britain's nuclear activities during and after World War II; Mark L. E. Oliphant and Lord Penney, 'John Douglas Cockcroft. 1897–1967', *Biographical Memoirs of Fellows of the Royal Society* 14 (1968): 139–88.
6. John Cockcroft, 'Nobel Banquet Speech', 1951, https://www.nobelprize.org/nobel_prizes/physics/laureates/1951/cockcroft-speech.html, accessed 4 April 2017.
7. Kragh, 'Electrons and Ether'; additionally, in particle physics, Dirac's aether has been adapted in an active line of research in quantum theory: see J.D. Bjorken, 'A Dynamical Origin for the Electromagnetic Field', *Annals of Physics* 24 (1963): 174–87; Yoichiro Nambu, 'Quantum Electrodynamics in Nonlinear Gauge', *Progress of Theoretical Physics Supplement* E68 (1968), 248–53, p. 190; Don Colladay and V. Alan Kostelecký, 'Lorentz-Violating Extension of the Standard Model', *Physical Review D* 58 (1998): 116002; in experiment, David Mattingly, 'Modern Tests of Lorentz Invariance', *Living Reviews in Relativity* 8 (2005): 5–84; and Stefano Liberati, 'Tests of Lorentz Invariance: A 2013 Update', *Classical and Quantum Gravity* 30 (2013), 133001.

8. 'Notes on Science: Ether, Abolished by Einstein, Restored', *New York Times*, 9 Dec 1951.
9. Rosenfeld's Marxism was on full display. See Anja Skaar Jacobsen, 'Léon Rosenfeld's Marxist Defense of Complementarity', *Historical Studies in the Physical and Biological Sciences* 37 Supplement (2007): 3–34.
10. Rosenfeld sent Dirac a clipping 'With Kind Regards', P. A. M. Dirac Papers, series 2, box 12, folder 2, Florida State University Archives, Tallahassee, FL. This was Rosenfeld's third Saturday article for the *Guardian*, coming after 'View of the Universe', 3 March 1950, and 'The Dome of Discovery', 5 December 1951.
11. Alistair Sponsel, 'Constructing a "Revolution in Science": The Campaign to Promote a Favourable Reception for the 1919 Solar Eclipse Experiments', *The British Journal for the History of Science* 35 (2002): 439–67; Daniel Kennefick, 'Not Only Because of Theory: Dyson, Eddington, and the Competing Myths of the 1919 Eclipse Expedition', in Christoph Lehner, Jürgen Renn and Matthias Schemmel, eds., *Einstein and the Changing Worldviews of Physics*, Einstein Studies 12 (Boston: Birkhäuser, 2012): 201–32.
12. C. D. Broad, *Scientific Thought* (London: Kegan Paul, Trench, Trumber & Co., 1923); absolutists, p. 115; 'stagnant ether', pp. 116, 118, 120; aesthetics, p. 203.
13. P. A. M. Dirac, 'Recollections of an Exciting Era', in C. Weiner, ed., *History of Twentieth Century Physics*, Proceedings of the International School of Physics 'Enrico Fermi', Varenna 1972 (New York: Academic Press, 1977): 109–46; Cunningham: R. H. Dalitz, 'Another Side to Paul Dirac', in Behram N. Kursunoglu and Eugene Paul Wigner, eds., *Reminiscences about a Great Physicist: Paul Adrien Maurice Dirac* (Cambridge: Cambridge University Press, 1987): 69–92, p. 72; see also Jagdish Mehra and Helmut Rechenberg, 'Quanta and Molecules: The Quantum Theory of Planck, Einstein and Nernst', in *The Historical Development of Quantum Theory, Volume 1, Part 1* (New York: Springer, 1982), 23–154, and Farmelo, *The Strangest Man*, pp. 39–43.
14. See Andrew Warwick, *Masters of Theory: Cambridge and the Rise of Mathematical Physics* (Chicago: University of Chicago Press, 2003), 443–500.
15. Arthur Stanley Eddington, *Space, Time and Gravitation: An Outline of the General Relativity Theory* (Cambridge: Cambridge University Press, 1920), p. 187.
16. Eddington, *Space, Time and Gravitation*, p. 16.
17. This was also Einstein's position *c.*1920; see Albert Einstein, 'Ether and the Theory of Relativity', in *The Collected Papers of Albert Einstein: Volume 7: The Berlin Years: Writings, 1918–1921*, ed. Michael Janssen, Robert Schulmann, József Illy, Christoph Lehner and Diana Kormos Buchwald (Princeton University Press, 2002), 305–23.
18. Overall, Eddington was a transcendental idealist; see Thomas Ryckman, '"World Building": Structuralism and Transcendental Idealism in Eddington', in *The Reign of Relativity: Philosophy in Physics 1915–1925* (New York: Oxford University Press, 2005), 177–217.
19. Helge Kragh, 'The Steady-State Alternative', in Kragh, *Cosmology and Controversy*, 142–201.
20. Ebenezer Cunningham, 'More Books on Relativity', *Nature* 109 (1922): 770–2, p. 772.
21. Warwick, *Masters of Theory*, 399–500.
22. Ebenezer Cunningham, *Relativity, The Electron Theory, and Gravitation* (London: Longmans, Green, 2nd edn, 1921), 'foredoomed', p. 4, metaphysics, pp. 30–1, 'concealed', pp. 4–5.
23. Stanley Goldberg, 'In Defense of Ether: The British Response to Einstein's Special Theory of Relativity, 1905-1911', *Historical Studies in the Physical Sciences* 2 (1970): 89–125, pp. 113–17.
24. Dirac Papers, series 2, box 32, folder 3.

25. S. C. Coutinho, 'Whittaker's Analytical Dynamics: A Biography', *Archive for History of Exact Sciences* 68 (2014): 355–407.
26. Whittaker was well known for his history of the aether: see Bruce J. Hunt, '[Review] Edmund Whittaker. A History of the Theories of Aether and Electricity. Introduction by Arthur I. Miller', *Isis* 79 (1988): 515–16.
27. Edmund Taylor Whittaker, 'The Influence of Gravitation on Electromagnetic Phenomena', *Journal of the London Mathematical Society* S1–3 (1928): 137–44, p. 137; Daniel M. Siegel, 'Thomson, Maxwell, and the Universal Ether in Victorian Physics', in G. N. Cantor and M. J. S. Hodge, eds., *Conceptions of Ether: Studies in the History of Ether Theories, 1740–1900* (Cambridge: Cambridge University Press, 1981), 239–68.
28. Whittaker, 'The Influence of Gravitation on Electromagnetic Phenomena', p. 137; Whittaker viewed Lorentz's aether as space endowed with properties: see Edmund Taylor Whittaker, *A History of the Theories of Aether and Electricity: From the Age of Descartes to the Close of the Nineteenth Century* (London: Longmans, 1910), pp. 419ff.
29. Whittaker, 'The Influence of Gravitation on Electromagnetic Phenomena', p. 138; the same quote was used in James Clerk Maxwell, *A Treatise on Electricity and Magnetism*, ed. by W. D. Niven, vol. 2 (New York: Dover Publications, 3rd edn, [1891] 1954), p. 432, footnote.
30. This is the opposite of the strategy of distinguishing 'modern' from 'classical' physics; see Richard Staley, 'On the Co-Creation of Classical and Modern Physics', in *Einstein's Generation: The Origins of the Relativity Revolution* (Chicago: University of Chicago Press, 2008), 347–396, and Graeme Gooday and Daniel Mitchell, 'Rethinking "Classical Physics"', in Jed Z. Buchwald and Robert Fox, eds., *The Oxford Handbook of the History of Physics* (Oxford: Oxford University Press, 2013), 721–64.
31. Einstein, 'Ether and the Theory of Relativity'.
32. Whittaker, 'The Influence of Gravitation on Electromagnetic Phenomena', pp. 139–40.
33. Jean Eisenstaedt, *'Trajectoires et impasses de la solution de Schwarzschild'*, *Archive for History of Exact Sciences* 37 (1987): 275–357, p. 305.
34. NB 'pre-relativity', not 'classical'. Whittaker, 'The Influence of Gravitation on Electromagnetic Phenomena', p. 140.
35. Joseph Larmor, 'The Exploded Ether', *Nature* 115 (1925): 419; Oliver Lodge, 'Ether and Matter and Relativity', *Nature* 115 (1925): 419; Oliver Lodge, 'The Ether and Relativity', *Nature* 126 (1930): 804–5; J. H. Jeans, 'The Ether and Relativity', *Nature* 126 (1930): 877; for earlier debates, see Baldwin, 'Keeping in the Race', p. 275ff, footnote 70.
36. E. A. Milne, 'Medicine for Philosophers', *Nature* 151 (1943): 62–4; see Kragh, 'The Steady-State Alternative'.
37. Lodge, p. 804.
38. Bruce J. Hunt, *The Maxwellians* (Ithaca, NY: Cornell University Press, 1991), p. 99; see also Siegel, 'Thomson, Maxwell, and the Universal Ether in Victorian Physics'.
39. From Dirac's first scientific publications to his appointment to the Lucasian chair of Mathematics at Cambridge.
40. P. A. M. Dirac, 'The Cosmological Constants', *Nature* 139 (1937): 323.
41. Helge Kragh, 'Cosmo-Physics in the Thirties: Towards a History of Dirac Cosmology', *Historical Studies in the Physical Sciences* 13 (1982): 69–108.
42. See Dirac, 'Recollections of an Exciting Era', p. 111, 134, Donald Franklin Moyer, 'Evaluations of Dirac's Electron, 1928–1932', *American Journal of Physics* 49 (1981): 1055–62, p. 1060, and Kragh, 'Electrons and Ether'.

43. Melinda Baldwin, *Making 'Nature': The History of a Scientific Journal* (Chicago: Chicago University Press, 2015), pp. 100, 130–2, 168; Helge Kragh, *Quantum Generations: A History of Physics in the Twentieth Century* (Princeton, NJ: Princeton University Press, 1999), pp. 20–6, 142–50.
44. P. A. M. Dirac, 'A Theory of Electrons and Protons', *Proceedings of the Royal Society of London A* 126 (1930): 360–5; P. A. M. Dirac, 'The Proton', *Nature* 126 (1930): 605–6.
45. P. A. M. Dirac, 'Does Conservation of Energy Hold in Atomic Processes?', *Nature* 137 (1936): 298–9; Kragh, *Dirac*, pp. 169–74.
46. There are only scanty archival materials to firmly indicate Dirac's motivations. Few records for *Nature* exist before the 1990s; see Baldwin, 'Keeping in the Race', footnote 6.
47. Kragh, *Dirac*, p. 200.
48. e.g. Queries 19 and 20 in Isaac Newton, *Opticks, or, A Treatise of the Reflections, Refractions, Inflections and Colours of Light* (London: William Innys, 4th edn, [1704] 1730), p. 324; also see James Clerk Maxwell, 'A Dynamical Theory of the Electromagnetic Field', *Philosophical Transactions of the Royal Society* 155 (1865): 459–512, p. 460. Dirac may have been influenced by the publication of the revised edition of Whittaker's *History*, although some printings of each of the 1910 and 1951 editions did not use 'æ'. I thank Tom Ryckman for suggesting this possibility.
49. Dirac and Whittaker made rhetorical use of the aether within their scientific communication. This use establishes a continuity among technical/scientific and non-technical/popular publications which each invoke the aether. For concerns about the historiographic unity of aether concepts, see Simon Schaffer, 'The Matter of Ether', *History of Science* 20 (1982): 297–303.
50. Cathryn Carson, 'Physics as Philosophy', in *Heisenberg in the Atomic Age: Science and the Public Sphere* (Washington, DC: German Historical Institute, 2010), 60–101.
51. P. A. M. Dirac, 'On the Theory of Quantum Mechanics', *Proceedings of the Royal Society of London A* 112 (1926): 661–77, p. 667.
52. Dirac, 'A Theory of Electrons and Protons'.
53. Helge Kragh, 'Quanta and Fields', in Kragh, *Dirac*, 118–150.
54. Quoted in Aaron Sidney Wright, 'A Beautiful Sea: P. A. M. Dirac's Epistemology and Ontology of the Vacuum', *Annals of Science* 73 (2016): 225–56, p. 234.
55. P. A. M. Dirac, 'Quantised Singularities in the Electromagnetic Field', *Proceedings of the Royal Society of London A* 133 (1931): 60–72; P. A. M. Dirac, 'Nobel Lecture: Theory of Electrons and Positrons' (Nobel Media AB, [1933] 2013), http://www.nobelprize.org/nobel_prizes/physics/laureates/1933/dirac-lecture.html, accessed 22 March 2018.
56. Donald Franklin Moyer, 'Vindications of Dirac's Electron, 1932–1934', *American Journal of Physics* 49 (1981): 1120–5; Xavier Roqué, 'The Manufacture of the Positron', *Studies in History and Philosophy of Science Part B* 28 (1997): 73–129.
57. Dirac Papers, series 2, box 2, folder 8. Tamm was awarded the Nobel Prize in 1958. See also A. B. Kojevnikov, ed., *Paul Dirac and Igor Tamm Correspondence; 1, 1928–1933* (Munich: Max-Planck-Institut für Physik, 1993).
58. Kragh, 'Quanta and Fields', pp. 107–17.
59. Dalida Monti, 'Dirac's Holes Model: From Proton to Positron', *Nuncius* 10 (1995): 99–130.
60. Silvan S. Schweber, 'The Birth of Quantum Field Theory', in *QED and the Men Who Made It: Dyson, Feynman, Schwinger, and Tomonaga* (Princeton: Princeton University Press, 1994), 1–75;

Silvan S. Schweber, 'The 1930s', in Schweber, *QED*, 76– 129; Silvan S. Schweber, 'The War and Its Aftermath', in Schweber, *QED*, 130–55; Silvan S. Schweber, 'Three Conferences: Shelter Islands, Pocono, and Oldstone', in Schweber, *QED*, 156–205; Laurie M. Brown, 'Some QED Myths-in-the-Making?', *Studies In History and Philosophy of Science Part B* 27 (1996): 81–90.

61. Alexander Rueger, 'Attitudes towards Infinities: Responses to Anomalies in Quantum Electrodynamics, 1927–1947', *Historical Studies in the Physical and Biological Sciences* 22 (1992): 309–37.
62. Kragh, *Dirac*, pp. 268–90.
63. Laurie M. Brown and Lillian Hoddeson, 'The Birth of Elementary Particle Physics: 1930–1950', in Laurie M. Brown and Lillian Hoddeson, eds., *The Birth of Particle Physics* (Cambridge: Cambridge University Press, 1983), 3–36.
64. Peter Galison, 'Particles and Theories', in *How Experiments End* (Chicago: University of Chicago Press, 1987), 75–134; and Andrew Pickering, 'The Old Physics: HEP, 1945–64', in *Constructing Quarks: A Sociological History of Particle Physics* (Chicago: University of Chicago Press, 1984), 46–84.
65. P. A. M. Dirac, 'Quantum Mechanics and the Aether', *The Scientific Monthly* 78 (1954): 142–6, p. 145.
66. Quoted in Schweber, *QED*, p. 85.
67. Kragh, *Dirac*, pp. 166–7; the major result was the calculation of the 'Lamb Shift'. Silvan S. Schweber, 'The Lamb Shift and the Magnetic Moment of the Electron', in Schweber, *QED*, 206–47.
68. Helge Kragh, 'The Principle of Mathematical Beauty, in Kragh, *Dirac*, 275–92.
69. Paul A. M. Dirac, 'Quantum Electrodynamics', *Communications of the Dublin Institute for Advanced Studies. Series A* 1 (1943): 1–36, p. 2.
70. Dirac, 'A New Classical Theory of Electrons', p. 291.
71. M. Born and L. Infeld, 'Foundations of the New Field Theory', *Proceedings of the Royal Society of London A* 144 (1934): 425–51; cf. P. A. M. Dirac, 'A Reformulation of the Born-Infeld Electrodynamics', *Proceedings of the Royal Society of London A* 257 (1960): 32–43.
72. Dirac, 'A New Classical Theory of Electrons', p. 291; see also P. A. M. Dirac, 'Classical Theory of Radiating Electrons', *Proceedings of the Royal Society of London A* 167 (1938): 148–69, pp. 148–9.
73. Dirac, 'Classical Theory of Radiating Electrons', p. 149.
74. P. A. M. Dirac, 'The Quantum Theory of the Electron', *Proceedings of the Royal Society of London A* 117 (1928): 610–24.
75. At least without introducing arbitrary new assumptions; see Dirac, 'Classical Theory of Radiating Electrons', p. 148.
76. Dirac, 'A New Classical Theory of Electrons', pp. 291, 293.
77. Dirac, 'A New Classical Theory of Electrons', pp. 293, 295, 296; emphasis added.
78. Dirac, 'Is There an Æther?', p. 906; see Andrew Pickering, 'Introduction', in Pickering, Constructing Quarks, 3–20.
79. As for Kuhn, each revolution undid the last; see Suman Seth, 'Crisis and the Construction of Modern Theoretical Physics', *The British Journal for the History of Science* 40 (2007): 25–51.
80. Dirac, 'Is There an Æther?', p. 906.
81. Dirac, 'Is There an Æther?', p. 906; Kragh, *Dirac*, p. 200.
82. Dirac, 'Is There an Æther?', p. 906.

83. Dirac, 'A New Classical Theory of Electrons', p. 293. The gauge-fixing term was $A_{\mu}A^{\mu} = k^2$, where $A_{\propto}$ is the electromagnetic potential, and *k* is a constant. Dirac defined an aether four-velocity $v_{\propto}$ and then made the identification $k^{-1}A_{\mu} = v_{\mu}$. See Dirac, 'Is There an Æther?', p. 906; cf. Bjorken, 'A Dynamical Origin for the Electromagnetic Field', and Nambu, 'Quantum Electrodynamics in Nonlinear Gauge'.
84. For Dirac: Abraham Pais, 'Playing with Equations, the Dirac Way', in Peter Goddard , ed., *Paul Dirac: The Man and His Work* (Cambridge: Cambridge University Press, 1998): 93–116; Olivier Darrigol, *From c-Numbers to q-Numbers: The Classical Analogy in the History of Quantum Theory* (Berkeley: University of California Press, 1992).
85. cf. Farmelo, *The Strangest Man*, p. 344.
86. Dirac, 'Is There an Æther?', pp. 906–7. Dirac first suggested a preferred direction in spacetime in P. A. M. Dirac, 'A New Basis for Cosmology', *Proceedings of the Royal Society of London A* 165 (1938): 199–208, pp. 199, 205; see Kragh, 'Cosmo-Physics'.
87. P. A. M. Dirac, 'The Lorentz Transformation and Absolute Time', *Physica* 19 (1953): 888–96; Dirac, 'Quantum Mechanics and the Aether'; P. A. M. Dirac, 'The Vacuum in Quantum Electrodynamics', *Il Nuovo Cimento* 6, 1 Supplement (1957): 322–39; Kragh, 'Electrons and Ether'.
88. Dirac Papers, series 2, box 27, folder 14. An audio recording of this lecture is available at http://www.mediatheque.lindau-nobel.org/videos/31423/electrons-and-the-vacuum-1956/meeting-1956, accessed 22 March 2018.
89. Cockcroft's talk is available at http://www.mediatheque.lindau-nobel.org/videos/33795/1956-scientific-and-technical-problems-in-the-development-of-nuclear-power, accessed 4 November 2017.
90. See Kragh, 'Electrons and Ether'; Bjorken, 'A Dynamical Origin for the Electromagnetic Field'; Nambu, 'Quantum Electrodynamics in Nonlinear Gauge'; Colladay and Kostelecký, 'Lorentz-Violating Extension of the Standard Model'; Mattingly, 'Modern Tests of Lorentz Invariance'; and Liberati, 'Tests of Lorentz Invariance: A 2013 Update'.

INDEX

Page numbers in italics refer to figures.